Handbook of Science and Technology

Handbook of Science and Technology

Edited by
Amber Hall

Willford Press
www.willfordpress.com

Published by Willford Press,
118-35 Queens Blvd., Suite 400,
Forest Hills, NY 11375, USA

ISBN: 978-1-68285-415-0

Cataloging-in-Publication Data

Handbook of science and technology / edited by Amber Hall.
p. cm.
Includes bibliographical references and index.
ISBN 978-1-68285-415-0
1. Science. 2. Technology. 3. Engineering. I. Hall, Amber.
Q158.5 .H36 2018
500--dc23

For information on all Willford Press publications
visit our website at www.willfordpress.com

Willford Press

Printed in China.

Contents

Preface VII

Chapter 1 **Bright Single-Photon Sources based on Anti-Reflection Coated Deterministic Quantum Dot Microlenses** 1
Peter Schnauber, Alexander Thoma, Christoph V. Heine, Alexander Schlehahn, Liron Gantz, Manuel Gschrey, Ronny Schmidt, Caspar Hopfmann, Benjamin Wohlfeil, Jan-Hindrick Schulze, André Strittmatter, Tobias Heindel, Sven Rodt, Ulrike Woggon, David Gershoni and Stephan Reitzenstein

Chapter 2 **Designing Closed-Loop Brain-Machine Interfaces using Model Predictive Control** 9
Gautam Kumar, Mayuresh V. Kothare, Nitish V. Thakor, Marc H. Schieber, Hongguang Pan, Baocang Ding and Weimin Zhong

Chapter 3 **Droplet-Assisted Laser Direct Nanoscale Writing on Silicon** 34
Yuan-Jen Chang, Chao-Hsuan Chang, Chao-Ching Ho, Jin-Chen Hsu and Chia-Lung Kuo

Chapter 4 **Magnetic Resonance Lithography with Nanometer Resolution** 44
Fahad AlGhannam, Philip Hemmer, Zeyang Liao and M. Suhail Zubairy

Chapter 5 **Membrane-Cryogenic Post-Combustion Carbon Capture of Flue Gases from NGCC** 54
Colin A. Scholes, Minh T. Ho and Dianne E. Wiley

Chapter 6 **Single Atoms Preparation using Light-Assisted Collisions** 64
Yin Hsien Fung, Pimonpan Sompet and Mikkel F. Andersen

Chapter 7 **Scale-Up Effects of CO_2 Capture by Methyldiethanolamine (MDEA) Solutions in Terms of Loading Capacity** 87
Samuel P. Santos, João F. Gomes and João C. Bordado

Chapter 8 **Dry Separation of Palm Kernel and Palm Shell using a Novel Five-Stage Winnowing Column System** 97
Rohaya Mohamed Halim, Ridzuan Ramli, Che Rahmat Che Mat, Choo Yuen May, Nasrin Abu Bakar and Nu'man Abdul Hadi

Chapter 9 **Terahertz and Cultural Heritage Science: Examination of Art and Archaeology** 112
Antonino Cosentino

Chapter 10 **Environmental Performance of Hypothetical Canadian Pre-Combustion Carbon Dioxide Capture Processes using Life-Cycle Techniques....125**
Lakkana Piewkhaow, Anastassia Manuilova, Christine W. Chan, Malcolm Wilson and Paitoon Tontiwachwuthikul

Chapter 11 **Fate and Transport of Fire-Born Particles in Porous Media....153**
Prabhakar Sharma, Fritjof Fagerlund, Ulrika Iverfeldt and Amanda Eskebaek

Chapter 12 **The Effects of Thermal Treatment and Steam Addition on Integrated CuO/CaO Chemical Looping Combustion for CO_2 Capture....166**
Alvaro Recio, Sip Chen Liew, Dennis Lu, Ryad Rahman, Arturo Macchi and Josephine M. Hill

Chapter 13 **Development of a CMOS Route for Electron Pumps to be used in Quantum Metrology....178**
Sylvain Barraud, Romain Lavieville, Louis Hutin, Heorhii Bohuslavskyi, Maud Vinet, Andrea Corna, Paul Clapera, Marc Sanquer and Xavier Jehl

Chapter 14 **Experimental Tests on Steel Plate-to-Plate Splices Bonded by C-FRPS Laminas with and without Wrapping....185**
Mario D'Aniello, Francesco Portioli and Raffaele Landolfo

Chapter 15 **A Preliminary Assessment of the Initial Compression Power Requirement in CO_2 Pipeline "Carbon Capture and Storage (CCS) Technologies"....196**
Abdussalam El-Suleiman, Nnamdi B. Anosike and Pericles Pilidis

Chapter 16 **Drilling of Copper using a Dual-Pulse Femtosecond Laser....205**
Chung-Wei Cheng and Jinn-Kuen Chen

Chapter 17 **Correlation Plenoptic Imaging with Entangled Photons....214**
Francesco V. Pepe, Francesco Di Lena, Augusto Garuccio, Giuliano Scarcelli and Milena D'Angelo

Permissions

List of Contributors

Index

Preface

Science and technology has made enormous growth in the past few decades. It has endless applications in the varied fields of health, education, transportation, etc. The importance and need of science and technology lies in its diverse benefits to society. This book attempts to understand the multiple branches that fall under the umbrella of science and technology and how such concepts have practical applications. In this book, using case studies and examples, constant effort has been made to make the understanding of the difficult concepts of science and technology as easy and informative as possible, for the readers.

Various studies have approached the subject by analyzing it with a single perspective, but the present book provides diverse methodologies and techniques to address this field. This book contains theories and applications needed for understanding the subject from different perspectives. The aim is to keep the readers informed about the progresses in the field; therefore, the contributions were carefully examined to compile novel researches by specialists from across the globe.

Indeed, the job of the editor is the most crucial and challenging in compiling all chapters into a single book. In the end, I would extend my sincere thanks to the chapter authors for their profound work. I am also thankful for the support provided by my family and colleagues during the compilation of this book.

Editor

1

Bright Single-Photon Sources Based on Anti-Reflection Coated Deterministic Quantum Dot Microlenses

Peter Schnauber [1], Alexander Thoma [1,*], Christoph V. Heine [1], Alexander Schlehahn [1], Liron Gantz [2], Manuel Gschrey [1], Ronny Schmidt [1], Caspar Hopfmann [1], Benjamin Wohlfeil [3], Jan-Hindrick Schulze [1], André Strittmatter [1], Tobias Heindel [1], Sven Rodt [1], Ulrike Woggon [4], David Gershoni [2] and Stephan Reitzenstein [1]

Academic Editor: Alexander V. Sergienko

1 Institut für Festkörperphysik, Technische Universität Berlin, Hardenbergstraße 36, 10623 Berlin, Germany; peter.schnauber@gmx.de (P.S.); maler_vincente@hotmail.de (C.V.H.); alexander.schlehahn@physik.tu-berlin.de (A.S.); MGschrey@mailbox.tu-berlin.de (M.G.); ronny.schmidt@tu-berlin.de (R.S.); caspar.hopfmann@tu-berlin.de (C.H.); j.schulze@tu-berlin.de (J.-H.S.); strittma@sol.physik.tu-berlin.de (A.S.); tobias.heindel@tu-berlin.de (T.H.); srodt@physik.tu-berlin.de (S.R.); stephan.reitzenstein@physik.tu-berlin.de (S.R.)

2 The Physics Department, Technion—Israel Institute of Technology, Haifa 32000, Israel; lironga@tx.technion.ac.il (L.G.); dg@physics.technion.ac.il (D.G.)

3 Zuse-Institut Berlin (ZIB), Takustraße 7, 14195 Berlin, Germany; wohlfeil@zib.de

4 Institut für Optik und Atomare Physik, Technische Universität Berlin, Straße des 17. Juni 135, 10623 Berlin, Germany; ulrike.woggon@tu-berlin.de

* Correspondence: alexander.thoma@tu-berlin.de

Abstract: We report on enhancing the photon-extraction efficiency (PEE) of deterministic quantum dot (QD) microlenses via anti-reflection (AR) coating. The AR-coating deposited on top of the curved microlens surface is composed of a thin layer of Ta_2O_5, and is found to effectively reduce back-reflection of light at the semiconductor-vacuum interface. A statistical analysis of spectroscopic data reveals, that the AR-coating improves the light out-coupling of respective microlenses by a factor of 1.57 ± 0.71, in quantitative agreement with numerical calculations. Taking the enhancement factor into account, we predict improved out-coupling of light with a PEE of up to 50%. The quantum nature of emission from QDs integrated into AR-coated microlenses is demonstrated via photon auto-correlation measurements revealing strong suppression of two-photon emission events with $g^{(2)}(0) = 0.05 \pm 0.02$. As such, these bright non-classical light sources are highly attractive with respect to applications in the field of quantum cryptography.

Keywords: quantum dots; deterministic devices; quantum optics; single-photon source; anti-reflection coating; extraction efficiency

1. Introduction

Single self-assembled semiconductor quantum dots (QD) are of particular interest regarding quantum communication schemes [1]. They have been proposed to act as efficient quantum-light sources emitting indistinguishable [2] and entangled photons [3], constituting essential building blocks towards the physical realization of quantum communication networks [4]. Recent experiments utilizing self-assembled QDs have demonstrated a record-high photon indistinguishability of 99.5% [5], while a fidelity for polarization-entangled photon pairs of 88% has been achieved exploiting strain-tunable optoelectronic devices [6]. However, critical to such applications and to proceed towards multi-node quantum networks, is to combine the individual benefits, e.g., the purity of single-photon emission or the entangled-state fidelity, within a single device concept compatible with a high photon-extraction efficiency (PEE) and based on deterministic device processing. Huge progress has been achieved with respect to the realization of high PEEs in recent years by exploiting cavity quantum electrodynamics (cQED) effects in optically and electrically driven photonic microcavity structures to enhance the interaction between a QD and a light field via the Purcell-effect [7–11]. Drawbacks arising from the use of cQED effects, however, are the rather small spectral bandwidth, due to the required large Q-factors of the corresponding microcavities, as well as the need for highly sophisticated growth and processing techniques, often hindering a scalable device fabrication.

A well-known alternative technique to enhance the in- or out-coupling efficiency of light in various types of light emitting structures, which is also used in an industrial scale, is to apply anti-reflection (AR) coatings, which reduce parasitic reflections at material surfaces [12]. So far, AR-coatings have been used in solar cells [13], light-emitting diodes [14], single-photon detectors [15] and magnetic field imaging of an ensemble of nitrogen-vacancy centers in diamond [16].

In this work, we present the application of an AR-coating in the field of nanophotonics by increasing the PEE of a single quantum emitter. In fact, we exploit the anti-reflective properties of a thin layer of Ta_2O_5 evaporated onto the curved surface of deterministic single-QD microlenses. Further, by demonstrating triggered single-photon emission with high single-photon purity and a high PEE using this novel approach, we merge the excellent quantum optical properties of single artificial atoms with industrial-scale fabrication techniques.

Our device concept is based on monolithic microlenses precisely aligned (with 24 nm accuracy [17]) to pre-registered single QDs, allowing for an increased PEE of single photons of up to 29% [18] and close to ideal quantum optical properties in terms of the multiphoton suppression and the photon indistinguishability [19]. An important advantage of the microlens concept is the broadband enhancement of the PEE (a bandwidth of ~50 nm has been calculated via finite-element method (FEM) simulations). This appealing feature offers the possibility to boost not only the emission of single photons, but also of bicolored polarization-entangled photon-pairs from the biexciton-exciton (XX-X) radiative cascade of QDs with diminishing fine-structure splitting [3,20] without the necessity for a complicated cavity structure. However, due to the refractive index contrast at the semiconductor-vacuum interface along the uncoated curved surface of the microlens, a significant portion of the incident light emitted by a quantum emitter is reflected back into the semiconductor, limiting the extraction-efficiency of the device. Considering a simplified model of a planar GaAs-vacuum interface and Fresnel's equations this amount is estimated to ~30% of the incident light. In order to circumvent this issue, we apply an anti-reflective layer of Ta_2O_5 to the QD microlens surface and provide a detailed statistical analysis of its impact with respect to the enhancement of the PEE and the quantum-optical properties.

2. Device Fabrication and Experimental Setup

To prove the impact of an AR-coating on the PEE of QDs integrated into microlenses, we utilize two different sample layouts. For a fundamental investigation of the AR-coating, we exploit a semiconductor structure with InGaAs QDs in the active layer grown on GaAs (001) substrate by means of metal-organic chemical vapor deposition. In the first step, 530 nm GaAs and 20 nm AlGaAs were deposited, followed by a 300 nm GaAs layer including a low-density layer of self-assembled Stranski-Krastanov InGaAs-QDs in the center. Next, again 20 nm AlGaAs were grown and finally capped by 230 nm of GaAs. To obtain maximum PEE in Sections 3.4 and 3.5 a second sample structure was grown. It includes a lower mirror based on a distributed Bragg reflector (DBR) consisting of 23 alternating $\lambda/4$-thick layers of AlGaAs (77.0 nm) and GaAs (65.7 nm). The DBR is followed by a GaAs spacer (65.0 nm), a low-density layer of Stranski-Krastanov self-assembled InGaAs-QDs, and a

400 nm thick GaAs capping layer [19]. This layout allows for the collection of QD-photons emitted into the lower hemisphere, thereby boosting the PEE of microlenses further.

The integration of pre-selected QDs into monolithic microlenses is based on three-dimensional (3D) in-situ electron-beam lithography (EBL) enabled by cathodoluminescence (CL) spectroscopy at cryogenic temperatures [21]. We utilize the EBL resist polymethyl-methacrylate (PMMA) which is spin-coated (thickness: 225 nm) onto the sample [22]. During the *in-situ* lithography process, the resist is gradually inverted by writing concentric circles with a carefully adjusted electron-dose centered at the positions of pre-selected QDs. Developing the resist at room temperature results in a 3D etch-mask of inverted, *i.e.*, cross-linked, PMMA with hemispheric profiles. In the subsequent plasma-etching step this mask is directly transferred into the semiconductor material to form monolithic microlenses. More details on the fabrication process can be found in Ref. [19].

In contrast to planar sample structures, for which total internal reflection limits the PEE to less than 1% [23], the microlens design leads to significantly higher PEEs because of its curved surface. However, the device is still suffering from partial reflection at the semiconductor-vacuum interface. In order to reduce the associated detrimental back-reflection, a 110 nm thick $\lambda/4$ layer of Ta_2O_5 was evaporated onto both sample structures via electron beam sputtering with 1 nm thickness accuracy. A schematic of the entire microlens structure is depicted in Figure 1a. To investigate the optical quality of AR-coated QD-microlenses and to evaluate their potential to act as bright quantum-light sources with enhanced PEE, we performed statistical spectroscopic studies using micro-photoluminescence (μPL) spectroscopy at 10 K. Single QDs were optically excited using a continuous wave (cw) diode laser emitting at a wavelength of 657 nm or alternatively a pulsed Ti:Sapphire laser (f = 80 MHz) at a wavelength of 840 nm operated in picosecond mode. Photoluminescence of single QD microlenses is collected by a microscope objective (numerical aperture (NA) = 0.4) and analyzed by a double-grating spectrometer (25 μeV spectral resolution) with an attached liquid-nitrogen-cooled Silicon-based charge-coupled-device camera. The non-classical photon-statistics of the QD emission is investigated via a fiber-based Hanbury-Brown and Twiss (HBT) setup with Silicon-based single-photon counting modules (SPCMs) with a combined timing resolution of 350 ps.

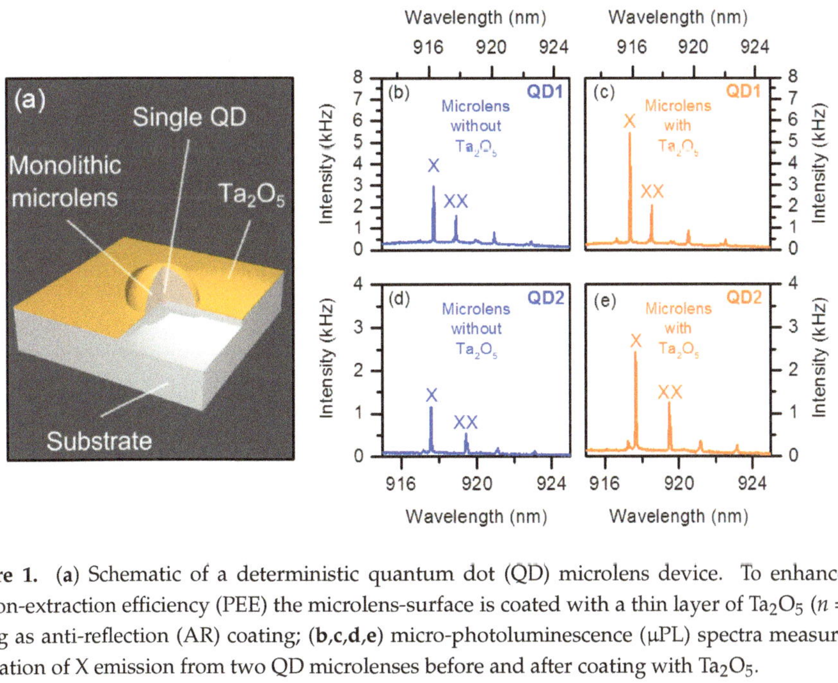

Figure 1. (**a**) Schematic of a deterministic quantum dot (QD) microlens device. To enhance the photon-extraction efficiency (PEE) the microlens-surface is coated with a thin layer of Ta_2O_5 (n = 2.1) acting as anti-reflection (AR) coating; (**b**,**c**,**d**,**e**) micro-photoluminescence (μPL) spectra measured at saturation of X emission from two QD microlenses before and after coating with Ta_2O_5.

3. Methods and Materials

3.1. Deposition of Ta_2O_5 onto the Microlens Surface

To reduce partial back reflection at the microlens surface resulting from the refractive index contrast between GaAs and vacuum, both samples were coated with a layer of a third material that fulfills the condition $n_{coating} \times d = \lambda/4$ with d being the thickness of the coating layer, $n_{coating}$ its refractive index and λ the wavelength of the photons. This discussion neglects the possible effect of the curved microlens surface which will be addressed in Section 3.3 dealing with numerical simulations. Optimum optical impedance matching requires $n_{coating} = (n_{vacuum} \times n_{GaAs})^{1/2} = 1.88$ [24]. Based on these considerations we deposited a thin film of d = 110 nm Ta_2O_5, with a close to ideal refractive index of n_{Ta2O5} = 2.1 at 930 nm, onto the samples. We confirm the deposition of a homogeneous layer of Ta_2O_5 even on the curved surface of the microlens (diameter of 2.4 μm) by scanning electron microscopy (SEM-) images (not shown here).

3.2. μPL Spectroscopy on AR-Coated Microlenses

We studied the impact of the AR-coating on the optical properties of deterministic QD-microlenses without backside mirror by μPL measurements at saturation pump power of the exciton. The latter ensures that we can rule out the influence of possibly different excitation efficiencies with and without AR-coating on the achieved PEE. Figure 1b,d present a μPL spectrum of QD1 and QD2 integrated into an uncoated microlens under above-band-gap excitation at 657 nm. Figure 1c,e show the spectrum of the corresponding QD after coating the microlens with Ta_2O_5. To compare the integrated μPL intensities of single QDs, we fitted the exciton emission line (X) with a Lorentzian profile. The effect of the AR-coating on the PEE is quantified by the ratio of the corresponding emission intensity after and before deposition of Ta_2O_5. In the present case this ratio is $I_{Int,Ratio,QD1} = 1.64 \pm 0.04$ and $I_{Int,Ratio,QD2} = 2.22 \pm 0.15$, which clearly demonstrates the positive impact of AR-coating on the extraction efficiency of the QD-microlens. Interestingly, the absolute spectral position of the emission lines observed in Figure 1 is blue-shifted by 1.28 nm for QD1 and 0.08 nm for QD2 after applying the AR-coating. A spectral shift has been observed consistently for all investigated AR-coated QD-microlenses. Thus, we can exclude that the blue-shift has a purely random character related, e.g., to varying charge configurations in the vicinity of the QD. We assign it to a modified quantum confined Stark shift arising from a different configuration of charge states at the semiconductor surface with and without AR coating. These surface states pin the Fermi level and thereby create an electric field, which in turn alters the emission energy of the QDs via the Stark-effect. We exclude that this effect affects significantly the outcoupling enhancement due to an increased electron-hole wavefunction overlap as we observe no correlation between the spectral shifts and the integrated intensity ratios.

We confirm the results obtained in the exemplary cases above by performing a statistical spectroscopic investigation on 19 QD-microlenses, where each structure was studied within a one-to-one comparison before and after applying the AR-coating. Figure 2 illustrates the statistical analysis performed on the brightest emission line in each spectrum in histogram representation. Here, (a) depicts the ratio of the integrated intensities $I_{Int,Ratio}$; (b) the ratio of the saturation pump powers $P_{Sat,Ratio}$ and (c) the spectral shifts $\Delta\lambda$. It reveals a mean value for the intensity ratio as high as $\overline{I_{Int,Ratio}} = 1.57 \pm 0.71$.

This finding confirms, that the observed increase in detected μPL intensity of the QD is due to an effective reduction of reflections by the AR-coating, which enhances the PEE. For the saturation power ratio a mean value of $\overline{P_{Int,Ratio}} = 0.42 \pm 0.27$ is extracted, implying that the AR-coating enhances on average also the excitation efficiency. The spectral shift is determined to $\overline{\Delta\lambda} = (1.30 \pm 0.84)$ nm.

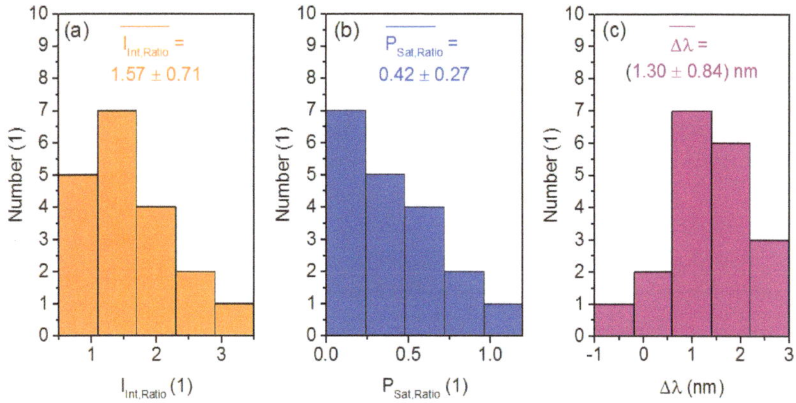

Figure 2. Statistical analysis performed on 19 QDs for (**a**) the ratio of the integrated intensities $I_{Int, Ratio}$; (**b**) the ratio of saturation power $P_{Int, Ratio}$; and (**c**) the spectral shift of the investigated QD emission line Δλ each before and after the coating with Ta_2O_5.

3.3. Simulation of Coated Microlenses Using Finite-Element Methods

Deeper insights into the outcoupling enhancement resulting from the AR-coating are obtained by performing numerical simulations of microlenses using a finite-element solver [25]. Here, a microlens structure was incorporated in a rotational symmetric (quasi two-dimensional) setting without backside reflector. Figure 3a displays the PEE as a function of the thickness d of the Ta_2O_5 coating for numerical apertures of the collection lens of 0.2, 0.4 and 1.0. The dependence of PEE(d) shows an oscillating behaviour resulting from interference effects caused by the coating. The first maximum at d_{sim} = 160 nm is close to the thickness of the AR-coating used in our experiment (d = 110 nm). Figure 3b shows the enhancement of the PEE *versus* AR-coating thickness d, as derived from Figure 3a by calculating PEE(d)/PEE(d = 0) for each curve. At d = 110 nm an enhancement of ~1.8 is extracted, in agreement with the experimental data.

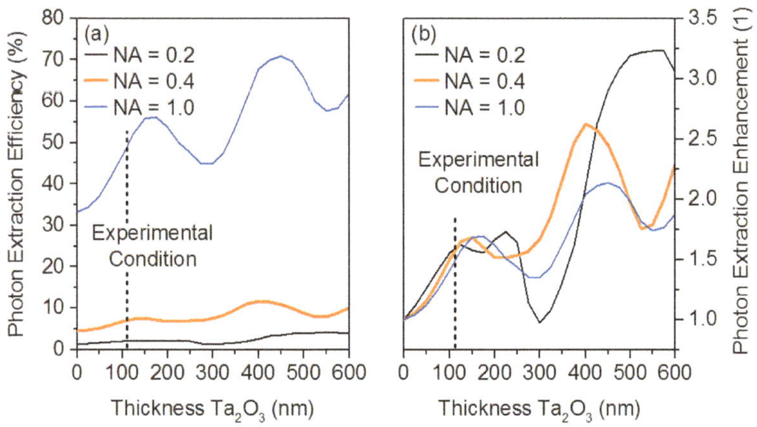

Figure 3. (**a**) Finite element method simulation of the PEE based on solving Maxwell's equations for a microlens without a backside distributed Bragg reflector (DBR) for numerical apertures of 0.2, 0.4 and 1.0 of the collecting optics as a function of Ta_2O_5 thickness; (**b**) Photon-extraction enhancement deduced from (**a**) as a function of Ta_2O_5 thickness.

3.4. *PEE of AR-Coated Deterministic QD Microlenses with DBR*

In addition to the bare lens design, we investigated the influence of a backside DBR on the outcoupling efficiency of AR-coated microlenses. Again, we performed a statistical μPL analysis of 14 QD-microlenses under saturation pump power of the exciton before and after applying the AR-coating. From the comparison we found a ratio of the integrated intensities of $\overline{I_{Int,Ratio}} = 2.19 \pm 1.34$, a ratio of the saturation pump power of $\overline{P_{Int,Ratio}} = 0.96 \pm 0.44$ and a spectral shift of $\overline{\Delta\lambda} = (1.18 \pm 0.65)$ nm. Besides the observation of a further increase in intensity as compared to Section 3.2, we are able to estimate the PEE for the given microlens structure. Taking the PEE of 25%–29% found for uncoated microlenses with backside DBR in Refs. [18,19] using a microscope objective with NA of 0.4 and multiplying it with the integrated intensity ratio from above, we predict a PEE of up to 50% for the AR-coated microlenses with DBR. This enhancement of PEE clearly highlights the benefits of suitable AR-coatings for QD-based single-photon sources that normally suffer from Fresnel-reflections at the semiconductor-vacuum interface.

3.5. *Triggered Single Photons from AR-Coated Deterministic QD Microlenses*

The quantum nature of emission from an AR-coated QD microlens with DBR is investigated via measurements of the photon-autocorrelation function $g^{(2)}(\tau)$ under pulsed non-resonant excitation at 840 nm. For this purpose the emission of the brightest QD state (X^+) was spectrally filtered by the monochromator and coupled into the HBT setup. Taking the combined detection rate of the SPCMs and the setup efficiency into account, we are able to deduce the photon flux F emitted into the first lens of the setup to F = (12.5 ± 0.6) MHz. The resulting coincidence histogram of $g^{(2)}(\tau)$ is presented in Figure 4. Triggered single-photon emission of the QD state is reflected in the strongly reduced number of coincidences at zero time delay ($\tau = 0$). At finite time delay, coincidence maxima occur at a periodicity of 12.5 ns, corresponding to the 80 MHz pulse repetition rate of the Ti:Sapphire laser. To quantitatively evaluate the suppression of two-photon emission events, we fitted the experimental data with equidistant peaks, each represented by the convolution of a two-sided exponential function (decay time constant $\tau_D = (0.83 \pm 0.01)$ ns) with a Gaussian accounting for the HBT's timing resolution (350 ps). Assuming a constant area A of the pulses at finite time delay, $g^{(2)}(0)$ is expressed by the ratio A_0/A, where A_0 corresponds to the zero-delay peak-area. The fitted model function (orange curve in Figure 4) reveals an antibunching value of $g^{(2)}(0) = 0.05 \pm 0.02$, which unambiguously proves the emission of triggered single photons of the AR-coating-enhanced emission of the QD-microlens.

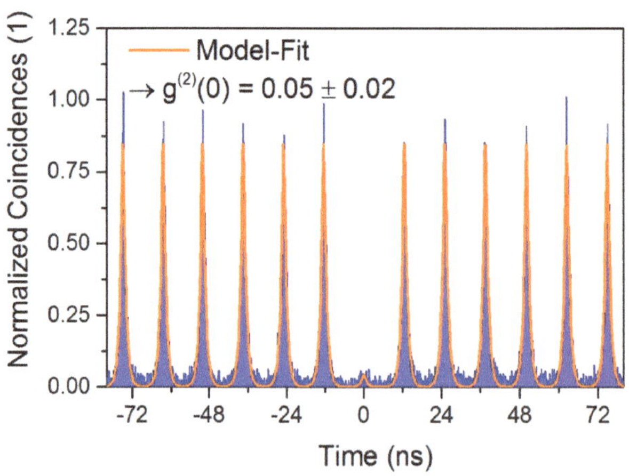

Figure 4. Photon-autocorrelation measurement of the X^+ emission-line under above-band pulsed excitation. Strong antibunching with $g^{(2)}(0) = 0.05 \pm 0.02$ is observed, clearly demonstrating triggered single-photon emission.

4. Conclusion

In summary, we have demonstrated a bright single-photon source by applying an AR-coating to monolithic microlenses containing single InGaAs-QDs. Depositing a Ta_2O_5 dielectric layer allows us to meet the $\lambda/(4 \times n_{coating})$-criterion to obtain destructive interference of the reflected rays. To our knowledge, this is the first demonstration of an AR-coated micro-structure surface in the single-quantum emitter regime reducing partial back reflections and thereby enhancing the PEE to an estimated value of 50%. This technique can be easily extended to an advanced coating design comprising graded-index stacked layers with gradually decreasing refractive index similar to surfaces covered with nanostructures to further optimize the optical impedance matching [14,26].

Acknowledgments: We gratefully acknowledge expert sample preparation and sample coating by C. Scharfenorth. This work was financially supported by the Deutsche Forschungsgemeinschaft (DFG) through SFB 787 "Semiconductor Nanophotonics: Materials, Models, Devices", the BMBF VIP-project QSOURCE (Grant-No.: 03V0630) and from the German-Israeli-Foundation for Scientific Research and Development (Grant-No.: 1148-77.14/2011).

Author Contributions: Peter Schnauber, Christoph Vincent Heine and Manuel Gschrey performed the CL lithography and Peter Schnauber, Christoph Vincent Heine, Alexander Thoma, Liron Gantz, Alexander Schlehahn, Caspar Hopfmann and Tobias Heindel carried out the μPL investigations. Christoph Vincent Heine Peter Schnauber and Alexander Thoma analyzed the data. Benjamin Wohlfeil conducted the numerical modelling. Ronny Schmidt processed the samples in the clean room. Jan-Hindrik Schulze and André Strittmatter grew the samples. Alexander Thoma wrote the manuscript with input from all authors, Sven Rodt, David Gershoni, Ulrike Woggon and Stephan Reitzenstein initiated the research and supervised the project. All authors have read and approved the final manuscript.

Conflicts of Interest: The authors declare no conflict of interest.

References

1. Kiraz, A.; Atatüre, M.; Imamoğlu, A. Quantum-dot single photon sources: Prospects for applications in linear quantum-information processing. *Phys. Rev. A* **2004**, *69*, 032305.
2. Santori, C.; Fattal, D.; Vučkovic, J.; Solomon, G.S.; Yamamoto, Y. Indistinguishable photons from a single photon device. *Nature* **2002**, *419*, 594–598. [PubMed]
3. Akopian, N.; Lindner, N.H.; Poem, E.; Berlatzky, Y.; Avron, J.; Gershoni, D.; Gerardot, B.D.; Petroff, P.M. Entangled Photon Pairs from Semiconductor Quantum Dots. *Phys. Rev. Lett.* **2006**, *96*, 130501. [CrossRef] [PubMed]
4. Gisin, N.; Thew, R. Quantum communication. *Nat. Photonics* **2007**, *1*, 165–171. [CrossRef]
5. He, Y.-M.; He, Y.; Wei, Y.-J.; Wu, D.; Atatüre, M.; Schneider, C.; Hoefling, S.; Kamp, M.; Lu, C.-Y.; Pan, J.-W. On-demand semiconductor single-photon source with near-unity indistinguishability. *Nat. Nano* **2013**, *8*, 217–220. [CrossRef] [PubMed]
6. Trotta, R.; Wildmann, J.S.; Zallo, E.; Schmidt, O.G.; Rastelli, A. Highly Entangled Photons from Hybrid Piezoelectric-Semiconductor Quantum Dot Devices. *Nano Lett.* **2014**, *14*, 3439–3444. [CrossRef] [PubMed]
7. Gérard, J.M.; Sermage, B.; Gayral, B.; Legrand, B.; Costard, E.; Thierry-Mieg, V. Enhanced Spontaneous Emission by Quantum Boxes in a Monolithic Optical Microcavity. *Phys. Rev. Lett.* **1998**, *81*, 1110–1113. [CrossRef]
8. Kiraz, A.; Michler, P.; Becher, C.; Gayral, B.; Imamoğlu, A.; Zhang, L.; Hu, E.; Schoenfeld, W.V.; Petroff, P.M. Cavity-quantum electrodynamics using a single InAs quantum dot in a microdisk structure. *Appl. Phys. Lett.* **2001**, *78*, 3932–3934. [CrossRef]
9. Ellis, D.J.P.; Bennett, A.J.; Dewhurst, S.J.; Nicoll, C.A.; Ritchie, D.A.; Shields, A.J. Cavity-enhanced radiative emission rate in a single-photon-emitting diode operating at 0.5 GHz. *New J. Phys.* **2008**, *10*, 043035. [CrossRef]
10. Dousse, A.; Suffczyński, J.; Beveratos, A.; Krebs, O.; Lemaître, A.; Sagnes, I.; Bloch, J.; Voisin, P.; Senellart, P. Ultrabright source of entangled photon pairs. *Nature* **2010**, *466*, 217–220. [CrossRef] [PubMed]
11. Heindel, T.; Schneider, C.; Lermer, M.; Kwon, S.H.; Braun, T.; Reitzenstein, S.; Höfling, S.; Kamp, M.; Forchel, A. Electrically driven quantum dot-micropillar single photon source with 34% overall efficiency. *Appl. Phys. Lett.* **2010**, *96*, 011107. [CrossRef]

12. Rayleigh, J.S. On the reflection of vibrations at the confines of two media between which the transition is gradual. *Proc. London Math. Soc.* **1880**, *11*, 51–56. [CrossRef]
13. Kuo, M.L.; Poxson, D.J.; Kim, Y.S.; Mont, F.W.; Kim, J.K.; Schubert, E.F.; Lin, S.Y. Realization of a near-perfect antireflection coating for silicon solar energy utilization. *Opt. Lett.* **2008**, *33*, 2527–2529. [CrossRef] [PubMed]
14. An, S.J.; Chae, J.H.; Yi, G.-C.; Park, G.H. Enhanced light output of GaN-based light-emitting diodes with ZnO nanorod arrays. *Appl. Phys. Lett.* **2008**, *92*, 121108. [CrossRef]
15. Rosfjord, K.; Yang, J.; Dauler, E.; Kerman, A.; Anant, V.; Voronov, B.; Gol'tsman, G.; Berggren, K. Nanowire single-photon detector with an integrated optical cavity and anti-reflection coating. *Opt. Express* **2006**, *14*, 527–534. [CrossRef]
16. Yeung, T.K.; le Sage, D.; Pham, L.M.; Stanwix, P.L.; Walsworth, R.L. Anti-reflection coating for nitrogen-vacancy optical measurements in diamond. *Appl. Phys. Lett.* **2012**, *100*, 251111. [CrossRef]
17. Gschrey, M.; Schmidt, R.; Schulze, J.-H.; Strittmatter, A.; Rodt, S.; Reitzenstein, S. Resolution and alignment accuracy of low-temperature *in situ* electron beam lithography for nanophotonic device fabrication. *J. Vac. Sci. Technol. B* **2015**, *33*, 021603. [CrossRef]
18. Schlehahn, A.; Gaafar, M.; Vaupel, M.; Gschrey, M.; Schnauber, P.; Schulze, J.-H.; Rodt, S.; Strittmatter, A.; Stolz, W.; Rahimi-Iman, A.; *et al.* Single-photon emission at a rate of 143 MHz from a deterministic quantum-dot microlens triggered by a mode-locked vertical-external-cavity surface-emitting laser. *Appl. Phys. Lett.* **2015**, *107*, 041105. [CrossRef]
19. Gschrey, M.; Thoma, A.; Schnauber, P.; Seifried, M.; Schmidt, R.; Wohlfeil, B.; Krüger, L.; Schulze, J.-H.; Heindel, T.; Burger, S.; *et al.* Highly indistinguishable photons from deterministic quantum-dot microlenses utilizing three-dimensional *in situ* electron-beam lithography. *Nat. Commun.* **2015**, *6*, 7662. [CrossRef] [PubMed]
20. Young, R.J.; Stevenson, R.M.; Atkinson, P.; Cooper, K.; Ritchie, D.A.; Shields, A.J. Improved fidelity of triggered entangled photons from single quantum dots. *New. J. Phys.* **2006**, *8*, 29. [CrossRef]
21. Gschrey, M.; Gericke, F.; Schüßler, A.; Schmidt, R.; Schulze, J.-H.; Heindel, T.; Rodt, S.; Strittmatter, A.; Reitzenstein, S. *In situ* electron-beam lithography of deterministic single-quantum-dot mesa-structures using low-temperature cathodoluminescence spectroscopy. *Appl. Phys. Lett.* **2013**, *102*, 251113. [CrossRef]
22. Gschrey, M.; Schmidt, R.; Kaganskiy, A.; Rodt, S.; Reitzenstein, S. Study of high-resolution electron-beam resists for applications in low-temperature lithography. *J. Vac. Sci. Technol. B* **2014**, *32*, 061601. [CrossRef]
23. Barnes, W.L.; Björk, G.; Gérard, J.M.; Jonsson, P.; Wasey, J.A.E.; Worthing, P.T.; Zwiller, V. Solid-state single photon sources: Light collection strategies. *Eur. Phys. J. D* **2002**, *18*, 197–210. [CrossRef]
24. Raut, H.K.; Ganesh, V.A.; Nair, A.S.; Ramakrishna, S. Anti-reflective coatings: A critical in-depth review. *Energy Environ. Sci.* **2011**, *4*, 3779–3804. [CrossRef]
25. JCMwave. JCMwave-Complete Finite Element Technology for Optical Simulations. Available online: http://www.jcmwave.com (accessed on 3 September 2015).
26. Huang, Y.-F.; Chattopadhyay, S.; Jen, Y.-J.; Peng, C.-Y.; Liu, T.-A.; Hsu, Y.-K.; Pan, C.-L.; Lo, H.-C.; Hsu, C.-H.; Chang, Y.-H.; *et al.* Improved broadband and quasi-omnidirectional anti-reflection properties with biomimetic silicon nanostructures. *Nat. Nano* **2007**, *2*, 770–774. [CrossRef] [PubMed]

2

Designing Closed-Loop Brain-Machine Interfaces Using Model Predictive Control †

Gautam Kumar [1], Mayuresh V. Kothare [2,*], Nitish V. Thakor [3], Marc H. Schieber [4], Hongguang Pan [5], Baocang Ding [6] and Weimin Zhong [7]

1 Electrical and Systems Engineering, Washington University in St. Louis, 1 Brookings Drive, St. Louis, MO 63130, USA; gautam.kumar@wustl.edu
2 Chemical and Biomolecular Engineering, Lehigh University, 111 Research Drive, Bethlehem, PA 18015, USA
3 Biomedical Engineering, Johns Hopkins University, Baltimore, MD 21218, USA; nitish@jhu.edu
4 Neurobiology and Anatomy, University of Rochester Medical Center, Rochester, NY 14642, USA; mhs@cvs.rochester.edu
5 School of Electrical and Control Engineering, Xi'an University of Science and Technology, Xi'an 710071, China; hongguangpan@163.com
6 Department of Automation, School of Electronic and Information Engineering, Xi'an Jiaotong University, Xi'an 710049, China; baocang.ding@gmail.com
7 Key Laboratory of Advanced Control and Optimization for Chemical Processes, Ministry of Education, East China University of Science and Technology, Shanghai 200237, China; wmzhong@ecust.edu.cn
* Correspondence: mayuresh.kothare@lehigh.edu
† This paper is an extended version of our paper published in 2013 American Control Conference, Washington, DC, USA, 17–19 June 2013 and 2015 American Control Conference, Chicago, IL, USA, 1–3 July 2015.

Academic Editor: Mikhail A. Lebedev

Abstract: Brain-machine interfaces (BMIs) are broadly defined as systems that establish direct communications between living brain tissue and external devices, such as artificial arms. By sensing and interpreting neuronal activities to actuate an external device, BMI-based neuroprostheses hold great promise in rehabilitating motor disabled subjects, such as amputees. In this paper, we develop a control-theoretic analysis of a BMI-based neuroprosthetic system for voluntary single joint reaching task in the absence of visual feedback. Using synthetic data obtained through the simulation of an experimentally validated psycho-physiological cortical circuit model, both the Wiener filter and the Kalman filter based linear decoders are developed. We analyze the performance of both decoders in the presence and in the absence of natural proprioceptive feedback information. By performing simulations, we show that the performance of both decoders degrades significantly in the absence of the natural proprioception. To recover the performance of these decoders, we propose two problems, namely tracking the desired position trajectory and tracking the firing rate trajectory of neurons which encode the proprioception, in the model predictive control framework to design optimal artificial sensory feedback. Our results indicate that while the position trajectory based design can only recover the position and velocity trajectories, the firing rate trajectory based design can recover the performance of the motor task along with the recovery of firing rates in other cortical regions. Finally, we extend our design by incorporating a network of spiking neurons and designing artificial sensory feedback in the form of a charged balanced biphasic stimulating current.

Keywords: closed-loop brain-machine interfaces; artificial sensory feedback; model predictive control

---

1. Introduction

Brain-machine interfaces (BMIs) [1,2] are broadly defined as systems that establish direct communications between living brain tissue and external devices such as artificial arm. The major components of these systems include measurements of cortical neuronal activity, extraction of task-relevant motor intention (decoder), and an encoder that feeds back the motor relevant sensory information back to the brain. Thus the brain, the BMI and the prosthetic device together act as a closed-loop BMI. Figure 1 shows a closed-loop BMI design (also known as brain-machine-brain interface (BMBI) [3]).

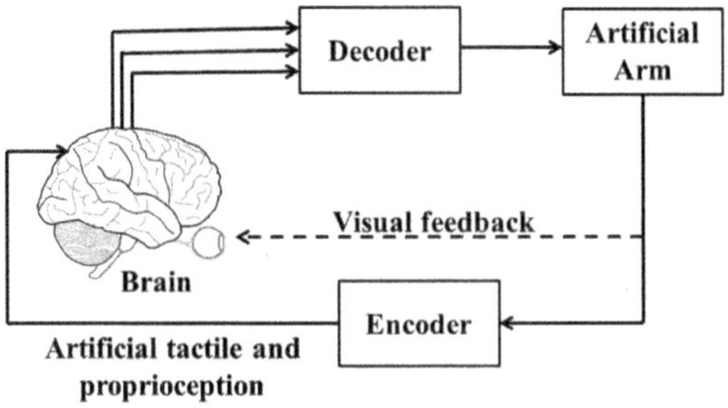

Figure 1. A closed-loop brain-machine interface (BMI).

In the last two decades, BMI based motor intended neural prosthetic systems have been studied extensively [4–11]. In most of these studies, healthy subjects are trained to perform a specific motor task such as reaching or grasping. Recorded data during the performance of the task are then used to develop a mathematical model called decoder. The decoder extracts the kinetic as well as the kinematic motor information from a continuously recorded firing activity of motor relevant cortical neurons. The performance of the decoder is typically measured by applying the decoded information to a prosthetic arm. The online movement based error correction during the reaching task is accomplished by the subject using the available visual feedback information in the absence of the natural proprioception. Therefore these BMIs are considered as partially closed-loop systems in their current formulations where the incorporation of artificial proprioception is neglected in their designs.

In the absence of tactile feedback, these BMIs can fail to differentiate visually similar textures. Similarly, in the absence of proprioception, these BMIs are unable to provide the natural sensation of the arm movement which are both experienced and used by healthy subjects in controlling their natural limb movements. It has been recognized in the BMIs community that the inclusion of sensory feedback from the actuated artificial limb in BMIs is necessary to improve the versatility of motor-based BMIs [12]. It has also recently been shown in [13] that kinesthetic feedback together with the visual feedback can significantly improve the BMI performance.

Recently, attempts have been made towards closing the BMI loop by incorporating artificial texture [3] and proprioception [14,15] information. In these studies, a intra-cortical micro-stimulation (ICMS) technique has been investigated as a promising approach in providing artificial sensation of motor tasks to the brain. The approach relies on a learning paradigm where the subject is trained to differentiate [3] or learn [14,15] artificial sensory feedback in a task-dependent context. Even though the approach is promising for developing future BMIs, the experimental trial and error approach in designing appropriate stimulating sensory input currents may change the natural functionality of the brain. Therefore, a systematic approach that uses optimal feedback control theory is highly desirable towards developing stimulation enhanced next generation BMIs. This approach provides flexibility in designing optimal stimulating sensory input currents and analyzing the closed-loop BMI under various feedback scenarios. It may also reduce the learning effort of the motor task by guiding the movement in the initial learning of the BMI motor task.

The intellectual merit of studying BMIs by taking control-theoretic approach is to exploit all the available degrees of freedom in developing the next generation of BMI-based feedback-enabled neuroprosthetic devices. The development of these feedback-enabled neuroprosthetic devices is necessary for making the prosthetic devices prone to error in decoding and targeting the intended action of the neurons. Moreover, issues such as prosthetic system stability, system reliability, impact of transmission loss, latency and time delays, impact of model complexity and uncertainty, optimality of the modeling and control framework, etc. are critical to the clinical deployment of next generation BMIs.

In recognition of these merits, systematic control-theoretic approaches have recently been taken to optimize the learning process in BMIs [7], rigorously analyze the BMI systems [16–18] and design artificial sensory feedback optimally [19,20] by developing a theoretical framework based on optimal feedback control (OFC) theory. For instance, the authors in [7] proposed a learning model to maximize the performance of closed-loop BMIs by leveraging tools such as inverse models and feedback control which are formally embedded in control theory. The authors in [18] developed a theoretical model of BMI experiments using optimal feedback control as a policy for brain control during BMI motor tasks. The framework incorporates visual and proprioceptive feedback in estimating states which are then used to compute optimal control inputs for stimulating neuronal models of the primary motor cortex and the premotor cortex neurons. Using this framework, the authors showed that the experimentally observed abrupt changes in neural modulations when switching to BMI control can be explained using optimal feedback control. In [16], the authors investigated the importance of visual and proprioceptive feedback in BMIs by using a framework of model predictive control in designing optimal stimulus for a single spiking neuron for a single joint movement task based closed-loop BMI. In [17], the authors proposed a stochastic optimal feedback controller as the closed-loop operation of the brain during a BMI performance. Using this framework, they analyzed the performance of open and closed-loop BMIs and explained key phenomenon in closed-loop BMI operation. In particular, they explained the experimentally observed parametric variations in closed-loop operation of BMIs such as the performance deterioration with increasing bin width and diminishing effect of decoder bias in closed-loop BMIs. In [19], the authors showed the first systematic approach in designing optimal artificial sensory feedback in closed-loop BMIs. In particular, the authors proposed an optimal design of feedback-enabled closed-loop BMI and designed artificial proprioceptive feedback using a model predictive controller in a single joint reaching task. The authors further extended their firing rate-based approach [19] to ICMS [20].

In this article, we theoretically demonstrate the recovery of closed-loop performance of a BMI for voluntary single joint extension task by designing an optimal artificial sensory feedback in the absence of the natural proprioceptive feedback pathways. Throughout our analysis, we exclude the treatment of visual feedback as well any form of cortical learning. A theoretical experiment is performed on a firing rate based cortical circuit model of a voluntary control of a single joint extension task to design a BMI. In particular, we design both the Wiener filter and the Kalman filter based decoders and compare their performances. We emphasize the degraded online performance of both decoders in the absence of the proprioceptive feedback. An optimal artificial sensory feedback in the framework of model predictive control is designed to compensate the loss of the natural proprioceptive feedback pathways. We extend our design by including a recurrent network of spiking neurons in the firing rate based neurophysiological cortical circuit model which allow us to design artificial sensory feedback in the form of a charge-balanced biphasic waveform of stimulating input current. We demonstrate the efficacy of our designs by performing simulations.

The remainder of this paper is organized as follows. In Section 2, we describe the neurophysiological cortical circuit model of a voluntary single joint extension task which we used to generate the synthetic data in Section 3. This is followed by the design of both the Wiener filter and the Kalman filter based linear decoders in Section 4. The performance analysis of the designed decoder in the presence and in the absence of the natural proprioception are described in Section 5. We formulate the model predictive control problems and design the artificial sensory feedback in Section 6. The paper ends with discussion.

2. Psycho-Physiological Cortical Circuit Model of Single Joint Movement

We used an average firing rate based psycho-physiological cortical circuit model, proposed by [21] and shown in Figure 2, for voluntary control of a single joint movement to design a closed-loop BMI. This minimal model captures the essential cortical pathways as well as the proprioceptive feedback pathways which are relevant during voluntary extension or flexion of a single joint such as elbow. Although the model excludes the treatment of visual feedback during the movement, the model has shown its capability in a qualitative reproduction of several experimentally observed results on voluntary control of a single joint movement. The details of the model and its connection with neurophysiology of a single joint voluntary movement can be found in [21].

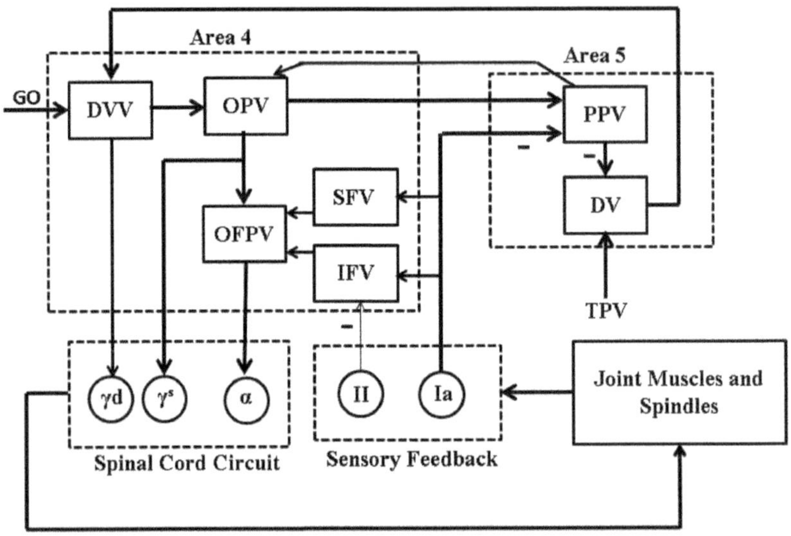

Figure 2. A psycho-physiological cortical circuit model for voluntary control of single joint movement: The diagram has been redrawn from Bullock *et al.* [21], Figure 1.1. Nomenclature (adopted from [21]): "GO" is a scalable gating signal; "DVV" is the desired velocity vector; "OPV" is the outflow position vector; "OFPV" is the outflow force and position vector; "SFV" is the static force vector; "IFV" is the inertial force vector; "PPV" is the perceived position vector; "DV" is the difference vector; "TPV" is the target position vector; "γ^d" and "γ^s" are dynamic and static gamma motoneurons respectively; "α" is alpha motoneuron; "Ia" and "II" are type Ia and II afferent fibers; – represents inhibitory feedback. The rest of the connections are excitatory.

Briefly, a population of area 5 (the parietal cortex) "DV" neurons computes the difference between the target and the perceived limb position vectors. The average firing activity of a population of these neurons is represented as

$$r_i(t) = \max\{T_i - x_i(t) + B^r, 0\}. \tag{1}$$

Here, $0 \leq r_i(t) \leq 1$ represents the average firing activity of a population of "DV" neurons associated with the agonist muscle i and shows a phasic behavior during the movement. Throughout the paper, we will denote the average firing activity of neurons associated with the agonist muscle i by the subscript i and the corresponding antagonist muscle by the subscript j. T_i is the target position vector ("TPV") command for the target position of the agonist muscle i. $x_i(t)$ is the average firing activity of a population of area 5 "PPV" neurons. These neurons continuously compute the present position of the agonist muscle i. B^r is the base firing activity of the "DV" neurons. Continuously computed difference vector information by the area 5 "DV" neurons is then scaled by a population of area 4 (the primary motor cortex (M1)) "DVV" neurons as

$$u_i(t) = \max\{g(t).(r_i(t) - r_j(t)) + B^u, 0\}. \tag{2}$$

Here, $u_i(t)$ is the average firing activity of a population of area 4 "DVV" neurons. B^u is the base firing activity of the "DVV" neurons. $g(t)$ is an internal "GO" signal which is assumed to be originated from the basal ganglia. "DVV" neurons fire only during the movement and thus their average firing activity shows a phasic-movement time (MT) behavior. The dynamics of the internal "GO" signal is modeled as

$$\frac{\mathrm{d}g^1(t)}{\mathrm{d}t} = \epsilon(-g^1(t) + (C - g^1(t))g^0), \tag{3a}$$

$$\frac{\mathrm{d}g^2(t)}{\mathrm{d}t} = \epsilon(-g^2(t) + (C - g^2(t))g^1(t)), \tag{3b}$$

$$g(t) = g^0 \frac{g^2(t)}{C}. \tag{3c}$$

Here, ϵ represents a slow integration rate and is treated as constant. C is a constant value at which the "GO" neurons saturate. The area 4 "OPV" neurons receive information from the area 4 "DVV" neurons as well as the area 5 "PPV" neurons and show tonic firing activity. The average firing activity of a population of "OPV" neurons is modeled as

$$\frac{\mathrm{d}y_i(t)}{\mathrm{d}t} = (1 - y_i(t))(\eta x_i(t) + \max\{u_i(t) - u_j(t), 0\}) - y_i(t)(\eta x_j(t) + \max\{u_j(t) - u_i(t), 0\}). \tag{4}$$

Here, η is a scaling factor. The average firing activity of a population of static ($\gamma_i^S(t)$) and dynamic ($\gamma_i^D(t)$) gamma motoneurons are modeled as

$$\gamma_i^S(t) = y_i(t), \tag{5a}$$

$$\gamma_i^D(t) = \rho \max\{u_i(t) - u_j(t), 0\}. \tag{5b}$$

Here, ρ is a scaling parameter. The average firing activity of the primary ("Ia") and the secondary ("II") muscle spindles afferents are modeled as

$$\begin{aligned} s_i^1(t) = S(\theta \max\{\gamma_i^S(t) - p_i(t), 0\} \\ + \phi \max\{\gamma_i^D(t) - \frac{\mathrm{d}p_i(t)}{\mathrm{d}t}, 0\}), \end{aligned} \tag{6a}$$

$$s_i^2(t) = S(\theta \max\{\gamma_i^S(t) - p_i(t), 0\}). \tag{6b}$$

Here, $s_i^1(t)$ and $s_i^2(t)$ are the primary and the secondary spindles afferents average firing activity respectively. p_i is the position of the agonist muscle i. θ is the sensitivity of the static nuclear bag and chain fibers. ϕ is the sensitivity of the dynamic nuclear bag fibers. The saturation of spindles afferents activity is given by the function $S(\omega) = \omega/(1 + 100\omega^2)$. The average firing activity $x_i(t)$ of a population of area 5 "PPV" neurons is modeled as

$$\begin{aligned} \frac{\mathrm{d}x_i(t)}{\mathrm{d}t} = (1 - x_i(t)) \max\{\Theta y_i(t) + s_j^1(t - \tau) - s_i^1(t - \tau), 0\} \\ - x_i(t) \max\{\Theta y_j(t) + s_i^1(t - \tau) - s_j^1(t - \tau), 0\}. \end{aligned} \tag{7}$$

Here, τ is the delay time of the spindles feedback. Θ is a constant gain. The average firing activity $q_i(t)$ of a population of area 4 "IFV" neurons is modeled as

$$q_i(t) = \lambda_i \max\{s_i^1(t-\tau) - s_i^2(t-\tau) - \Lambda, 0\}. \tag{8}$$

Here, Λ is a constant threshold. The average firing activity $f_i(t)$ of a population of area 4 "SFV" neurons is modeled as

$$\frac{\mathrm{d}f_i(t)}{\mathrm{d}t} = (1 - f_i(t))hs_i^1(t-\tau) - \psi f_i(t)(f_j(t) + s_j^1(t-\tau)). \tag{9}$$

Here, h is a constant gain which controls the strength and speed of an external load compensation. ψ is an inhibitory scaling parameter. The average firing activity $a_i(t)$ of a population of the area 4 "OFPV" neurons is modeled as

$$a_i(t) = y_i(t) + q_i(t) + f_i(t). \tag{10}$$

The average firing activity of these neurons shows a phasic-tonic behavior. The average firing activity $\alpha_i(t)$ of alpha motoneurons is modeled as

$$\alpha_i(t) = a_i(t) + \delta s_i^1(t), \tag{11}$$

where δ is a stretch reflex gain. The limb dynamics is described by

$$\begin{aligned}\frac{\mathrm{d}^2 p_i(t)}{\mathrm{d}t^2} &= \frac{1}{I}(M(c_i(t) - p_i(t)) - M(c_j(t) - p_j(t)) \\ &+ E_i - V\frac{\mathrm{d}p_i(t)}{\mathrm{d}t}).\end{aligned} \tag{12}$$

Here $p_i(t)$ is the position of the agonist muscle i within its range of origin-to-insertion distances. $p_j(t)$ is the position of the antagonist muscle such that $p_i(t) + p_j(t) = 1$. I is the moment of inertia of the limb. V is the joint viscosity. E_i is the external force applied to the joint. $M(c_i(t), p_i(t)) = \max\{c_i(t) - p_i(t), 0\}$ represents the total force generated by the agonist muscle i. $c_i(t)$ is the muscle contraction activity dynamics of which is given by

$$\frac{\mathrm{d}c_i(t)}{\mathrm{d}t} = \nu(-c_i(t) + \alpha_i(t)). \tag{13}$$

3. Synthetic Experimental Data Generation for Extension Task

In a typical non-human primate experiment, a monkey is trained to accomplish a given motor task such as reaching or grasping. After the training, spiking activity of single neurons are recorded through implanted multi-channel electrodes from various motor relevant cortical areas such as the primary motor cortex (M1), the premotor area (PMv, PMd), and the primary somatosensory area (S1). Simultaneously, kinetic and kinematic information such as joint torque, velocity and position of the real arm are measured to generate a data set to design a decoder.

In this work, we generated a synthetic experimental data set for voluntary control of a single joint extension task by simulating the system model (1)–(13) in MATLAB (Version R2011b, The MathWorks, Inc., Natick, MA, USA). The target position of the agonist muscle i was set to the desired one at $t = 0$. The "GO" signal was turned on at $t = 50$ ms. During the initial 50 ms, the system was at the priming state. The initial condition of variables was set to 0 except $x_i(0) = x_j(0) = 0.5$, $y_i(0) = y_j(0) = 0.5$, $p_i(0) = p_j(0) = 0.5$, $u_i(0) = u_j(0) = B^u$ and $r_i(0) = r_j(0) = B^r$. For the simulation, we used the following model parameters [21]: $I = 200$, $V = 10$, $\nu = 0.15$, $B^r = 0.1$, $B^u = 0.01$, $\Theta = 0.5$, $\theta = 0.5$, $\phi = 1$, $\eta = 0.7$, $\rho = 0.04$, $\lambda_1 = 150$, $\lambda_2 = 10$, $\Lambda = 0.001$, $\delta = 0.1$, $C = 25$, $\epsilon = 0.05$, $\psi = 4$, $h = 0.01$, $T_1 = 0.7$ and $\tau = 0$.

In BMI experiments, a trial is considered successful if the trained monkey accomplishes the specified motor task in a given time duration. Thus the accomplishment time of the task in successful trials is allowed to vary. In our case, the "GO" signal controls the velocity of the joint movement and thus the accomplishment time of a given task. Therefore we assumed that there is a trial-to-trial variability in the internal "GO" signal. To introduce the trial-to-trial variability in the "GO" signal, we modeled g^0 as a Gaussian distributed random variable with mean 0.75 and variance 0.0025. For a given trial, g^0 is constant. It should be noted that the "GO" signal has no effect on the accuracy of the movement.

We simulated the model and generated synthetic data for 1600 independent trials of the voluntary single joint extension task. In each of these trials, the simulation was performed for the duration of 1.45 s which includes a variable holding period at the target position after the accomplishment of the task. To generate a synthetic data set, we measured the average firing activity of a population of area 4 "DVV", "OPV", and "OFPV" agonist and the corresponding antagonist neurons sampled at every 10 ms. Simultaneously, we measured the agonist muscle position $p_i(t)$, the agonist muscle velocity $v_i(t) = dp_i(t)/dt$, and the total force difference between the agonist and the corresponding antagonist muscle $\Delta M(t) = M(c_i(t), p_i(t)) - M(c_j(t), p_j(t))$ at every 10 ms. With this, we created a data set of 233,600 samples by embedding recorded data from 1600 trials.

4. Wiener and Kalman Filters Based Decoder Designs

Neurophysiological as well as BMI experimental studies have shown the encoding of task-relevant kinetic as well as kinematic motor information in the spike trains of the cortical area 4 neurons. For a given motor task, this motor information is extracted from a continuously recorded spike train of the cortical area 4 neurons by developing a mathematical model called decoder. In the past, several linear and nonlinear decoders have been developed in BMI studies. Among them, the most popular are based on the Wiener filter [22], the Kalman filter and its variations [23–28], artificial neural networks [29] and recurrent neural networks [30].

We designed both the adapted Wiener filter [22] and the Kalman filter [28]) based decoder models (linear decoders) to extract the total force difference between the agonist and the corresponding antagonist muscle ($\Delta M(k)$), the agonist muscle position ($p_i(k)$), and the agonist muscle velocity ($v_i(k)$) from the recorded firing activity of the area 4 "DVV", "OPV", and "OFPV" neurons and performed comparisons between the performance of these two decoders. Here, $k = 1, 2, \cdots$ is a discrete sample time at which data were recorded for a given trial. $k = 1$ corresponds to $t = 0$ ms, $k = 2$ corresponds to $t = 10$ ms, and so on. We emphasize that these neurons have direct contribution to the spinal cord circuit of the neurophysiological system shown in Figure 2.

4.1. Wiener Filter Based Decoder Design

In a discrete-time adapted Wiener filter based decoder design [22], the relation between $\Delta M(k)$ and the average firing activity of area 4 neurons, *i.e.*, "DVV", "OPV", and "OFPV" neurons can be expressed as

$$\Delta M(k) = \mathbf{w}^T \mathbf{z}(k) + n_1(k). \tag{14}$$

Here, $\mathbf{w}$ is a $(L.N) \times 1$ weight vector. L is the number of delay elements. $(\cdot)^T$ is the transpose of a vector. $\mathbf{z}(k) = [z_1(k), z_1(k-1), \cdots, z_1(k-L+1), z_2(k), \cdots, z_N(k-L+1)]^T$. $z_m(k-l)$ represents the average firing activity of the population m delayed by l samples. For our system, $z_1 = y_i$, $z_2 = y_j$, $z_3 = u_i$, $z_4 = u_j$, $z_5 = a_i$, and $z_6 = a_j$. Thus, $N = 6$. We assumed the number of delay elements $L = 10$. Thus the weight vector $\mathbf{w}$ has a dimension of 60×1. We also assumed that there is no measurement noise in obtaining data *i.e.*, $n_1(k) = 0$. Similarly, the relation between $[p_i(k), v_i(k)]^T$ and the average firing activity of area 4 neurons *i.e.*, "DVV", "OPV", and "OFPV" neurons can be expressed as

$$[p_i(k), v_i(k)]^T = \mathbf{W}^T \mathbf{z}(k). \tag{15}$$

Here, $\mathbf{W}^T$ is a weight matrix of dimension 60×2. We again assumed that there is no measurement noise in obtaining data.

For consistency with BMI experiments, in this work, we used 220,000 samples of the recorded synthetic data to train the weight vector "**w**" and the weight matrix "**W**" of the designed decoders (Equations (14) and (15)). For this, we used the following normalized least mean squares algorithm [22]:

$$\mathbf{w}(k+1) = \mathbf{w}(k) + \frac{\eta}{\beta + ||\mathbf{z}(k)||^2}\mathbf{z}(k)e(k). \tag{16a}$$

$$\mathbf{W}(k+1) = \mathbf{W}(k) + \frac{\eta}{\beta + ||\mathbf{z}(k)||^2}\mathbf{z}(k)(\mathbf{e}(k))^T. \tag{16b}$$

Here, η and β are constants. $||\cdot||$ represents the Euclidean norm. In Equation (16a), $e(k)$ represents a scalar error between the recorded $\Delta M(k)$ and the estimated value through Equation (14). In Equation (16b), $\mathbf{e}(k)$ represents the error vector between the recorded $[p_i(k), v_i(k)]^T$ and the estimated value through Equation (15). For our study, we set $\eta = 0.01$ and $\beta = 1$. After the training, we froze the weight vector "**w**" and the weight matrix "**W**" to the final adapted value. Then we used the rest of 13,600 samples to validate the performance of both decoders.

Figure 3A,C shows the offline performance of the adapted Wiener filter based decoder in decoding the joint position $p_i(k)$ and the joint velocity $v_i(k)$ respectively, as defined by Equation (15), on the test data for 1000 samples. Figure 3E shows the offline performance of the adapted Wiener filter based decoder in decoding the force difference between the agonist and the corresponding antagonist muscle, $\Delta M(k)$, as defined by Equation (14), on the test data for 1000 samples.

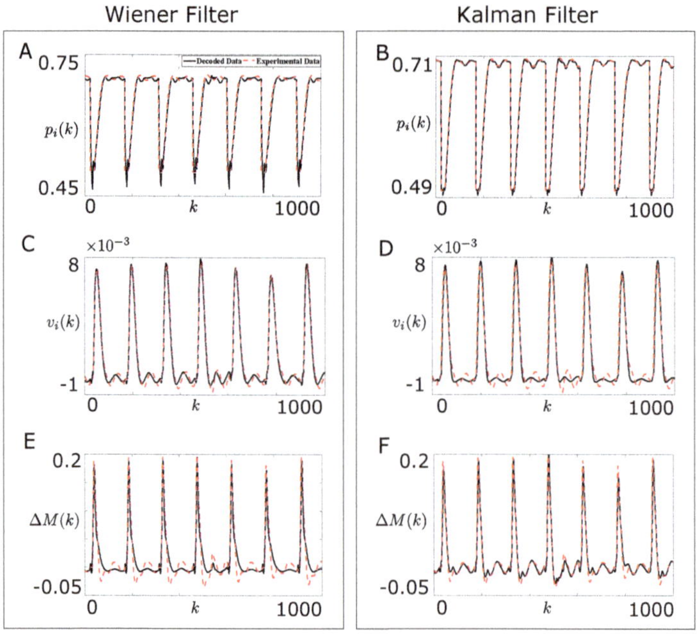

Figure 3. Comparison of the offline performances of the Wiener filter and the Kalman filter based decoders for the single joint reaching task on a sample part of the test data. (**A**,**C**,**E**) show the comparison between the experimental (dotted line, red) and the decoded (solid line, blue) joint position $p_i(k)$, joint velocity, $v_i(k)$ and force difference between the agonist and the corresponding antagonist muscle, $\Delta M(k)$ respectively for the Wiener filter based decoder; (**B**,**D**,**F**) show the comparison between the experimental (dotted line, red) and the decoded (solid line, blue) joint position $p_i(k)$, joint velocity, $v_i(k)$ and force difference between the agonist and the corresponding antagonist muscle, $\Delta M(k)$ respectively for the Kalman filter based decoder.

4.2. Kalman Filter Based Decoder Design

The following set of dynamical equations describes the design of the Kalman filter based decoder [28]:

$$\hat{x}(k \mid k-1) = A\hat{x}(k-1), \tag{17a}$$

$$\hat{P}(k \mid k-1) = A\hat{P}(k-1)A^T + R, \tag{17b}$$

$$\hat{x}(k) = \hat{x}(k \mid k-1) + K_k(\mathbf{z}(k) - C\hat{x}(k \mid k-1)), \tag{17c}$$

$$\hat{P}(k) = (I - K_k C)\hat{P}(k \mid k-1), \tag{17d}$$

$$K_k = \hat{P}(k \mid k-1)C^T(C\hat{P}(k \mid k-1)C^T + Q)^{-1}. \tag{17e}$$

Here, $\hat{x}(k \mid k-1)$ and $\hat{x}(k)$ represent *a priori* and *a posteriori* estimate of the state vector $x(k)$ of dimension $n \times 1$ at time k respectively. $\hat{P}(k \mid k-1)$ and $\hat{P}(k)$ are the estimate of *a priori* and *a posteriori* covariance matrix respectively. $\mathbf{z}(k)$ is the observation (firing rate) vector of dimension $r \times 1$. K_k is the Kalman gain and I is an identity matrix. $A \in \mathbb{R}^{n \times n}$ is the state matrix and is given by $A = X_2 X_1^T (X_1 X_1^T)^{-1}$. $C \in \mathbb{R}^{r \times n}$ represents the observation matrix and is given by $C = ZX^T(XX^T)^{-1}$. $R = \frac{1}{D-1}(X_2 - AX_1)(X_2 - AX_1)^T$ and $Q = \frac{1}{D}(Z - CX)(Z - CX)^T$ are covariance matrices of Gaussian noise sources with mean zero to the state and the observation vectors respectively. The structure of matrices Z, X, X_1, and X_2 is given by

$$Z = \begin{pmatrix} z_{1,1} & \cdots & z_{1,D} \\ \vdots & \ddots & \vdots \\ z_{r,1} & \cdots & z_{r,D} \end{pmatrix} \tag{18a}$$

$$X = \begin{pmatrix} x_{1,1} & \cdots & x_{1,D} \\ \vdots & \ddots & \vdots \\ x_{n,1} & \cdots & x_{n,D} \end{pmatrix} \tag{18b}$$

$$X_1 = \begin{pmatrix} x_{1,1} & \cdots & x_{1,D-1} \\ \vdots & \ddots & \vdots \\ x_{n,1} & \cdots & x_{n,D-1} \end{pmatrix} \tag{18c}$$

$$X_2 = \begin{pmatrix} x_{1,2} & \cdots & x_{1,D} \\ \vdots & \ddots & \vdots \\ x_{n,2} & \cdots & x_{n,D} \end{pmatrix} \tag{18d}$$

Here, $z_{i,j}$ represents the jth firing rate data of the ith neuron. $x_{i,j}$ represents the jth data of the ith state.

For our system, $x(k) \equiv \Delta M(k)$ $(n = 1)$ if the total force difference between the agonist and the corresponding antagonist muscle $(\Delta M(k))$ is extracted, and $x(k) \equiv [p_i(k), v_i(k)]^T$ $(n = 2)$ if the position and the velocity of the agonist muscle are extracted from the average firing activity of area 4 neurons *i.e.*, "DVV", "OPV", and "OFPV" neurons. $r = 6$ and $D = 220{,}000$. $\mathbf{z}(k) = [y_i(k), y_j(k), u_i(k), u_j(k), a_i(k), a_j(k)]^T$.

For consistency with the Wiener filter based decoder design, we used 220,000 samples of the recorded synthetic data to compute A, C, R, and Q for both $x(k) \equiv \Delta M(k)$ and $x(k) \equiv [p_i(k), v_i(k)]^T$. Then we used the rest of 13,600 samples to validate the performance of the decoder for both cases.

Figure 3F shows the offline performance of the Kalman filter based decoder on the test data for 1000 samples when $x(k) \equiv \Delta M(k)$ and Figure 3B,D shows the offline performance of the Kalman filter based decoder on the test data for 1000 samples when $x(k) \equiv [p_i(k), v_i(k)]^T$. Clearly, the Kalman filter based decoder performed better than the Wiener filter based decoder on the test data.

4.3. Comparison of Designed Decoders

We compared the performance of the designed decoders (*i.e.*, the Wiener filter and the Kalman filter) in an open-loop BMI system design shown in Figure 4.

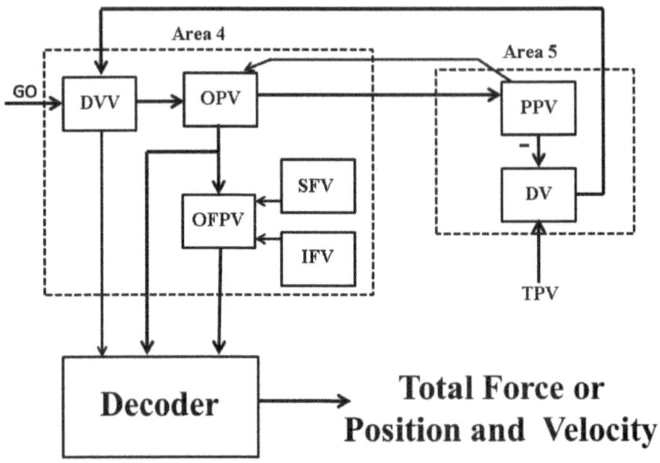

Figure 4. An open-loop BMI system design.

As shown in Figure 4, the average firing activity of area 4 neurons *i.e.*, "DVV", "OPV", and "OFPV" neurons are used by the decoder to extract either the total force ($\Delta M(k)$) or the position ($p_i(k)$) and the velocity ($v_i(k)$) of the agonist muscle. To compare the performance of both decoders (*i.e.*, the Wiener filter and the Kalman filter based decoders), We simulated (1)–(10) along with the particular decoder model (*i.e.*, (14)–(16) for the Wiener filter and (17) and (18) for the Kalman filter) and $s_i^1(t-\tau) = s_i^2(t-\tau) = s_j^1(t-\tau) = s_j^2(t-\tau) = 0$ and $g^0 = 0.75$. Remaining model parameters are the same as given in Section 3.

Figure 5A,B compare the open-loop performance of the Wiener filter based decoder and the Kalman filter based decoder with the performance of the closed-loop real system shown in Figure 2 when the extracted information from both decoders is the position ($p_i(t)$) and the velocity ($v_i(t)$) of the agonist muscle in real time t respectively. As shown clearly in these figures, the performance of both decoders degrades substantially in the absence of the natural proprioception information. Moreover, the Kalman filter performed better over the Wiener filter in decoding position compared to the velocity of the agonist muscle.

Figure 5C compares the open-loop performance of the Wiener filter based decoder and the Kalman filter based decoder with the performance of the closed-loop real system shown in Figure 2 when the extracted information from both the decoders is the total force ($\Delta M(t)$) in real time t. As shown in Figure 5C, the performance of both decoders degrades substantially in the absence of the natural proprioception information.

Without the loss of generality, in the rest of the paper, we considered the Wiener filter based decoder design with $\Delta M(t)$ as the extracted information to design closed-loop BMIs. Note that the described approach can be implemented to other forms of decoders as well.

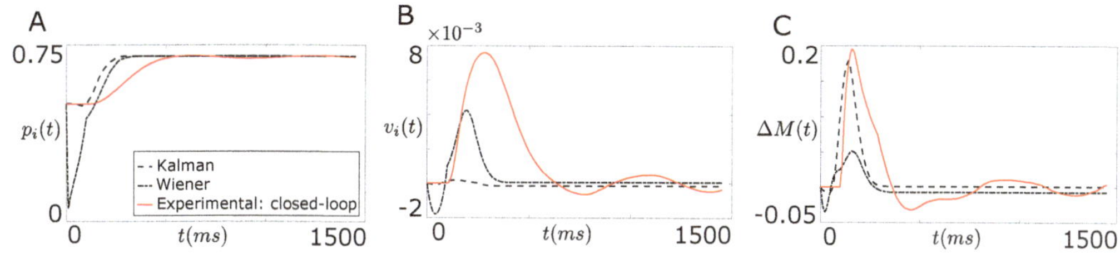

Figure 5. Comparison of the open-loop performances of the Wiener filter based decoder and the Kalman filter based decoder with the performance of the closed-loop real system shown in Figure 2. (**A**) compares the open-loop decoder performance and the performance of the closed-loop real system when the extracted information from both decoders was the joint position, $p_i(t)$; (**B**) shows the open-loop decoder performance and the performance of the closed-loop real system when the extracted information from both decoders was the joint velocity, $v_i(t)$; (**C**) compares the open-loop decoder performance and the performance of the closed-loop real system when the extracted information from both decoders was the force difference between the agonist and the corresponding antagonist muscle, $\Delta M(t)$.

5. Need of a Closed-Loop BMI

We studied the online performance of the Wiener filter based decoder with $\Delta M(k)$ as the extracted information in the presence and in the absence of the natural proprioceptive feedback information when the decoder interacts with the dynamics of the muscle as shown in Figure 6.

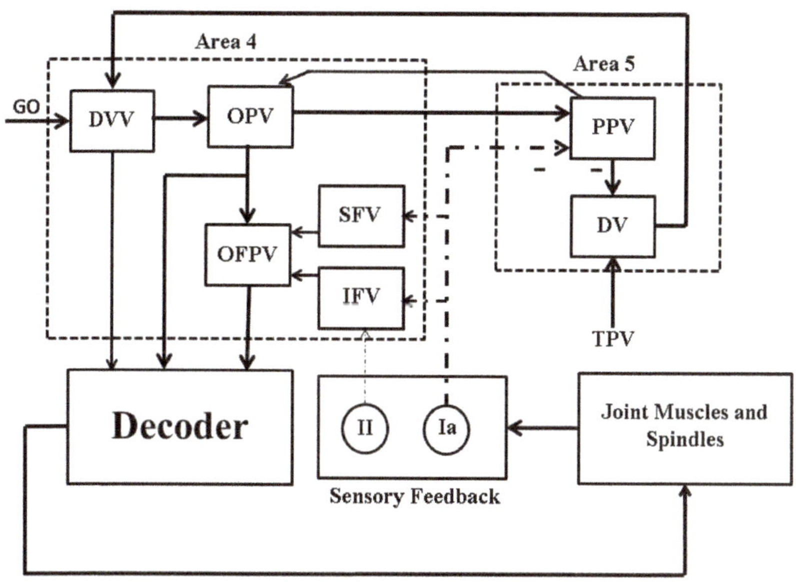

Figure 6. A closed-loop BMI system design using the natural proprioceptive feedback information (sensory feedback).

To make a realistic comparison of the performance of the decoder with the neurophysiological (psycho-physiological) system, we first carried out our investigation by studying the performance of the neurophysiological system shown in Figure 2 in the presence and in the absence of the natural proprioceptive feedback *i.e.*, the sensory feedback. For both cases, we simulated (1)–(13) with $g^0 = 0.75$. The remaining model parameters are the same as given in Section 3 for both cases except $\theta = 0$ and

$\rho = 0$ in the case of no proprioception. This means that the primary ("Ia") and the secondary ("II") muscle spindles afferents become inactive (see (6)) in the absence of proprioception.

Figure 7A shows the position trajectory of the agonist muscle i ($p_i(t)$) in the presence and in the absence of the natural proprioception for the neurophysiological system.

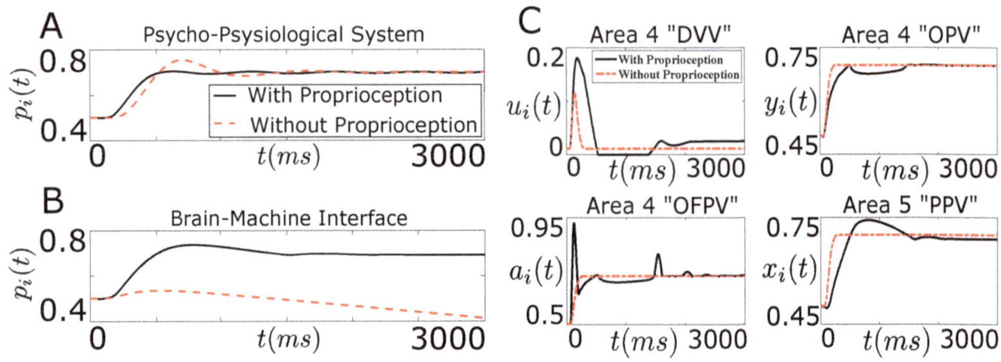

Figure 7. Degradation in the performance of BMI in the absence of proprioception. (**A**,**B**) show the position trajectory of the agonist muscle i ($p_i(t)$) as a function of time (t) in the presence (solid line) and in the absence (dotted line) of the natural proprioceptive feedback information. (**A**) shows the position trajectory for the neurophysiological system ("Psycho-Physiological System") while (**B**) shows the position trajectory for the designed BMI ("Brain-Machine Interface"). The desired position target (T_i) for the agonist muscle i in (**A**,**B**) is 0.7; (**C**) shows the average firing activity of a population of agonist area 4 "DVV" ($u_i(t)$), area 4 "OPV" ($y_i(t)$), area 4 "OFPV" ($a_i(t)$) and area 5 "PPV" ($x_i(t)$) neurons in the presence (solid line) and in the absence (dotted line) of the natural proprioceptive feedback information for the designed BMI.

As shown in Figure 7A, the desired position of the agonist muscle i has been achieved in both cases for the neurophysiological system. The result is consistent with a prior neurophysiological experiment where it was shown that a trained monkey (in the absence of visual feedback) can reach the desired target position in the presence and in the absence of proprioception [31].

Next we studied the performance of the closed-loop BMI (decoder) (in the presence of the natural proprioceptive feedback information) and the open-loop BMI (decoder) (in the absence of the natural proprioceptive feedback information) shown in Figure 6. For this, we simulated (1)–(10), (12) and (14). Here we assumed that the limb dynamics is the same for the neurophysiological system and the BMI. For the open-loop BMI, we again set $\theta = 0$ and $\rho = 0$.

Figure 7B shows the position trajectory of the agonist muscle i ($p_i(t)$) in the presence and in the absence of the natural proprioception for the closed-loop and the open-loop BMI.

It is clear from Figure 7B that the decoder performance degrades substantially when the decoder, trained with the closed-loop data, is applied on the open-loop system. Since the decoder was trained with the closed-loop firing activity of the area 4 "OPV", "DVV" and "OFPV" neurons, the firing activity of these neurons must have changed significantly in the absence of the natural proprioception feedback information. To see this, we plotted the firing activity of these neurons in the presence and in the absence of the natural proprioception feedback information for the BMI design shown in Figure 6. Figure 7C shows the firing activity of the area 4 "OPV", "DVV" and "OFPV" neurons and the area 5 "PPV" neurons in the presence and in the absence of the natural proprioceptive feedback information.

As shown in Figure 7C, the firing activity of cortical neurons deviates significantly from the closed-loop activity in the absence of the natural proprioceptive feedback information. Since the weights of the designed decoder were not adapted to accommodate these significant deviations in the firing activity of the area 4 neurons, the decoder performance degrades substantially in the absence of the natural proprioceptive feedback information. These results clearly show that there is a necessity for designing an artificial proprioceptive feedback to regain the closed-loop performance of the designed decoder in the absence of the natural proprioceptive feedback.

6. Artificial Proprioceptive Feedback Design

As shown in Figure 2, the area 5 "PPV" neurons receive the position feedback information through the primary (Ia) muscle spindles afferents. These neurons then use this proprioceptive feedback information to compute the present position vector command. In the absence of the natural proprioceptive feedback pathways, this feedback information is lost. In order to compensate the lost feedback information of the area 5 "PPV" neurons, we design an artificial sensory feedback in a model predictive control framework. The goal is to recover the closed-loop performance of the decoder (the Wiener filter based decoder with $\Delta M(k)$ as the extracted information) by providing the designed optimal artificial sensory feedback to the "PPV" neurons in the absence of the natural proprioceptive feedback pathways. Note that we are not designing artificial feedback to compensate the loss of sensory feedback to the area 4 "IFV" and "SFV" neurons in this study. Thus in the absence of the natural sensory feedback, these neurons remain inactive during our analysis.

6.1. Model Predictive Control

Model predictive control (MPC) is an optimal control strategy that explicitly incorporates a dynamic model of the system as well as constraints in determining control actions. At each time k, the system measurements are obtained and a model of the system is used to predict future outputs of the system $O_{k+l+1|k}$, $l = 0, 1, 2, \cdots, N_p - 1$ as a function of current and future control moves $I_{k+l|k}$, $l = 0, 1, 2, \cdots, N_c - 1$. How far ahead in the future the predictions are computed is called the prediction horizon N_p and how far ahead the control moves are computed is called the control horizon N_c. Figure 8 illustrates the idea of prediction and control horizon in a model-based receding horizon control strategy.

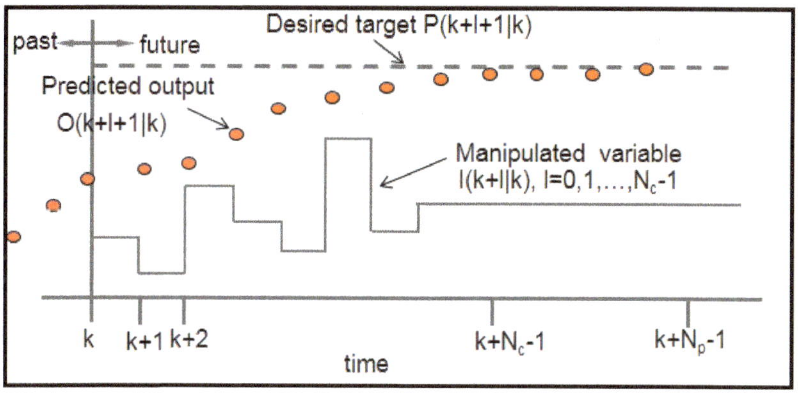

Figure 8. Prediction and controller move optimality in MPC.

Using the predictions from the model, the N_c control moves $I_{k+l|k}$, $l = 0, 1, \cdots, N_c - 1$ are optimally computed by minimizing a cost function $\mathbb{J}_k$ over the prediction horizon N_p subject to constraints on the control inputs as well as any other constraints on the internal states and outputs of the system as follows:

$$\min_{I_{k+l|k}, l=0,1,\cdots,N_c-1} \mathbb{J}_k \tag{19}$$

subjects to constraints on control inputs and the system. A typical quadratic objective cost function may be of the form

$$\mathbb{J}_k = \sum_{l=0}^{N_p-1} \{[O_{k+l+1|k} - P_{k+l+1|k}]^T Q_1 [O_{k+l+1|k} - P_{k+l+1|k}]\} + \sum_{l=0}^{N_c-1} \{I_{k+l|k}^T R_1 I_{k+l|k}\}. \quad (20)$$

Here, $P_{k+l+1|k}$ is the output to be tracked. The matrices Q_1 and R_1 are positive semidefinite and positive definite weighting matrices respectively which determine the relative importance of the tracking objective versus control effort. Only the first optimally computed move $I_{k|k}$ is implemented out of m computed optimal moves at time k. At the next time $k+1$, new system measurements are obtained and the optimization problem is solved again with the new measurements. Thus, the control and prediction horizon recede by one step as time moves ahead by one step. The measurements at each sampling time provide feedback for rejecting inter-sample disturbances, model uncertainty and noises, all of which cause the model predictions to be different from the true system output.

6.2. *Firing Rate Based Closed-Loop BMI Design*

In order to recover the closed-loop (natural) performance of the decoder, we formulate two control problems in this section. In the first problem, we design an optimal artificial sensory feedback to stimulate the population of area 5 "PPV" neurons such that the position trajectory of the agonist muscle i matches the position trajectory obtained in the presence of the natural proprioception during the reaching task. We call this "Problem 1". Although we are not treating the visual feedback in this work, this position trajectory tracking problem can be considered equivalent to the BMI experiments where the visual feedback is used by the subject to make online corrections in the trajectory during the reaching task in the absence of proprioception. In the second problem, the goal is to match the average firing activity of the agonist population of "PPV" neurons to its natural firing activity by designing an optimal artificial feedback to stimulate the agonist population of the area 5 "PPV" neurons. We call this "Problem 2". This firing rate trajectory tracking problem can be considered equivalent to the BMI experiments where the primary somatosensory area (S1) neurons are stimulated artificially to restore the natural proprioception information. Figure 9A,B show the design of a closed-loop BMI operation during the reaching task for problem 1 and problem 2 respectively.

As shown in Figure 9A,B, for both problems we use the MPC strategy (described above) to design the optimal artificial sensory feedback and formulate the following control problem for the systems shown in Figure 9A,B:

$$\min_{I(k|k),I(k+1|k),\cdots,I(k+N_c-1|k)} \mathbb{J}_p(k) \quad (21a)$$

such that

$$I(k+l \mid k) \in [-0.5, 0.5] \;\; for \;\; 0 \leq l \leq N_c - 1, \quad (21b)$$

$$I(k+l \mid k) = 0 \;\; for \;\; N_c \leq l \leq N_p - 1. \quad (21c)$$

Here, $I(k+l \mid k)$ for $l = 0, 1, \cdots, N_c - 1$ is the designed artificial sensory input. $\mathbb{J}_p(k) = \sum_{l=0}^{N_p-1}(O(k+l+1 \mid k) - P(k+l+1 \mid k))^2$ is the cost function. Note that $O(k+l+1 \mid k)$ and $P(k+l+1 \mid k)$ are scalars in our case with $Q_1 = R_1 = 1$ (see Equation (20)). In case of Problem 1 (see Figure 9A), the measured output of the system $O(k \mid k)$ at a given time k is the position of the agonist muscle i, *i.e.*, $p_i(k \mid k)$. $P(\cdot)$ represents the desired position trajectory. In case of Problem 2 (see Figure 9B), the measured output of the system $O(k \mid k)$ at a given time k is the average firing activity of the area 5 "PPV" neurons associated with the agonist muscle i, *i.e.*, $x_i(k \mid k)$. $P(\cdot)$ represents the desired average firing activity of the area 5 "PPV" neurons.

To solve the control problem (21), we first compute the desired position and firing activity trajectory for Problem 1 and 2 respectively. For this, we simulate (1)–(5), (6a), (7), (10), (12) and (14). It should be noted that we have included only the natural sensory feedback (Ia) to the area 5 "PPV" neurons through Equation (6a) for computing the desired trajectory. Thus in the absence of natural proprioception to the area 4 "IFV" and "SFV" neurons, $q_i(t) = f_i(t) = 0$ in (10). Next we compute $p_i(k+m+1 \mid k)$ and $x_i(k+m+1 \mid k)$ for $m = 0, 1, 2, \cdots, N_p - 1$ for Problem 1 and 2 respectively. For this, we use a model of the system given by (1)–(4), (10), (12) and (14) along with the following modified firing activity dynamics of the "PPV" neurons:

$$\begin{aligned}\frac{dx_i(t)}{dt} &= (1 - x_i(t))\max\{\Theta y_i(t) - I(k+l \mid k), 0\} \\ &\quad - x_i(t)\max\{\Theta y_j(t) + I(k+l \mid k), 0\}.\end{aligned} \tag{22}$$

Here, $I(k+l \mid k)$ is constant during t and $t+10$ ms *i.e.*, between the sample time. Note that the average firing activity of the primary ("Ia") muscle spindles afferents is set to 0 in (22) (see (7) for the comparison).

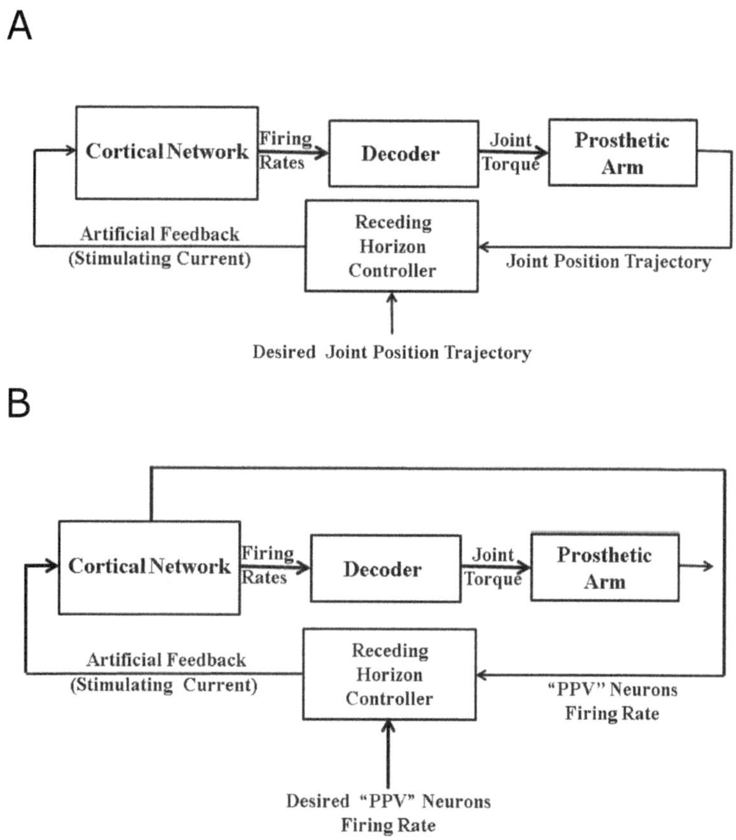

Figure 9. Model Predictive Control (MPC) based closed-loop BMI for Problem 1 and Problem 2. (**A**) shows the design for "Problem 1". Here the model predictive controller designs the "Artificial Feedback" to stimulate "PPV" neurons such that the system output ("Single Joint Position" trajectory) mimics the "Desired Joint Position" trajectory; (**B**) shows the design for "Problem 2". Here the model predictive controller designs the "Artificial Feedback" to stimulate "PPV" neurons such that the system output ("PPV Neurons Firing Rate") mimics the "Desired PPV Neurons Firing Rate".

With this, we solve the optimization problem (21) numerically in MATLAB for both problems. We use the MATLAB optimization function "fmincon" with the sequential quadratic programming "sqp" algorithm. For both problems, we set $N_c = 5$ and $N_p = 30$. Figure 10A shows the performance of the controller in tracking the desired position trajectory of the agonist muscle i for "Problem 1".

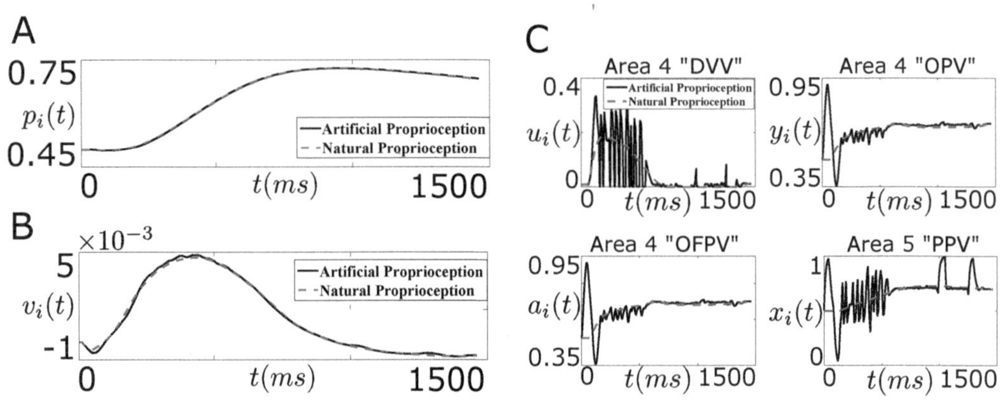

Figure 10. Performance of the designed closed-loop BMI in tracking the desired position trajectory ("Problem 1", Figure 9A). (**A**) shows the position trajectory of the agonist muscle i in the presence of the designed artificial sensory feedback (solid line, blue) and the natural sensory feedback (dotted line, red); (**B**) shows the velocity trajectory of the agonist muscle i in the presence of the designed artificial sensory feedback (solid line, blue) and the natural sensory feedback (dotted line, red); (**C**) shows the average firing activity of a population of the cortical area 4 "DVV" neurons ($u_i(t)$), "OPV" neurons ($y_i(t)$), "OFPV" neurons ($a_i(t)$), and the cortical area 5 "PPV" neurons ($x_i(t)$) in the presence of the artificial sensory feedback (solid line, blue) and the natural sensory feedback (dotted line, red).

As shown in Figure 10A, the controller performs well in tracking the desired position trajectory. Also the stimulation of the area 5 "PPV" neurons by the designed optimal artificial sensory feedback recovers the closed-loop velocity trajectory, as shown in Figure 10B. Thus the designed optimal artificial sensory feedback is sufficient in recovering the closed-loop performance of the decoder in the absence of the natural proprioceptive feedback pathways. Next we wonder if the designed optimal artificial sensory feedback in "Problem 1" also recovers the natural average firing activity of the cortical neurons.

Figure 10C shows the average firing activity of the cortical area 4 and 5 neurons during the reaching task. As shown in this figure, the average firing activity of the cortical area 4 and 5 neurons during the artificial stimulation differs significantly from the natural one. This shows that although the artificial stimulation of the "PPV" neurons through the design of "Problem 1" recovers the closed-loop performance of the decoder, it fails to recover the natural firing activity of the cortical neurons.

Next we study the performance of the designed controller for "Problem 2" (see Figure 9B). Remember that the objective of the controller here is to track the natural average firing activity of the area 5 "PPV" neurons by designing optimal artificial sensory input to the "PPV" neurons. Figure 11A shows the average firing activity of the cortical area 4 and 5 neurons.

As shown in the bottom right plot of Figure 11A, the designed controller performs well in tracking the natural firing activity of the area 5 "PPV" neurons. Moreover, the stimulation results in recovering the natural firing activity of the area 4 cortical neurons.

Figure 11B,C show the position and velocity trajectory of the agonist muscle i during the movement. As shown in these figures, the decoder recovers the closed-loop (natural) performance in this case. This shows that the designed controller in "Problem 2" not only recovers the closed-loop performance of the decoder but also recovers the natural firing activity of the cortical neurons through the optimal artificial stimulation.

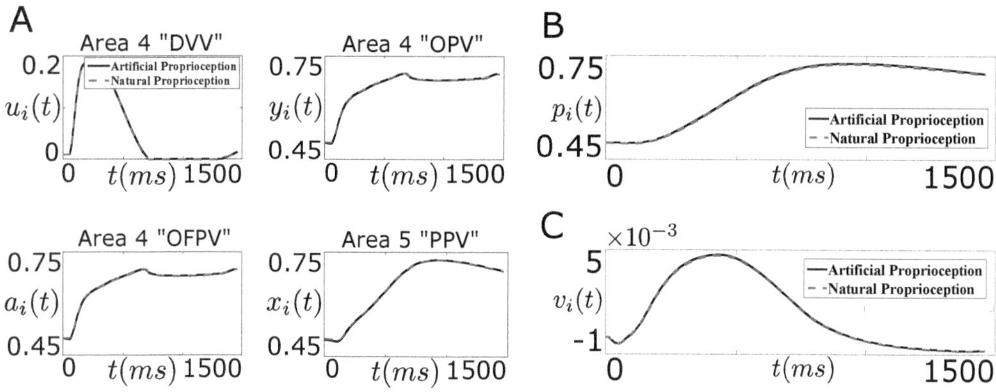

Figure 11. Performance of the designed closed-loop BMI in tracking the desired firing rate trajectory ("Problem 2", Figure 9B). (**A**) shows the average firing activity of the cortical area 4 "DVV" neurons ($u_i(t)$), "OPV" neurons ($y_i(t)$), "OFPV" neurons ($a_i(t)$), and the cortical area 5 "PPV" neurons ($x_i(t)$) in the presence of artificial sensory feedback (solid line, blue) and the natural sensory feedback (dotted line, red); (**B**) shows the position trajectory of the agonist muscle i ($p_i(t)$) in the presence of artificial sensory feedback (solid line, blue) and the natural sensory feedback (dotted line, red); (**C**) shows the velocity trajectory of the agonist muscle i ($v_i(t)$) in the presence of artificial sensory feedback (solid line, blue) and the natural sensory feedback (dotted line, red).

6.3. Intracortical Micro-Stimulation Based Closed-Loop BMI Design

In the previous section, we focused on designing optimal artificial proprioception stimuli in the form of average firing rates. However in experimental BMIs studies, intracortical micro-stimulation (ICMS) based currents in a charge-balanced biphasic waveform are typically used to stimulate cortical sensory neurons and thus to provide artificial sensory feedback during closed-loop operation of BMIs [3,14,32].

Therefore, we modified the MPC-based closed-loop BMI design shown in Figure 9A which allowed us to use such waveform of currents in the present firing activity based framework, as shown in Figure 12.

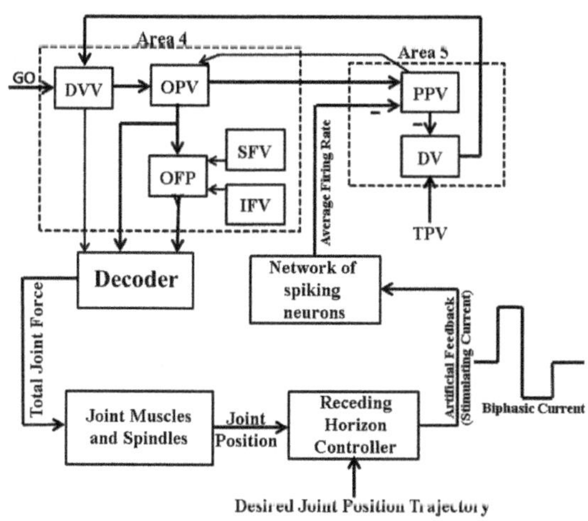

Figure 12. Receding horizon controller based closed-loop BMI design I. Here the receding horizon controller designs the "Artificial Feedback" stimulating current in a charge-balanced biphasic waveform to stimulate "Network of Spiking Neurons" such that the system output ("Single Joint Position" trajectory) mimics the "Desired Joint Position" trajectory.

We modeled the "Network of spiking neurons" in Figure 12 by a recurrent network of the Integrate-and-Fire neurons with dynamics [33]

$$\tau_k \frac{dv_k(t)}{dt} = -v_k(t) + R_s I_k(t) \tag{23a}$$

$$I_k(t) = I_k^s(t) + I^E(t) \tag{23b}$$

$$I_k^s(t) = -g_k(t)(v_k(t) - E) \tag{23c}$$

$$g_k(t) = \sum_{l \neq k=1}^{N_k} \sum_f w_l K(t - t_l^f) \tag{23d}$$

$$K(t - t_l^f) = \frac{q_l}{\tau_l}(t - t_l^f)exp(-\frac{t - t_l^f}{\tau_l})\Theta(t - t_l^f), \tag{23e}$$

where $v_k(t)$ is the membrane potential of neuron k at time t, τ_k is the membrane time constant of neuron k and R_s is the membrane resistance. $I_k(t)$ is the total input current to the neuron k and is the sum of the synaptic current $I_k^s(t)$ and the stimulating current $I^E(t)$. $v_k(t)$ is reset to v_r whenever $v_k(t)$ exceeds a constant firing threshold v_{th} and an action potential is assumed at this time. $g_k(t) \geq 0$ is the excitatory conductance and E is the excitatory membrane reversal potential. N_k is the total number of presynaptic neurons to the neuron k, w_l is the weight of the synapse from the lth neuron to the post-synaptic neuron k, t_l^f is the time of the fth incoming action potential from the lth presynaptic neuron, $K(t - t_l^f)$ is the stereotypical time course of postsynaptic conductance following presynaptic spikes, q_l is the maximum conductance transmitted by the lth neuron, and $\Theta(t - t_l^f)$ is the heavy-side function.

We considered a fully recurrent network of 3 excitatory neurons to represent both the agonist population and the antagonist population of neurons. Network parameters for both populations are provided in Appendix A.1. We computed the average firing rate from the spiking activity of both agonist and antagonist populations as the ratio of the number of spikes in a time window of $T_s = 30$ ms and the total number of neurons in individual population, *i.e.*, 3.

We used a rectangular charge-balanced biphasic waveform shown in Figure 13 to design the stimulating current. Here a_1, a_2, d_1, d_2, d_3 and d_4 characterize a single biphasic pulse of the current $I^E(t)$. The symmetric form of this current minimizes tissue damage, prevents irreversible electrode corrosion and avoids release of toxic metal oxides [34]. This form of current is also the accepted practice in various microstimulation applications such as vestibular prosthesis [35], cochlear implant [36], FES [37,38], DBS [39], retinal prosthesis [40], and BMIs [3,14].

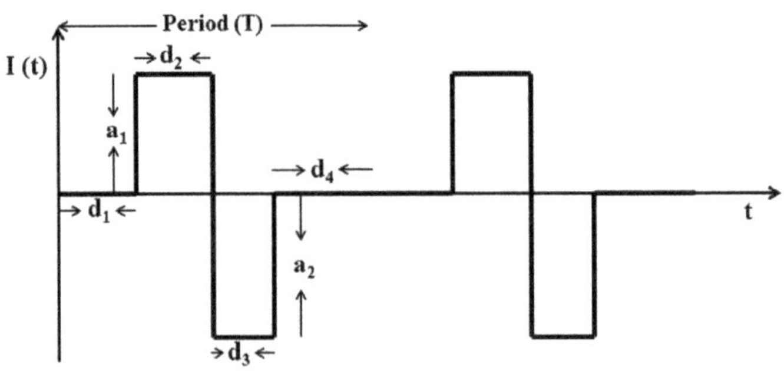

Figure 13. Charge-balanced Intra-cortical Micro-stimulation (ICMS) current in a biphasic waveform. Here the net current $\int_0^T I(t)dt = 0$ for $T = \sum_{i=1}^4 d_i$.

We formulated the following optimal control problem in the MPC framework to recover the closed-loop (natural) performance of the decoder:

$$\min_{\boldsymbol{\beta}(k|k),\boldsymbol{\beta}(k+1|k),\cdots,\boldsymbol{\beta}(k+N_c-1|k)} \mathbb{J}_p(k), \tag{24a}$$

such that

$$a_1(k+l \mid k) \in [0, A_{max}] \;\; for \;\; 0 \leq l \leq N_c - 1, \tag{24b}$$

$$a_2(k+l \mid k) \in [A_{min}, 0] \;\; for \;\; 0 \leq l \leq N_c - 1, \tag{24c}$$

$$d_i(k+l \mid k) \geq 0 \;\; for \;\; i = 1,2,3, \;\; 0 \leq l \leq N_c - 1, \tag{24d}$$

$$n_t(k+l \mid k) \in \{1,2,3,\cdots\} \;\; for \;\; 0 \leq l \leq N_c - 1, \tag{24e}$$

$$n_t(k+l \mid k) \sum_{i=1}^{4} d_i(k+l \mid k) = T_s \;\; for \;\; 0 \leq l \leq N_c - 1, \tag{24f}$$

$$d_2(k+l \mid k) a_1(k+l \mid k) = d_3(k+l \mid k) a_2(k+l \mid k) \;\; for \;\; 0 \leq l \leq N_c - 1. \tag{24g}$$

Here, $\boldsymbol{\beta}(k+l|k) = [a_1(k+l \mid k), a_2(k+l \mid k), d_1(k+l \mid k), d_2(k+l \mid k), d_3(k+l \mid k), n_t(k+l \mid k)]'$ for $l = 0,1,2,\cdots,N_c - 1$. $(\cdot)'$ is the transpose of a vector. $\mathbb{J}_p(k) = \sum_{m=0}^{N_p-1}(O(k+m+1 \mid k) - P(k+m+1 \mid k))^2$ is the cost function. N_c and N_p are the control and the prediction horizon respectively. $a_1(k+l \mid k)$, $a_2(k+l \mid k)$, $d_1(k+l \mid k)$, $d_2(k+l \mid k)$, and $d_3(k+l \mid k)$ for $l = 0,1,\cdots,N_c - 1$ characterize a single biphasic pulse of current $I^E(t)$ shown in Figure 13. T_s is the decoder sample time. $n_t(k+l \mid k)$ for $l = 0,1,\cdots,N_c - 1$ is the number of biphasic pulse in the sample time T_s.

In this work, we set $A_{max} = 10{,}000$, $A_{min} = -10{,}000$, $T_s = 30$ ms, $n_t(k+l \mid k) = 1$, $N_p = 30$, and $N_c = 15$. Using the constraints (24f) and (24g) and $n_t(k+l \mid k) = 1$, our optimization variables reduced to $\boldsymbol{\beta}(k+l|k) = [a_1(k+l \mid k), d_1(k+l \mid k), d_2(k+l \mid k), d_3(k+l \mid k)]'$ for $l = 0,1,2,\cdots,N_c - 1$. We used a modified particle swarm optimization (PSO) algorithm within the parallel computing toolbox of MATLAB to solve the optimization problem (24) (see Appendix A.2 for details). Figure 14A,B show the performance of the designed controller (24) in tracking the desired position and velocity trajectories of the agonist muscle i.

As shown in Figure 14A, the controller performs well in tracking the desired position trajectory (although not perfect). Also the stimulation of the area 5 "PPV" neurons by the designed optimal artificial sensory feedback in the form of biphasic waveform recovers the closed-loop velocity trajectory, as shown in Figure 14B. We note that the controller performance is poor compared to the firing rate design shown in Figure 10.

Figure 14C shows the average firing activity of the cortical area 4 and 5 neurons during the reaching task. As shown in this figure, the average firing activity of the cortical area 4 and 5 neurons during the artificial stimulation differs significantly from the natural as we also observed in Figure 10C. This shows that although the artificial stimulation of the "PPV" neurons through the design of ICMS current (approximately) recovers the closed-loop performance of the decoder, it fails to recover the natural firing activity of the cortical neurons. Figure 14D shows the firing rate trajectory of the agonist and antagonist population of the spiking network.

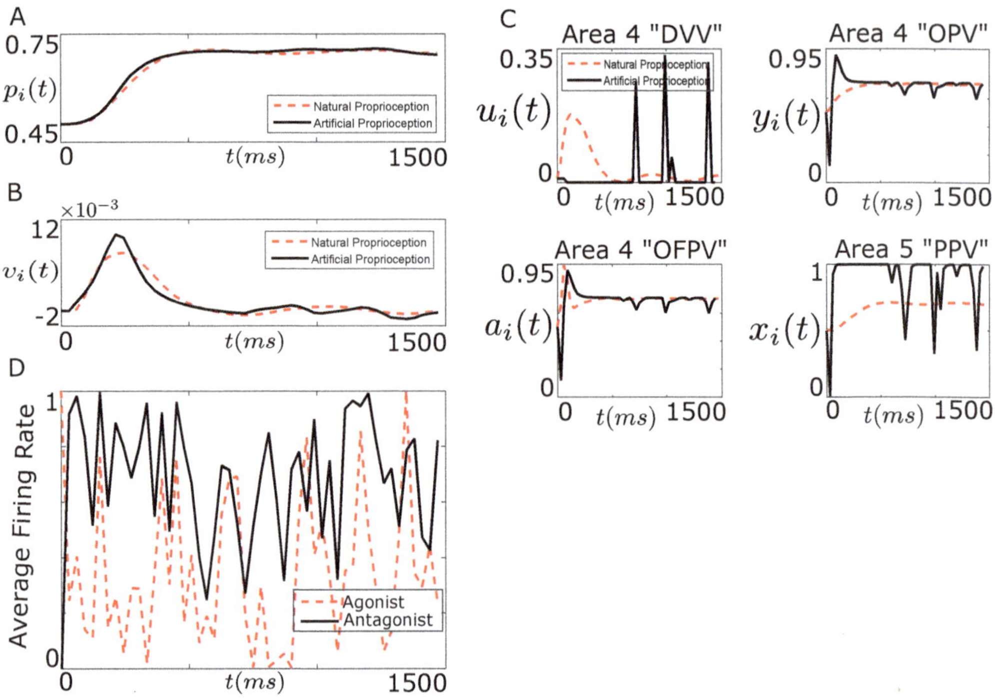

Figure 14. Performance of the designed closed-loop BMI in tracking the desired position trajectory (Figure 12). (**A**) shows the position trajectory of the agonist muscle i in the presence of the designed artificial sensory feedback (solid line, blue) and the natural sensory feedback (dotted line, red); (**B**) shows the velocity trajectory of the agonist muscle i in the presence of the designed artificial sensory feedback (solid line, blue) and the natural sensory feedback (dotted line, red); (**C**) shows the average firing activity of a population of the cortical area 4 "DVV" neurons ($u_i(t)$), "OPV" neurons ($y_i(t)$), "OFPV" neurons ($a_i(t)$), and the cortical area 5 "PPV" neurons ($x_i(t)$) in the presence of the artificial sensory feedback (solid line, blue) and the natural sensory feedback (dotted line, red); (**D**) Firing trajectory of the agonist and the antagonist population of the spiking network.

7. Discussion

7.1. Tracking Firing Rate of Neurons which Encode Sensory Information Recovers the Natural Performance of BMI

In this article, we have designed optimal artificial sensory feedback in a control-theoretic framework to recover the closed-loop performance of a brain-machine interface (BMI) during voluntary single joint extension task. This is the first systematic attempt to incorporate artificial proprioception in BMIs towards stimulation enhanced next generation BMIs. Our approach differs from recent closed-loop BMI system modeling approaches where the focus has mainly been on the analysis of closed-loop BMIs within optimal feedback control framework [17,18]. Two control problems, namely, the position trajectory tracking problem and the cortical sensory neurons average firing rate tracking problem, have been investigated towards designing an optimal artificial sensory feedback for the BMI in the receding horizon control framework. Our position trajectory tracking problem designs the artificial proprioceptive feedback which differs from the natural proprioceptive feedback and can potentially be learned by the BMI user as shown in [3,15]. The cortical sensory neurons average firing rate tracking problem designs the natural proprioceptive feedback through external stimulations [14]. Our results shows that tracking the natural firing activity of the cortical sensory neurons using an external stimulating controller is the appropriate approach towards recovering the natural performance of the motor task. Such an approach requires a complete understanding of how our brain encodes the sensory information. In the absence of a complete understanding of neural encoding, the position trajectory problem provides a more practical way to incorporate artificial sensory feedback in BMIs and thus to close the BMI-loop as also recognized in [15].

7.2. Generalization beyond Tracking Problems

We formulated trajectory tracking control problems to design optimal artificial sensory feedback wherein we assumed that the desired trajectory (position or firing rate) is known. Obtaining such an information a priori may not be feasible in BMI motor tasks. Nevertheless, our proposed framework, based on optimal feedback control, can be generalized to design the necessary sensory feedback and thus close the BMI-loop using the available information such as the desired end-point of the arm effector and various realistic constraints. For example, one can formulate a control problem in the model predictive control framework to minimize the error between the current position and the desired position along with penalizing the derivative of the total force to stabilize the position while ensuring the smoothness of the movement. Such control problems within the optimal control framework have been studied in the context of sensorimotor control [18,41].

7.3. Limitations

In this article, we used an experimentally validated psycho-neurophysiological cortical circuit model of a single joint extension/flexion task to design a BMI. The model provided a basis to investigate the role of proprioceptive feedback in single joint tasks and to design optimal artificial sensory feedback in a BMI using the model predictive control framework. However, this model limited our investigation to a single sensory modality as it does not include visual feedback and learning. For example, we have not investigated how our design will get affected in the presence of visual feedback and any form of cortical learning of a motor task. Since our designed controller is adaptive in nature, the designed artificial sensory feedback will adapt with the learning. It may also be possible that the design controller can help in the initial learning of the BMI task by guiding the movement and thus can reduce the learning effort of the motor task.

Acknowledgments: Financial support from the US National Science Foundation, Cyber Enabled Discovery and Innovation (CDI) program, is gratefully acknowledged.

Author Contributions: Gautam Kumar, Mayuresh V. Kothare, Nitish V. Thakor and Marc H. Schieber formulated the problem. Gautam Kumar, Mayuresh V. Kothare, Hongguang Pan, Baocang Ding and Weimin Zhong developed the solution. Gautam Kumar, Mayuresh V. Kothare, Nitish V. Thakor, Marc H. Schieber, Hongguang Pan, Baocang Ding and Weimin Zhong wrote the paper. All authors have read and approved the final manuscript.

Conflicts of Interest: The authors declare no conflict of interest.

Appendix A

Appendix A.1 Model Parameters for Recurrent Spiking Networks (Equation (23))

We used the following parameters to model the "Network of Spiking Neurons": $N_k = 2$, $\tau_k = 10$ ms, $R_s = 0.04$, $v_r = -65$ mV, $v_{th} = -45$ mV, $E = 0$. For the agonist population, q_l and τ_l for $l = 1,2,3$ were set to $\{16.9610, 17.7975, 16.2787\}$, and $\{5.1071, 7.5474, 7.8020\}$, respectively; the weight matrix w_l was set to $\begin{pmatrix} 0 & 0.9572 & 0.1419 \\ 0.1576 & 0 & 0.4218 \\ 0.9706 & 0.8003 & 0 \end{pmatrix}$. For the antagonist population, q_l and τ_l for $l = 1,2,3$ were set to $\{17.8244, 13.7881, 17.8530\}$ and $\{5.8355, 6.6406, 7.8725\}$, respectively; the weight matrix w_l was set to $\begin{pmatrix} 0 & 0.1419 & 0.7922 \\ 0.4854 & 0 & 0.9595 \\ 0.8083 & 0.9157 & 0 \end{pmatrix}$. It should be noted here that the total number of neurons in the network for each population is $N_k + 1$.

Appendix A.2 Particle Swarm Optimization Algorithm

We solved the optimization problem (24) using a particle swarm optimization (PSO) algorithm [42–44]. Briefly, an individual particle S' ($S' \in \{1, 2, \cdots, S'_n\}$) in the PSO is composed of three vectors:

1. Position in the D_n-dimensional search space (constraints set) $X_{S'} = \{X_{S'1}, X_{S'2}, \cdots, X_{S'D_n}\}$;
2. Best position that it has individually found $P_{S'} = \{P_{S'1}, P_{S'2}, \cdots, P_{S'D_n}\}$;
3. Velocity $V_{S'} = \{V_{S'1}, V_{S'2}, \cdots, V_{S'D_n}\}$.

All the particles are originally initialized in a uniform random manner throughout the search space. The velocity is also initialized randomly. These particles then move throughout the search space and the PSO updates the entire swarm at each iteration k' by updating the velocity and position of each particle in each dimension. We have used the following update rule:

$$V_{S'D}(k'+1) = \omega(k')V_{S'D}(k') + C_p\varepsilon_1(k')(P_{S'D}(k') - X_{S'D}(k')) + C_p\varepsilon_2(k')(P_{gD}(k') - X_{S'D}(k')), \quad (A1)$$

$$X_{S'D}(k'+1) = X_{S'D}(k') + V_{S'D}(k'+1). \quad (A2)$$

Here C_p is a constant. We set $C_p = 2.0$. ε_1 and ε_2 are independent random numbers which are uniquely generated in $(0, 1)$ at every update for each individual dimension $D = 1, 2, \cdots, D_n$. $P_g(k') = \{P_{g1}(k'), P_{g2}(k'), \cdots, P_{gD_n}(k')\}$ is the global best position at iteration k'.

As shown in Figure 13, we have 6 control inputs to be designed $(a_1, a_2, d_1, d_2, d_3, d_4)$. Using constraints (24f) and (24g), we reduce the number of control inputs to be designed to 4. In particular, we design a_1, d_1, d_2 and d_3. In other words, the optimization problem (24) has $4 \times N_c$ independence decision variables. Thus, the particle position $X_{S'D}$ is composed of a_1, d_1, d_2, d_3. The particle fitness function is the same as the cost function $\mathbb{J}_p(k)$ at each sample time. $\omega(k')$ in each step is updated using

$$\omega(k') = 0.2 + k'\frac{1.8 - 0.2}{K'_{\max}}. \quad (A3)$$

Here, $K'_{\max}$ is the maximum number of iteration.

Throughout our work, we consider only one biphasic waveform in each sample time $T_s = 30$ ms. To satisfy the constraints (24e), (24f) and (24g) while updating the velocity and position of each particle through (A1) and (A2), we design some additional constraints which are imposed during the process of particle generation as follows (Algorithm 1):

Algorithm 1

1: if $k' = 1$, randomly generate the particle position $X_{S'} = (a_1, w_1, w_2, w_3)$ for each particle S' satisfying (24b) and (24c); if $k' > 1$, update the particle position a_1, d_1, d_2, d_3 by (A1) and (A2);
2: check a_1, if a_1 satisfies constraint (24b), hold it, if not, let a_1 equal the nearer bound;
3: check d_2. Using (24f) and (24g), obtain $d_3 \geq \frac{a_1 d_2}{A_{max}}$, $d_3 \leq T_s - d_2$ and $d_2 \geq 0$. This implies $d_2 \in [0 \;\; \frac{T_s \times A_{max}}{A_{max} + a_1}]$. Since d_2 is an integer, $d_2 \in [0 \;\; int(\frac{T_s \times A_{max}}{A_{max} + a_1})]$. Here, $int(\cdot)$ denotes rounding up. If w_2 satisfies this constraint, let $d_2 = int(w_2)$; if not, let dw_2 equal the nearer bound;
4: check d_3. $d_3 \in [int(\frac{a_1 d_2}{A_{max}}) \;\; T_s - d_2]$. If d_2 satisfies this constraint, let $d_3 = int(d_3)$; if not, let d_3 equal the nearer bound;
5: check if $d_1 \in [0 \;\; T_s - d_2 - d_3]$. If d_1 satisfies this constraint, hold it, if not, let d_1 equal the nearer bound.

We implement the following Algorithm 2 to solve the optimization problem (24):

Algorithm 2 Offline stage

1: given the parameters of the system model and the network of spiking neurons, compute the desired position trajectory $R(\cdot)$;
2: provide A_{max}, A_{min} and T_s of the biphasic waveform of charge-balanced ICMS currents;
3: set the particle number S'_n and iterations $K'_{\max}$;
4: provide the simulation steps $K_{\max}$.

Online stage, for each $k > 0$:

1: compute the particle position $X_{S'}$ $(S' = 1, 2, \cdots, S'_n)$ such that the constraints (24b)–(24g) are satisfied via Algorithm 1;
2: if $k' = 1$, calculate $O(k \mid k), O(k+1 \mid k), \cdots, O(k + N_p - 1 \mid k)$ and the fitness function $\mathbb{J}_p(k)$ for each particle $X_{S'}$. Update $P_{S'}(k')$ and $P_g(k')$, then go to step 5; else, go to step 3;
3: update $\omega(k')$ using (A3) and the particle position $X_{S'}(k')$ using (A1) and (A2);
4: choose the control and the prediction horizon as N_p and N_c respectively. Calculate $O(k \mid k), O(k + 1 \mid k), \cdots, O(k + N_p - 1 \mid k)$ and the fitness function $\mathbb{J}_p$ for each particle X_S, and update $P'_S(k')$ and $P_g(k')$;
5: if $k' = K'_{\max}$, let the control inputs $u(k)$ equal $P_g(k')$, and go to step 6; else, let $k' = k' + 1$ and go to step 2;
6: if $k > K_{\max}$, go to step 7; else, let $k = k + 1$ and go to step 1;
7: impose control inputs $u(k)$, get the system outputs and terminate the program.

For our simulations, we set $S_n = 96$, $K'_{\max} = 30$, $K_{\max} = 49$. We used $N_p = 6$ and $N_c = 3$ $(k = 1, 2, \cdots, 16)$ till $t = 480$ ms and $N_p = 30$ and $N_c = 15$ $(k = 17, 18, \cdots, 49)$ for $t > 480$ ms.

References

1. Nicolelis, M.A.L.; Lebedev, M.A. Principles of neural ensemble physiology underlying the operation of brain-machine interfaces. *Nat. Rev.* **2009**, *10*, 530–540.
2. Moritz, C.T.; Perlmutter, S.I.; Fetz, E.E. Direct control of paralyzed muscles by cortical neurons. *Nature* **2008**, *486*, 639–643.
3. Doherty, J.E.; Lebedev, M.A.; Ifft, P.J.; Zhuang, K.Z.; Shokur, S.; Bleuler, H.; Nicolelis, M.A.L. Active tactile exploration using a brain-machine-brain interface. *Nature* **2011**, *479*, 228–231.
4. Hochberg, L.R.; Serruya, M.D.; Friehs, G.M.; Mukand, J.A.; Saleh, M.; Caplan, A.H.; Branner, A.; Chen, D.; Penn, R.D.; Donoghue, J.P. Neuronal ensemble control of prosthetic devices by a human with tetraplegia. *Nature* **2006**, *442*, 164–171.
5. Velliste, M.; Perel, S.; Spalding, M.C.; Whitford, A.S.; Schwartz, A.B. Cortical control of a prosthetic arm for self-feeding. *Nature* **2008**, *453*, 1098–1101.
6. Koyoma, S.; Chase, S.M.; Whitford, A.S.; Velliste, M.; Schwartz, A.B.; Kass, R.E. Comparison of brain-computer interface decoding algorithms in open-loop and closed-loop control. *J. Comput. Neurosci.* **2010**, *29*, 73–87.
7. Héliot, R.; Ganguly, K.; Jimenez, J.; Carmena, J.M. Learning in closed-loop brain–machine interfaces: Modeling and experimental validation. *IEEE Trans. Syst. Man Cybern. Part B Cybern.* **2010**, *40*, 1387–1397.
8. Cunnigham, J.P.; Nuyujukian, P.; Gilja, V.; Chestek, C.A.; Ryu, S.I.; Shenoy, K.V. A closed-loop human simulator for investigating the role of feedback-control in Brain Machine Interfaces. *J. Neurophysiol.* **2011**, *105*, 1932 1949.
9. Moritz, C.T.; Fetz, E.E. Volitional control of single cortical neurons in a brain-machine interface. *J. Neural Eng.* **2011**, *8*, 025017.
10. Hochberg, L.R.; Bacher, D.; Jarosiewicz, B.; Masse, N.Y.; Simeral, J.D.; Vogel, J.; Haddadin, S.; Liu, J.; Cash, S.S.; van der Smagt, P.; *et al.* Reach and grasp by people with tetraplegia using a neurally controlled robotic arm. *Nature* **2012**, *485*, 372–375.

11. Ethier, C.; Oby, E.R.; Bauman, M.J.; Miller, L.E. Restoration of grasp following paralysis through brain-controlled stimulation of muscles. *Nature* **2012**, *485*, 368–371.
12. Scherberger, H. Neural control of motor prostheses. *Curr. Opin. Neurobiol.* **2009**, *19*, 629–633.
13. Suminski, A.J.; Tkach, D.C.; Fagg, A.H.; Hatsopoulos, N.G. Incorporating feedback from multiple sensory modalities enhances brain-machine interface control. *J. Neurosci.* **2010**, *30*, 16777–16787.
14. Weber, D.J.; London, B.M.; Hokanson, J.A.; Ayers, C.A.; Gaunt, R.A.; Torres, R.R.; Zaaimi, B.; Miller, L.E. Limb-state information encoded by peripheral and central somatosensory neurons: Implications for an afferent interface. *IEEE Trans. Neural Syst. Rehabil. Eng.* **2011**, *19*, 501–513.
15. Dadarlat, M.C.; O'Doherty, J.E.; Sabes, P.N. A learning-based approach to artificial sensory feedback leads to optimal integration. *Nat. Neurosci.* **2015**, *18*, 138–144.
16. Kumar, G.; Aggarwal, V.; Thakor, N.V.; Schieber, M.H.; Kothare, M.V. An optimal control problem in closed-loop neuroprostheses. In Proceedings of the 2011 50th IEEE Conference on Decision and Control and European Control Conference (CDC-ECC), Orlando, FL, USA, 12–15 December 2011; pp. 53–58.
17. Lagang, M.; Srinivasan, L. Stochastic optimal control as a theory of brain-machine interface operation. *Neural Comput.* **2013**, *25*, 374–417.
18. Benyamini, M.; Zacksenhouse, M. Optimal feedback control successfully explains changes in neural modulations during experiments with brain-machine interfaces. *Front. Syst. Neurosci.* **2015**, *9*, 71, doi:10.3389/fnsys.2015.00071.
19. Kumar, G.; Schieber, M.H.; Thakor, N.V.; Kothare, M.V. Designing closed-loop brain-machine interfaces using optimal receding horizon control. In Proceedings of the 2013 American Control Conference, Washington, DC, USA, 17–19 June 2013; pp. 5029–5034.
20. Pan, H.; Ding, B.; Zhong, W.; Kumar, G.; Kothare, M.V. Designing Closed-loop Brain-Machine Interfaces with Network of Spiking Neurons Using MPC Strategy. In Proceedings of the 2015 American Control Conference, Chicago, IL, USA, 1–3 July 2015; pp. 2543–2548.
21. Bullock, D.; Cisek, P.; Grossberg, S. Cortical Networks for Control of Voluntary Arm Movements under Variable Force Conditions. *Cerebral Cortex* **1998**, *8*, 48–62.
22. Kim, S.P.; Sanchez, J.C.; Rao, Y.N.; Erdogmus, D.; Carmena, J.M.; Lebedev, M.A.; Nicolelis, M.A.L.; Principe, J.C. A comparison of optimal MIMO linear and nonlinear models for brain-machine interfaces. *J. Neural Eng.* **2006**, *3*, 145–161.
23. Wu, W.; Mumford, D.; Gao, Y.; Bienenstock, E.; Donoghue, J.P. Modeling and decoding motor cortical activity using a switching Kalman filter. *IEEE Trans. Biomed. Eng.* **2004**, *51*, 933–942.
24. Wu, W.; Black, M.J.; Mumford, D.; Gao, Y.; Bienenstock, E.; Donoghue, J.P. Bayesian population decoding of motor cortical activity using a Kalman filter. *Neural Comput.* **2006**, *18*, 80–118.
25. Mulliken, G.H.; Musallam, S.; Anderson, R.A. Decoding trajectories from posterior parietal cortex ensembles. *J. Neurosci.* **2008**, *28*, 12913–12926.
26. Li, Z.; O'Doherty, J.E.; Hanson, T.L.; Lebedev, M.A.; Henriquez, C.S.; Nicolelis, M.A.L. Unscented Kalman Filter for Brain-Machine Interfaces. *PLoS ONE* **2009**, *4*, e6243, doi:10.1371/journal.pone.0006243.
27. Dangi, S.; Gowda, S.; Héliot, R.; Carmena, J.M. Adaptive Kalman filtering for closed-loop Brain-Machine Interface systems. In Proceedings of the 2011 5th International IEEE/EMBS Conference on Neural Engineering (NER), Cancun, Mexico, 27 April–1 May 2011; pp. 609–612.
28. Gilja, V.; Nuyujukian, P.; Chestek, C.A.; Cunningham, J.P.; Yu, B.M.; Fan, J.M.; Churchland, M.K.; Kaufman, M.T.; Kao, J.C.; Ryu, S.I.; *et al.* A high-performance neural prosthesis enabled by control algorithm design. *Nature Neurosci.* **2012**, *15*, 1752–1757.
29. Aggarwal, V.; Acharya, S.; Tenore, F.; Shin, H.C.; Cummings, R.E.; Schieber, M.H.; Thakor, N.V. Asynchronous decoding of dexterous finger movements using M1 neurons. *IEEE Trans. Neural Syst. Rehabil. Eng.* **2008**, *16*, 3–14.
30. Sussillo, D.; Nuyujukian, P.; Fan, J.M.; Kao, J.C.; Stavisky, S.D.; Ryu, S.I.; Shenoy, K.V. A recurrent neural networks for closed-loop intracortical brain-machine interface decoders. *J. Neural Eng.* **2012**, *9*, 026027.
31. Bizzi, E.; Accornero, N.; Chapple, W.; Hogan, W. Posture control and trajectory formation during arm movement. *J. Neurosci.* **1984**, *4*, 2738–2744.
32. Hanson, T.L.; Omarsson, B.; O'Doherty, J.E.; Peikon, I.A.; Lebedev, M.A.; Nicolelis, M.A.L. High-side digitally current controlled biphasic bipolar microstimulator. *IEEE Trans. Neural Syst. Rehabil. Eng.* **2012**, *20*, 331–340.

33. Gerstner, W.; Kistler, W.M. *Spiking Neuron Models: Single Neurons, Populations, Plasticity*; Cambridge University Press: Cambridge, UK, 2002.
34. Merrill, D.R.; Bikson, M.; Jeffreys, J.G.R. Electrical stimulation of excitation tissue: Design of efficacious and safe protocols. *J. Neurosci. Methods* **2005**, *141*, 171–198.
35. Davidovics, N.S.; Fridman, G.Y.; Chiang, B.; Santina, C.C.D. Effects of biphasic current pulse frequency, amplitude, duration, and interphase gap on eye movement responses to prosthetic electrical stimulation of the vestibular nerve. *IEEE Trans. Neural Syst. Rehabil. Eng.* **2011**, *19*, 84–94.
36. Cogan, S. Neural stimulation and recording electrodes. *Annu. Rev. Biomed. Eng.* **2008**, *10*, 275–309.
37. Rakos, M.; Freudenschuss, B.; Girsch, W.; Hofer, C.; Kaus, J.; Meiners, T.; Paternostro, T.; Mayr, W. Electromyogram-controlled functional electrical stimulation for treatment of the paralyzed upper extremity. *Artif. Organs* **1999**, *23*, 466–469.
38. Micera, S.; Keller, T.; Lawrence, M.; Morari, M.; Popovic, D.B. Wearable neural prostheses. *IEEE Eng. Med. Biol. Mag.* **2010**, *29*, 64–69.
39. Butson, C.R.; McIntyre, C.C. Differences among implanted pulse generator waveforms cause variations in the neural response to deep brain stimulation. *Clin. Neurophysiol.* **2007**, *118*, 1889–1894.
40. Watanabe, T.; Kikuchi, H.; Fukushima, T.; Tomita, H.; Sugano, E.; Kurino, H.; Tanaka, T.; Tamai, M.; Koyanagi, M. Novel retinal prosthesis system with three dimensionally stacked LSI chip. In Proceedings of the 36th European Solid-State Device Research Conference, Montreux, Switzerland, 19–21 September 2006; pp. 327–330.
41. Todorov, E. Optimality principles in sensorimotor control. *Nat. Neurosci.* **2004**, *7*, 907–915.
42. Bratton, D.; Kennedy, J. Defining a standard for particle swarm optimization. In Proceedings of the Swarm Intelligence Symposium, Honolulu, HI, USA, 1–5 April 2007; pp. 120–127.
43. Wei-Min, Z.; Shao-Jun, L.; Feng, Q. θ-PSO: A new strategy of particle swarm optimization. *J. Zhejiang Univ. Sci. A* **2008**, *9*, 786–790.
44. Shi, Y.; Eberhart, R. A modified particle swarm optimizer. In Proceedings of the 1998 IEEE International Conference on Evolutionary Computation, IEEE World Congress on Computational Intelligence, Anchorage, AK, USA, 4–9 May 1998; pp. 69–73.

3

Droplet-Assisted Laser Direct Nanoscale Writing on Silicon †

Yuan-Jen Chang [1,*,‡], Chao-Hsuan Chang [1,‡], Chao-Ching Ho [2], Jin-Chen Hsu [1] and Chia-Lung Kuo [1]

1 Department of Mechanical Engineering, National Yunlin University of Science and Technology, Douliou, Yunlin 640, Taiwan; u9711012@yuntech.edu.tw (C.-H.C.); hsujc@yuntech.edu.tw (J.-C.H.); kuocl@yuntech.edu.tw (C.-L.K.)

2 Department of Mechanical Engineering, National Taipei University of Technology, Taipei 106, Taiwan; hochao@mail.ntut.edu.tw

* Correspondence: changy@yuntech.edu.tw

† This is an extended version of conference paper published in IEEE-NEMS 2014.

‡ These authors contributed equally to this work.

Academic Editor: Manoj Gupta

Abstract: Nano-structuring using laser direct writing technology has shown great potential for industrial applications. A novel application of water droplets to this technology is proposed in this paper. With a hydrophobic layer and a controlled substrate temperature, a layer of randomly distributed water droplets with a high contact angle is formed on the substrate. These liquid droplets can be used as lenses to enhance the laser intensity at the bottom of the droplets. As a result, nanoscale holes can be fabricated on the substrate by controlling the laser energy density. We successfully fabricated holes with a diameter of 600 nm at a substrate temperature of 12 °C and a power density of 1.2×10^8 W/cm^2 in our experiments. We also found that the hole diameter was around a ninth of the water droplet diameter. Meanwhile, the machined holes are not affected much by the focal length of the lens, but a hole with less than 100 nm in diameter at the center was observed.

Keywords: direct writing; laser; droplet; nanoscale

---

1. Introduction

Nanoscale manufacturing technology has been widely applied in the areas of integrated circuits [1], hard disks [2], flexible electronics [3], and optoelectronics [4,5]. However, the cost of conventional optical nanolithography is extremely high. The next-generation nanofabrication techniques still have disadvantages, such as the low throughput of e-beam lithography, and the reliance of the mold fabrication of nanoimprinting on conventional nanolithography. One laser direct writing nanolithography technology is achieved by near-field optics, and three major methods have been developed recently. Laser nanolithography uses a semiconductor, metallic nanoparticles, or a tip-based microscope to enhance the laser intensity directly beneath the particles or tips [6]; however, they have their own technical barriers. In tip-based near-field laser nanolithography, the tip is easily damaged [7], and the throughput is still low [8]. By applying micro/nanoscale spherical particles, the flexibility of the pattern design is limited and the particles are one-time use only. Although different incident angles of the laser increase the time of use of each micro/nanoparticle, the reliability and the uniformity still need to be improved [9,10]. Another method to apply near-field laser direct writing lithography is optical trapping technology, in which particles are moved around on the surface of the substrates [11,12]. The lifetime of the particles is extended by keeping them a tiny distance from the substrate. However, with optical trapping technology, the parallel control of multiple particles is complicated. A novel application of liquid droplets to enhance the laser intensity was investigated in this study. The droplets could be formed and removed easily. Moreover, the droplets instantly cooled down the ablation spots and helped remove the debris during the process. In this study, our intention was to demonstrate the potential of using liquid droplets in laser direct writing nanofabrication. With the condensation of water on silicon in air, the droplets could be randomly distributed on the silicon substrate; therefore, the machined holes were randomly distributed as well. By carefully controlling the laser power density, the debris around the machined holes could be minimized.

2. Principle

The liquid droplet acts as a lens to refract the laser beam, as shown in Figure 1. With a higher refractive index than air, the laser beam is focused onto the substrate. The curvature of the droplet depends on both the contact angle of the droplet, θ, as well as the diameter of the droplet. With a larger curvature, the radius of the beam focused onto the substrate is smaller. In order to minimize the radius of the beam, a larger droplet contact angle and a smaller volume of the droplet are preferable. When the laser beam is focused onto the substrate, laser ablation occurs if the energy density reaches the threshold ablation energy density of the silicon. Taking advantage of the Gaussian distribution of a laser beam, the threshold energy density can be adjusted to be reached in a small area at the center part of the laser beam. Thereby, a nanoscale silicon hole by laser ablation is obtained.

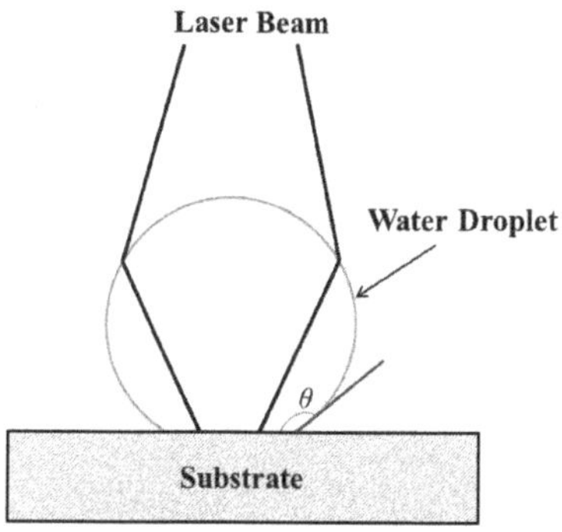

Figure 1. Schematic view of laser direct nano-structuring technology.

3. Materials and Methods

A schematic of the experimental setup is shown in Figure 2. Silicon (Hsin-Chu, Taiwan) coated with a hydrophobic layer was the substrate for patterning. The hydrophobic layer, 1H-1H-2H-2H-Perfluorooctyl-trichlorosilane, was deposited on the silicon substrate by vapor deposition. The substrate was then placed on a thermoelectric cooler (TEC), and the temperature of the silicon was manually controlled with a DC power supply. A pulsed Nd-YAG laser (Lotis TII, Japan) with a wavelength of 532 nm was utilized due to the high absorption by silicon [13] and the low absorption by water [14]. The pulse duration was 6 ns while the output beam size was 6.3 mm in diameter. After being reflected by the first mirror, the laser beam passed through three neutral density filters with different optical density (OD) numbers, 0.2, 0.3, and 1.0, in a series. The laser beam was positively defocused with respect to the substrate, i.e. the focal point was above the substrate and the water droplets. The focal length of the lens was 83.8 mm. The TEC and the silicon substrate were placed on a block of aluminum with a thin layer of thermal paste for heat dissipation. The whole setup was then placed on the stage of a Sodick AP1L Micro Precision Electrical Dischage Machining (EDM) machine (Sodick, Japan). A zoom lens with a charge-coupled device (CCD) was set to take an *in situ* image of droplets while the silicon substrate was cooling. The image was taken first, and the substrate was moved to the designated position under the lens within 6 s afterwards. A picture of the setup is shown in Figure 3.

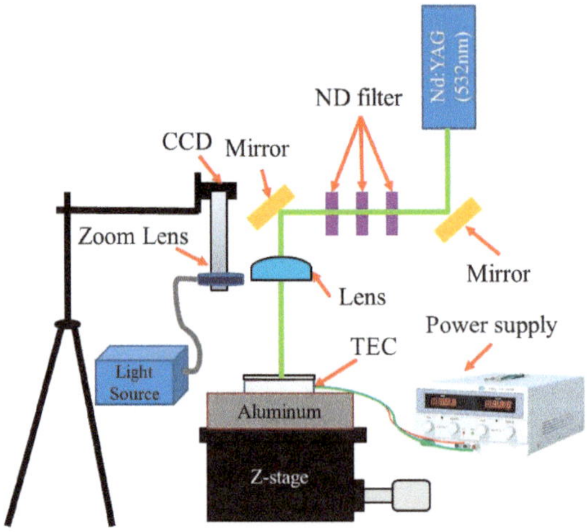

Figure 2. Schematic view of the experimental setup.

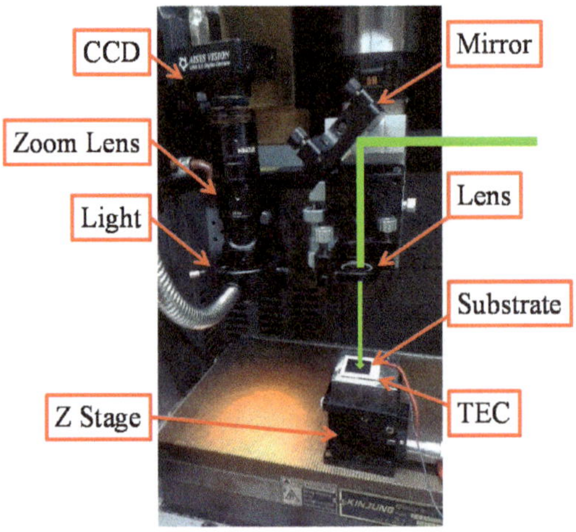

Figure 3. Picture of the experimental setup.

4. Results and Discussion

When the substrate temperature was below 12 °C, the water droplets were randomly condensed on the silicon, while the relative humidity was in the normal range of 55% to 45%. We measured the size of droplets from 1 to 7 min after reaching the designated substrate temperature. The observed results are shown in Figure 4. The size of the droplets increased with time and decreased with temperature. We further reduced the condensation time to 20 s and 30 s at both 10 °C and 12 °C. A picture of the condensed water droplets distributed randomly on the silicon is shown in Figure 5. The diameter in pixels was measured manually for each droplet with the scale bar for obtaining size distribution, as shown in Figure 6. Five different positions on the substrate were measured, including the center and the four corners of the substrate. The droplet diameter distributions are shown in Figure 7. The range of the droplet diameter was 13 μm to 22 μm, depending on the temperature and the condensing time. The variation in diameter at each temperature was around 4 μm. The results also showed that the droplet diameter distribution at 10 °C for 30 s was similar to that at 12 °C for 20 s. The contact angle of the droplet on the hydrophobic surface was measured to be ~126° by a contact angle analyzer (Phoenix 150, SEO, Korea), as shown in Figure 8.

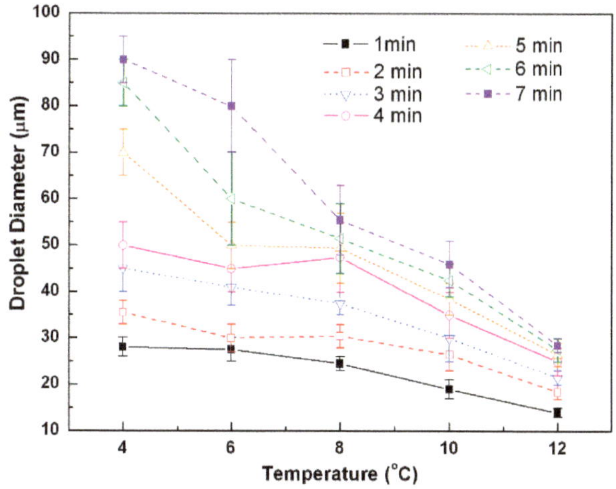

Figure 4. The diameter of droplets on silicon changed with substrate temperature and with the time after reaching the designated temperatures.

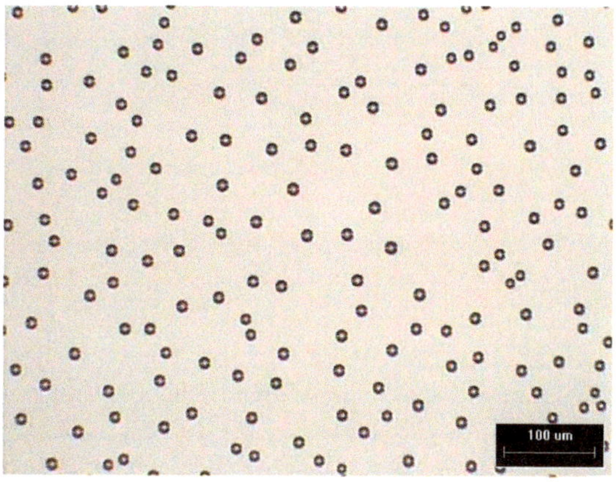

Figure 5. OM picture of condensed water droplet on silicon at 12 °C for 20 s.

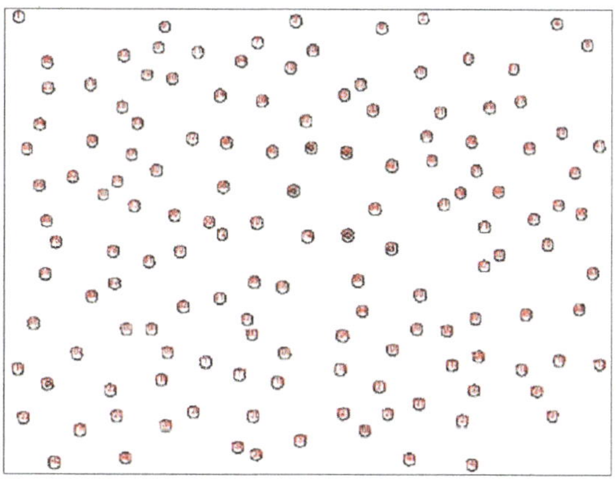

Figure 6. Binary image of condensed water droplet on silicon at 12 °C for 20 s.

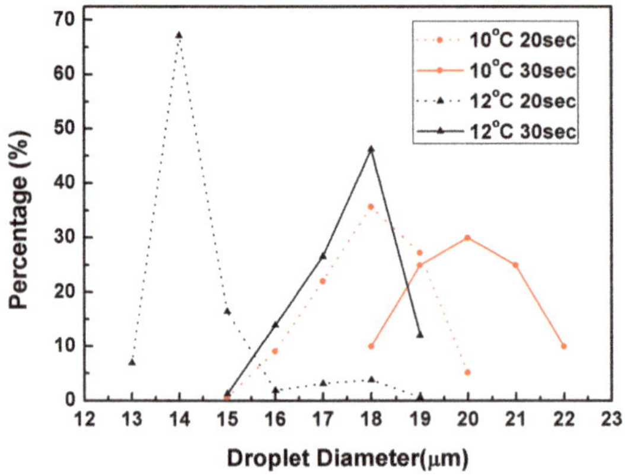

Figure 7. Size distribution of droplets at 10 °C and 12 °C.

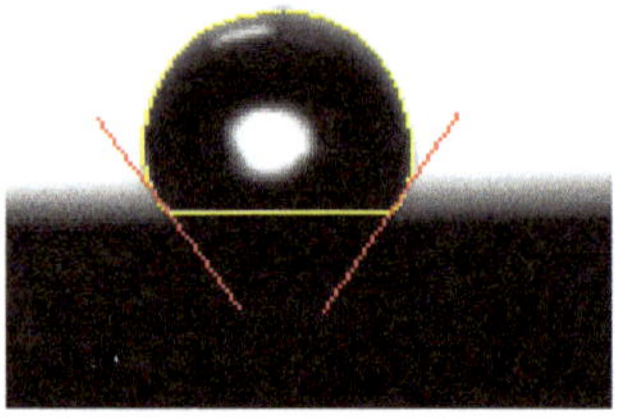

Figure 8. Contact angle measurement result.

The laser radiation power after passing through the neutral density filters was 41 mW, with a pumping energy of 23 Joules. Four different power densities were used in our experiments: 1.9×10^8, 1.6×10^8, 1.4×10^8, and 1.2×10^8 W/cm^2, and were achieved by adjusting the defocusing distance. The scanning electron microscope (SEM) pictures of the machined holes at different power densities with a substrate temperature of 12 °C are shown in Figure 9. From the pictures, we found that the machined area was increased with laser intensity. It can be seen that the holes still exist in the lower left part of the machined area with the lowest power density. This indicates the power density distribution in the laser beam. Higher laser intensity produced the melted silicon sputtered around the hole, while lower laser intensity fabricated circular melting zones, as shown in Figure 9. In Figure 9a, different shapes of holes represent the local power density distribution of the laser beam. With the lowest power density of 1.2×10^8 W/cm^2 in our experiments, it can be seen in an enlarged image of Figure 9d that the hole shapes were all circular with less silicon debris surrounding the holes, as shown in Figure 10. With the local power density slightly over the silicon damage threshold, the minimum hole size could be obtained.

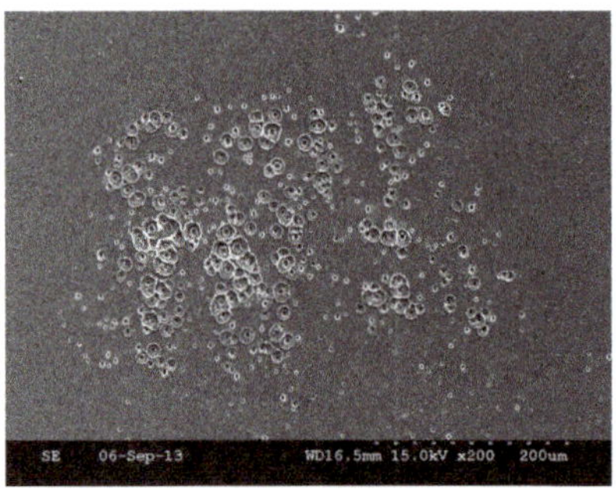

(**a**) 1.9×10^8 W/cm^2

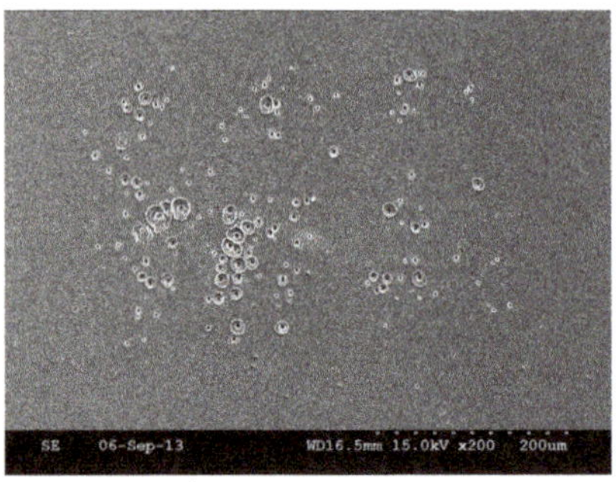

(**b**) 1.6×10^8 W/cm^2

Figure 9. *Cont.*

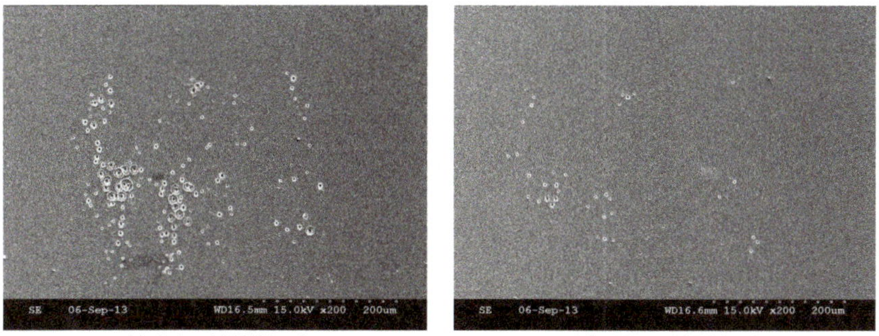

(c) 1.4 × 10⁸ W/cm² (d) 1.2 × 10⁸ W/cm²

Figure 9. SEM pictures of machined holes at different laser power densities (12 °C for 20 s).

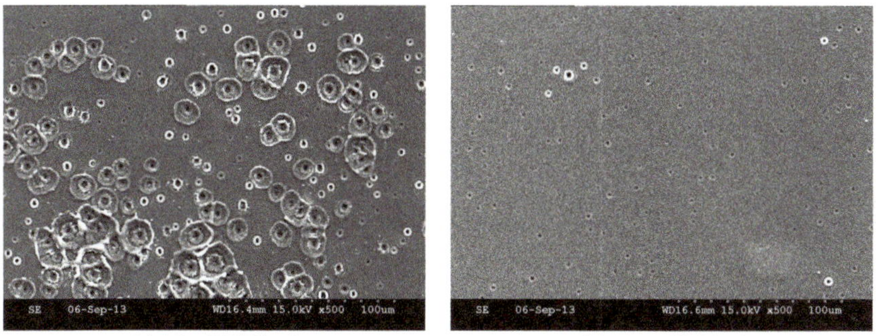

(a) 1.9 × 10⁸ W/cm² (b) 1.2 × 10⁸ W/cm²

Figure 10. Enlarged SEM pictures of machined holes with 12 °C for 20 s.

Since the machined holes were all perfectly circular with a power density of 1.2 × 10⁸ W/cm², we measured the width of the melting area and the diameter of the holes machined at this specific power density. An example of the measured hole is shown in Figure 11. The size distribution of the melting zone and the diameter of the machined holes were measured, and the results are shown in Figures 12 and 13 respectively. The diameter of the melting zone, *i.e.*, the outer diameter of the melting silicon area, was 3 μm to 7.2 μm. The diameter range of the machined hole at the center was 0.6 μm to 4 μm. Lower temperatures and longer condensing times produced larger machined holes. The results also showed that the melting areas of the machined holes were all around 30% of the diameter of the water droplets at the corresponding temperatures and condensation times, as shown in Figure 14. This showed that the power density area in a focused Guassian beam through the liquid lens possibly corresponded to the silicon damage threshold.

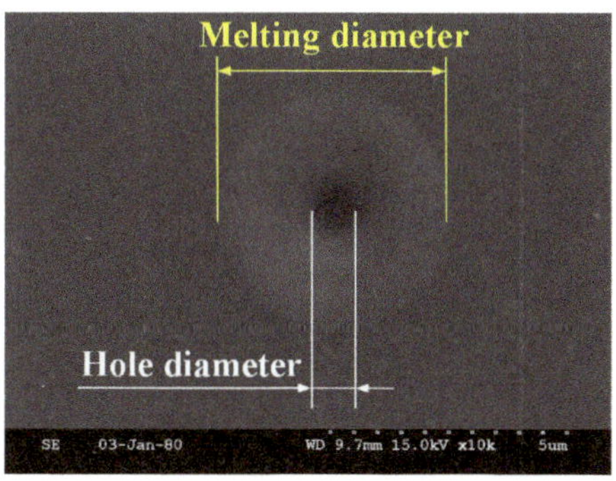

Figure 11. SEM picture of a machined hole with power density of 1.2 × 10⁸ W/cm² at 12 °C for 20 s.

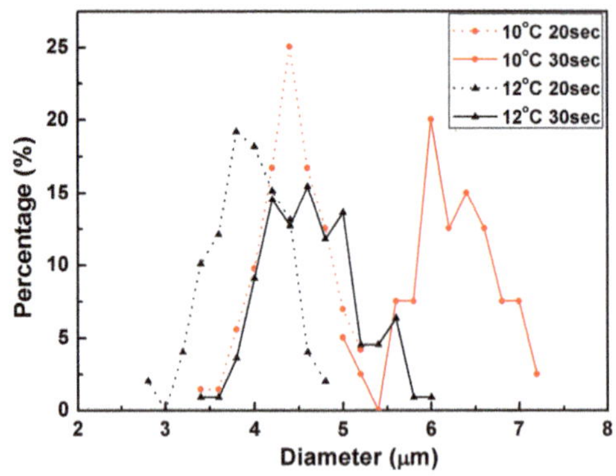

Figure 12. Size distribution of melting width of machined holes at 10 °C and 12 °C with a power density of 1.2×10^8 W/cm^2.

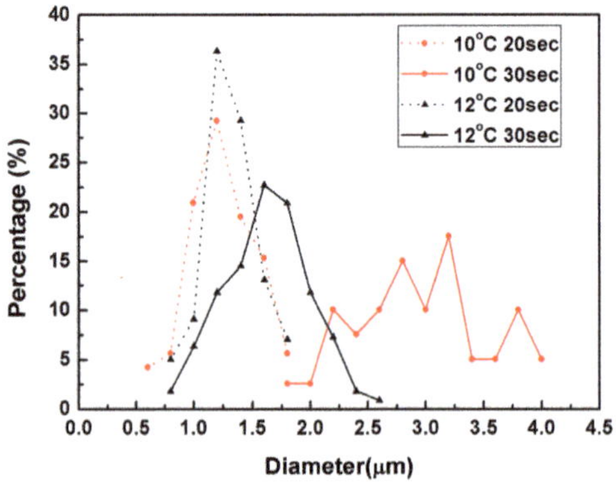

Figure 13. Diameter distribution of machined holes at 10 °C and 12 °C with a power density of 1.2×10^8 W/cm^2.

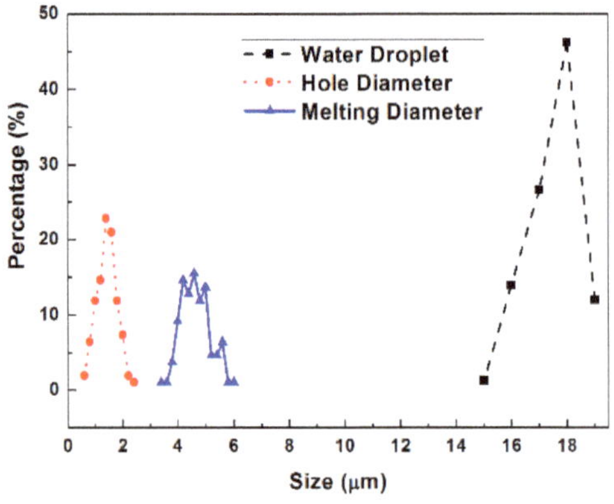

Figure 14. Comparison of droplet size, hole diameter and melting diameter of machined holes at 12 °C condensing for 30 s.

We also compared the hole diameters with water droplet size at different designated temperatures, and the results are shown in Figure 15. The power density was 1.88×10^8 W/cm^2. In these experiments, we fabricated the holes on the substrate after 1 min when the designated temperature of silicon was reached. The temperature ranged from 6 °C to 12 °C. The size of the holes were measured with an

SEM. We observed that the higher the substrate temperature that we applied, the smaller the hole diameter that we obtained. The results also show that the smaller droplet size generates smaller holes. Smaller droplets provide larger curvature and results in smaller spot by refraction in the droplet. We also observed that the hole diameters were around a third of the diameter of droplets.

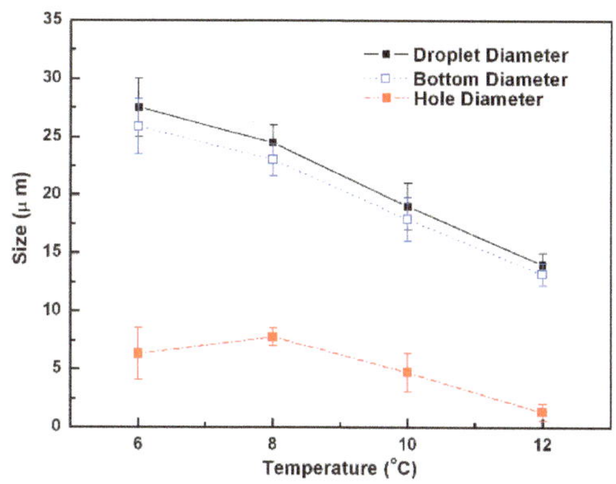

Figure 15. Comparisons among droplet diameters, bottom diameters of droplets, and fabricated hole diameters, with the substrate temperature ranging from 6 °C to 12 °C.

Moreover, we compared the fabricated holes with different focal lengths, including 56 mm, 83.8 mm, and 111.8 mm. The results are shown in Figure 16. The power density used in these experiments was 1.5 × 10^8 W/cm^2. From the SEM pictures, ultra-small holes were found at the center of machined holes. The size of the holes was less than 100 nm. We suggest that it was due to the concentrated energy of a Gaussian laser beam.

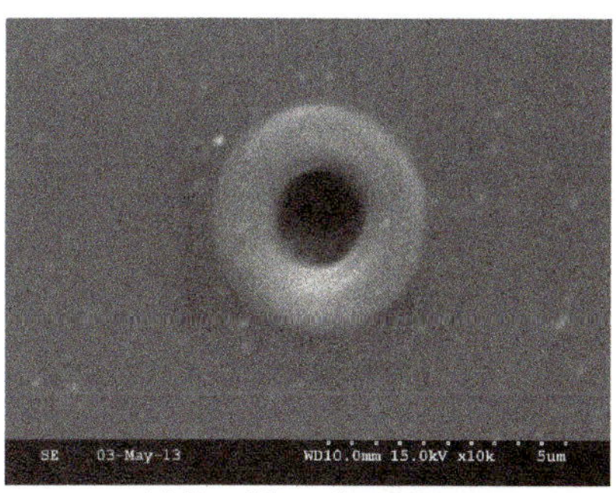

(**a**) Focal length = 56mm

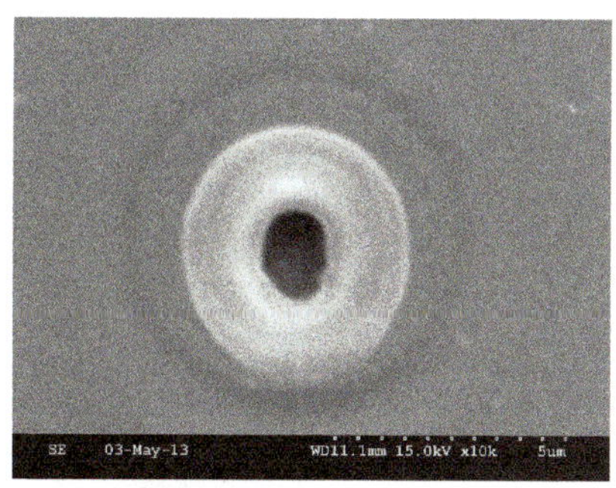

(**b**) Focal length = 83.8 mm

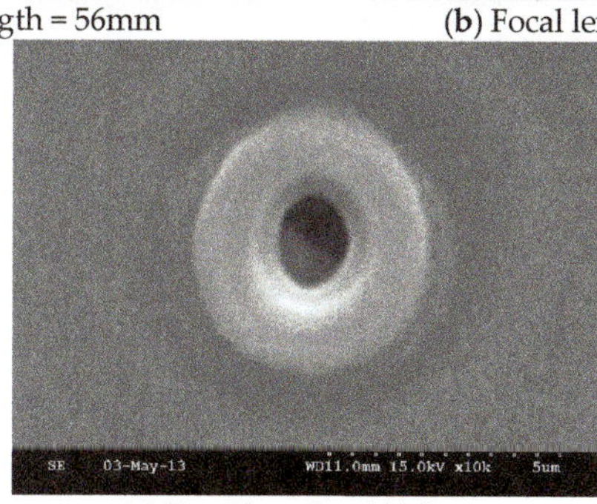

(**c**) Focal length = 111.8 mm

Figure 16. SEM picture of machined holes with different focal length (scale bar = 5 μm).

We reduced the power density to 1.0×10^8 W/cm^2, and the machined holes with different focal lengths are shown in Figure 17. It was clearly found that the ultra-small hole at the center disappeared with the focal lengths of 83.8 mm and 111.8 mm. Only the machined hole with the focal length of 56 mm still has the ultra-small hole at the center. The aspect ratios of machined holes, defined as the ratio of diameter to depth, were measured with these three different lenses and the results are shown in Figure 18. It shows shorter focal length produced holes with higher aspect ratio. Meanwhile, the aspect ratio decreased with the power density. A jump with the focal length of 111.8 mm might be due to the instability of the laser pulse.

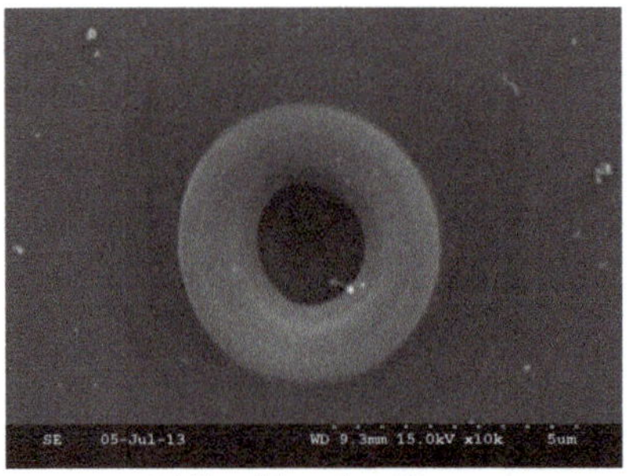

(**a**) Focal length = 56mm

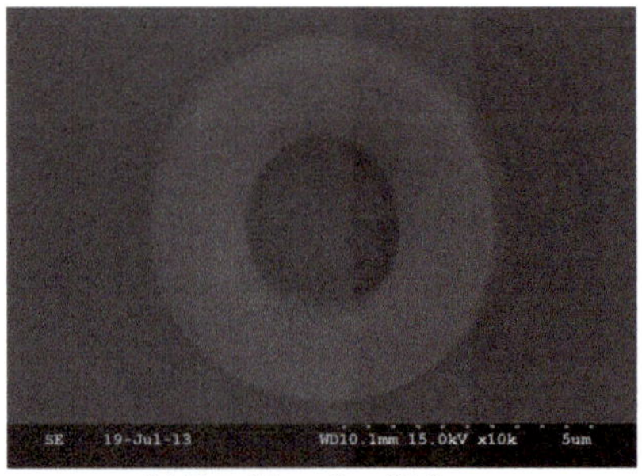

(**b**) Focal length = 83.8 mm

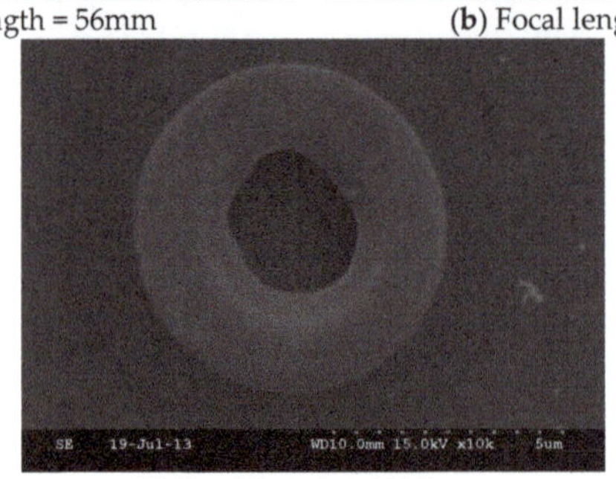

(**c**) Focal length = 111.8 mm

Figure 17. SEM picture of machined holes with a power density of 1.0×10^8 W/cm^2 (Scale bar = 5 μm).

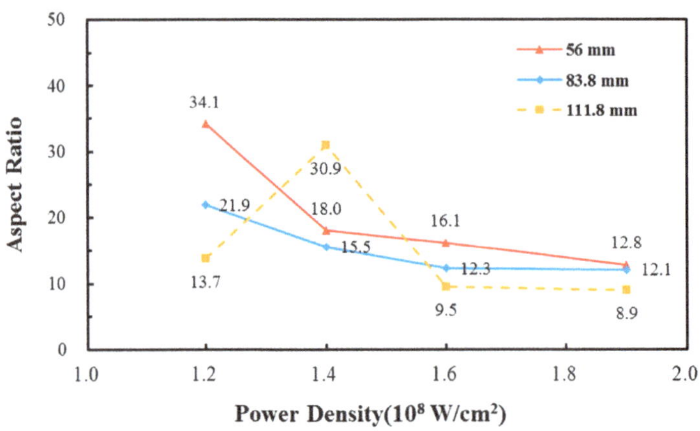

Figure 18. Aspect ratio of holes at different power density with three different focal length.

5. Conclusions

In conclusion, we have demonstrated that water droplets can be used to achieve laser direct writing nanofabrication. With a droplet diameter of ~13 μm, a hole with a diameter of 600 nm was successfully fabricated at a substrate temperature of 12 °C for 20 s. We also found that the hole diameter is around a third of the water droplet diameter. Therefore, smaller droplets produce smaller holes. The machined holes were not affected much by the focal length of the lens, but a hole with less than 100 nm in diameter at the center was observed. The aspect ratio decreased with the power density as well as the focal length of the lens. To reduce the power density further, sub-100 nm holes could possibly be fabricated.

Acknowledgments: The authors would like to thank the National Science Council (NSC102-2221-E-224-015-) for their financial support, and Y.C. Lee's Lab in the Department of Mechanical Engineering at National Cheng Kung University, Taiwan, for their equipment in supporting the experiments herein.

Author Contributions: Yuan-Jen Chang conceived and designed the experiments; Chao-Hsuan Chang performed the experiments; Chao-Ching Ho, Jin-Chen Hsu and Chia-Lung Kuo analyzed the data; Yuan-Jen Chang wrote the paper. All authors have read and approved the final manuscript.

Conflicts of Interest: The authors declare no conflict of interest.

References

1. Forrest, S.R. The path to ubiquitous and low-cost organic electronic applications on plastic. *Nature* **2004**, *428*, 911–918. [CrossRef] [PubMed]
2. Sbiaa, R.; Piramanayagam, S.N. Patterned Media Towards Nano-bit Magnetic Recoarding: Fabrication and Challenges. *Recent Pat. Nanotech.* **2007**, *1*, 29–40. [CrossRef]
3. Jeans, A.; Almanza-Workman, M.; Cobene, R.; Elder, R.; Garcia, R.; Gomez-Pancorbo, F.; Jackson, W.; Jam, M.; Kim, H.-J.; Kwon, O.; *et al.* Advanceds in roll-to-roll imprint lithography for display applications. *Proc. SPIE* **2010**, *7637*. [CrossRef]
4. Madden, S.; Han, T.; Bulla, D.; Davies, B.L. Low loss chalcogenide glass waveguides fabricated by thermal nanoimprint lithography. In Proceedings of Optical Fiber Communication Conference (OFC), San Diego, CA, USA, 11–15 March 2010; pp. 1–3.
5. Liu, A.; Jones, R.; Liao, L.; Samara-Rubio, D.; Rubin, D.; Cohen, O.; Nicolaescu, R.; Paniccia, M. A high-speed silicon optical modulator based on a metal-oxide-semiconductor capacitor. *Nature* **2004**, *427*, 615–618. [CrossRef] [PubMed]
6. Li, L.; Hong, M.; Schmidt, M.; Zhong, M.; Malshe, A.; Huis in'tVeld, B.; Kovalenko, V. Laser nano-manufacturing–State of the art and challenges. *CIRP Ann. Manuf. Technol.* **2011**, *60*, 735–755. [CrossRef]
7. Lu, Y.F.; Hu, B.; Mai, Z.H.; Wang, W.J.; Chim, W.K.; Chong, T.C. Laser-scanning probe microscope based nanoprocessing of electronics materials. *Jpn. J. Appl. Phys.* **2001**, *40*, 4395–4398. [CrossRef]
8. Murphy-DuBay, N.; Wang, L.; Xu, X. Nanolithography using high transmission nanoscale ridge aperture probe. *Appl. Phys. A* **2008**, *93*, 881–884. [CrossRef]
9. Guo, W.; Wang, Z.B.; Li, L.; Whitehead, D.J.; Luk'yanchuk, B.S.; Liu, Z. Near-field laser parallel nanofabrication of arbitrary-shaped patterns. *Appl. Phys. Lett.* **2007**, *90*, 243101. [CrossRef]
10. Li, L.; Guo, W.; Wang, Z.B.; Liu, Z.; Whitehead, D.J.; Luk'yanchuk, B.S. Large-area laser nano-texturing with user-defined patterns. *J. Micromech. Microeng.* **2009**, *19*, 054002. [CrossRef]
11. McLeod, E.; Arnold, C.B. Subwavelength direct-write nanopatterning using optically trapped microspheres. *Nat. Nanotech.* **2008**, *3*, 413–417. [CrossRef] [PubMed]
12. Leitz, K.-H.; Quentin, U.; Alexeev, I.; Schmidt, M. Process investigations of optical trap assisted direct-write microsphere near-field nanostructuring. *CIRP Ann. Manf. Technol.* **2012**, *61*, 207–210. [CrossRef]
13. Santbergen, R.; van Zolingen, R.J.C. The absorption factor of crystalline silicon PV cells: A numerical and experimental study. *Sol. Energy Mater. Sol. Cells* **2008**, *92*, 432–444. [CrossRef]
14. Pope, R.M.; Fry, E.S. Absorption spectrum (380–700 nm) of pure water. II. integrating cavity measurements. *Appl. Opt.* **1997**, *36*, 8710–8723. [CrossRef] [PubMed]

4

Magnetic Resonance Lithography with Nanometer Resolution

Fahad AlGhannam [1,2,†], Philip Hemmer [3,†], Zeyang Liao [1,†] and M. Suhail Zubairy [1,*]

1 Institute for Quantum Science and Engineering (IQSE) and Department of Physics and Astronomy, Texas A&M University, College Station, TX 77843-4242, USA; falghannam@tamu.edu (F.A.); zeyangliao@physics.tamu.edu(Z.L.)
2 The National Center for Applied Physics, KACST, P.O.Box 6086, Riyadh 11442, Saudi Arabia
3 Department of Electrical & Computer Engineering, Texas A&M University, College Station, TX 77843-4242, USA; prhemmer@exchange.tamu.edu
* Correspondence: zubairy@physics.tamu.edu
† These authors contributed equally to this work.

Academic Editors: Ronald E. Meyers and Yanhua Shih

Abstract: We propose an approach for super-resolution optical lithography which is based on the inverse of magnetic resonance imaging (MRI). The technique uses atomic coherence in an ensemble of spin systems whose final state population can be optically detected. In principle, our method is capable of producing arbitrary one and two dimensional high-resolution patterns with high contrast.

Keywords: superresolution; subwavelength lithography; NMR imaging; optically detected magnetic resonance

PACS: 42.50.St; 76.60.Pc; 76.70.Hb

---

1. Introduction

Optical lithography is widely used to print circuit images onto substrates because of its inherent parallelism and the fact that its non-ionizing illumination avoids substrate damage. However, the resolution of traditional optical lithography is restricted by the Rayleigh diffraction limit [1–4]. Due to this limit, we have to use shorter wavelengths such as extreme ultraviolet or X-ray to print smaller patterns [5,6]. However, insulating materials such as silicon dioxide, when exposed to photons with energy greater than the band gap, release free electrons which subsequently can cause adverse charging which can lead to damage [7]. It is therefore interesting and useful to invent an optical lithography scheme that can overcome the diffraction limit.

In the past two decades, a number of methods have been proposed including multiphoton absorption [8–10], quantum entanglement [11–13], dopperlon [14–16], and dark states [17]. However, these methods either require higher order multiphoton absorption, higher order quantum entangled states or multi atomic levels to write patterns with higher order resolution. Thus, improving resolution more than 10 times the optical wavelength with these methods is difficult in practice. In 2010, we proposed a novel scheme to overcome the diffraction limit in optical lithography by inducing Rabi oscillations between two energy levels of the medium [18–20]. The extension from lower resolution to higher resolution is straightforward where we just need to increase the power of the excitation laser or the interaction time. However, effectively printing arbitrary one dimensional (1D) and two dimensional (2D) patterns with low background is difficult in this method. Donut beam lithography [21,22],which is the inverse of the ground state depletion (GSD) super-resolution technique [23,24], can write arbitrary patterns with resolution beyond the diffraction limit, but it requires point-by-point scanning instead of full frame exposure and is very inefficient since most of the laser light is actually used to suppress resist exposure. Using frequency-encoded light masks in neutral atom lithography it is possible to print arbitrary sub-diffraction-limited features [25]. However, in atom lithography with very high atom collimation is required for high resolution and this is difficult to achieve with sufficient throughput to be useful for most lithography applications.

In this paper, we propose an approach based on the MRI super-resolution technique [26]. Our method can be conveniently called magnetic resonance lithography (MRL) [27]. This method can be highly parallel, where the parallelism results from the fact that coherent excitation of two-level spin-flip transitions is employed. In this method, the extension from lower resolution to higher resolution is also straightforward where we just need to apply larger magnetic gradient and narrower bandwidth of microwave. Furthermore, it is also possible to print arbitrary 1D and 2D patterns with low background.

This paper is organized as follows. In Section 2, we discuss the basic physics of MRL and propose the pulse sequence to print arbitrary 1D and 2D patterns. In Section 3, we numerically demonstrate how to print an arbitrary 1D and 2D patterns into the spin states and the photoresist. In Section 4, we discuss some possible materials for MRL. Finally, we summarize the results.

2. Basic Principle and Proposed Experiment

MRL requires a paramagnetic system with high spin transition quality factors and high contrast optical readout of the spin state. In addition, to write an arbitrary 2D pattern, the system should have three magnetic sublevels. The proposed schematic setup is shown in Figure 1a where the paramagnetic system is imbeded near the surface of the substrate. We can create the magnetic gradient fields using conductive metal wires fabricated on the substrate surface by means of a conventional photolithography process. The energy levels of the paramagnetic system is shown in Figure 1b. The energy scale of the lower three magnetic sublevels are in the microwave region (1–100 GHz or 5–500 μV). The energy scale between the ground magnetic sublevels and the higher excited state is in the optical region which is of the order of eV. A static magnetic field B_0 is applied from the perpendicular direction to uniformly split the magnetic sublevels $m = 0$ and $m = \pm 1$ shown in Figure 1b. A microwave pulse with special spectral components is also applied from the perpendicular direction to excite the spin transitions $m = 0 \rightleftharpoons m = \pm 1$. After preparing the required pattern in the spin states, we may use a laser to excite the spins in state $m = 0$ into a fluorescing excited state. A layer of photoresist covers the substrate's surface to capture the energy from the fluorescing excited state and the spin pattern can then be transfered to the photoresist.

To see how MRL can achieve high resolution, let us first consider how to write a narrow line in a spin pattern. If we apply a gradient magnetic field, the Zeeman shift of the $m = +/-1$ sublevels is a function of position. A narrowband microwave π pulse is applied to excite the spins from $m = 0$ to one of the $m = +/-1$ sublevels. Only spins with the proper Zeeman shift are resonant with the microwave and the others are not appreciably excited. This is similar to the "resonant z-slice" in MRI [26]. The absolute resolution limit of this technique is given by [28]

$$\Delta x = \frac{\Gamma}{d\Delta_{ZM}/dx} \tag{1}$$

where Γ is the spin transition linewidth and $d\Delta_{ZM}/dx$ is the spatial gradient of Zeeman shift ($1/T_2$ in [28] is the spin transition linewidth Γ here and γG in [28] is the spatial gradient of the Zeeman energy shift in Equation (1)). The width of the line is not limited by the wavelength but by the spin transition linewidth and the gradient of the magnetic field. In principle, we can achieve arbitrary resolution by increasing the gradient of the magnetic field. For example, if the dephasing time is about 10 μs [29] and the gradient of the magnetic field is 100 Gauss/μm, the resolution can be a few nanometers.

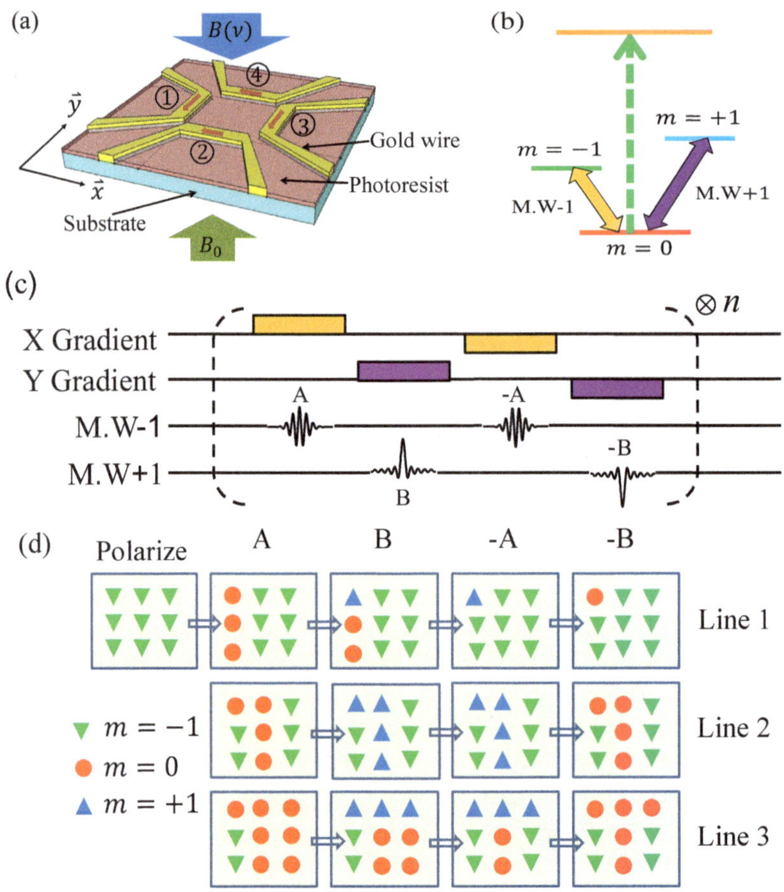

Figure 1. (Color online) (**a**) Schematic setup for magnetic resonance lithography. The spins are embedded innear the surface of the substrate. We first write the patterns into the spins and then transfer the spins pattern to the photoresist via the Forster resonance energy transfer, B_0 is static magnetic field applied to split the magnetic sublevles $m = \pm 1$ while $B(\nu)$ is microwave applied to couple the transition $m = 0$ to $m = \pm 1$; (**b**) Energy levels for magnetic resonance lithography (MRL). The spin sublevels are coupled by two microwave fields. The green laser is applied to excite the spins to the fluorescence excited state (greed dashed line); (**c**) The pulse sequence to write an arbitrary 2D binary pattern; (**d**) An example showing how to write a letter "T" without crosstalk between lines.

We can also address many spatial locations in parallel as in conventional MRI where the narrowband microwave is replaced by a pulse with more complex spectral contents [27]. For example, if we want to print an arbitrary 1D pattern characterized by $f(x)$ whose value is normalized into 0 to 1 range, we should have excitation probability given by [30]

$$f(x) = \frac{1}{2}\{1 - \cos[\Omega_R(x)\tau_p]\} \tag{2}$$

where τ_p is the pulse duration and $\Omega_R(x) = \mu_{ab}\mathcal{E}(x)/\hbar$ is the resonant part of the Rabi frequency seen by atoms at the position x. Here, $\mathcal{E}(x)$ is the amplitude of the microwave frequency component which is resonant to the transition frequency $\omega(x)$ at position x, μ_{ab} is the transition dipole moment between the two relevant energy levels, and $\hbar$ is the Planck's constant. The transition frequency at position x is given by $\omega(x) = \omega_0 + g\mu_B B(x)/\hbar$ where ω_0 is the zero field splitting, $g\mu_B = 2.8$ MHz/Gauss, and

$B(x)$ is the magnetic field at position x. The pulse in the time domain is the Fourier transformation of the required spectrum which in turn is related to position because of the magnetic gradient.

2D patterns can also be printed in this setup where all three energy sublevels are involved and both X and Y magnetic gradients should be applied. The proposed pulse sequence is shown in Figure 1c. For example, we want to write a letter "T" (Figure 1d). First, all the spins are polarized to the state $|-1\rangle$ (green down triangles). This is the initial state. To write column 1, the X-gradient magnetic field is turned on and a narrowband microwave π pulse (A), whose frequency is tuned to be resonant with spins in column 1, rotates spins in this column from $|-1\rangle$ to $|0\rangle$ state. Then the X-gradient is turned off and the Y-gradient is turned on and another microwave pulse (B) with the desired spectrum rotates the spin in the first row from $|0\rangle$ to $|+1\rangle$ (blue up triangles) state. To write another column without crosstalk, we need to turn on the negative X-gradient magnetic field again (and turn off the Y-gradient) and then apply the negative of pulse A to rotate unwanted spins (red circle dots) in column 1 back to the state $|-1\rangle$ (green down triangles). Finally, we again turn on the negative Y-gradient (with X-gradient off) and the negative of pulse B is applied to rotate the spins in state $|+1\rangle$ back to state $|0\rangle$. A 1D pattern is then formed in the column 1 on the $|0\rangle$ state. After that we can move to the next column and repeat the above procedures. As seen in the figure, the desired 2D pattern "T" can then be encoded in the $|0\rangle$ state. In the ideal case, one spin state such as $m = 0$ can be made to fluoresce with excitation laser, while other spin states are dark. An optical readout can then excite the spins in the state $m = 0$ into a fluorescing excited state. If the photoresist is within the near-field region of the spins, the fluorescence exposes the photonresist by Föster resonance energy transfer (FRET) [31,32]. To avoid the direct exposure of the photo-resist by the excitation light, we require that the paramagnetic material has quite different absorption and emission frequency bandwidth. For example the negatively charged Nitrogen Vacancy (NV) color center in diamond can be excited by a green light but emit a red light. We can then choose a photo-resist that is only sensitive at the wavelength of the emission light but not sensitive at the wavelength of the excitation light. Usually, the absorption band of a photoresist is determined by the photoinitiator. With commercially available photoinitiators, it is possible to formulate a photoresist that is only sensitive to a narrowband of the optical spectrum, for example the H-Nu Blue 640 and 660 which is red but not green sensitive [33].

The FRET efficiency is given by $e = [1 + (R/R_0)^6]^{-1}$, where R_0 is the critical FRET radius and R is the distance between the emitter and the receiver. To capture the high resolution features of the patteren, the photoresist needs to be within the FRET radius from the paramagnetic system. This could be achieved in several ways depending on the nature of the paramagnetic system. For example, for paramagnetic defects in crystals, this could be achieved by shallow implantation or engineered growth techniques. A more practical approach, however, would be to coat the substrate surface with a photoresist that contains paramagnetic molecules or nanoparticles with paramagnetic defects. In this way, patterns with higher aspect ratio can be achieved.

3. Numerical Simulation

In the following, we apply a numerical simulation to write a 1D and a 2D pattern onto the $|0\rangle$ state using MRL. The dynamics of the system is govern by the master equation

$$\dot{\rho}(x,y,t) = -\frac{i}{\hbar}[H(x,y,t), \rho(x,y,t)] - \mathcal{L}_\rho \tag{3}$$

where ρ is 3×3 density matrix for the three energy sub-levels, H is the Hamiltonian given by $H(x,y,t) = \hbar\Delta_1(x,t)|-1\rangle\langle -1| + \hbar\Delta_2(x,t)|+1\rangle\langle +1| + \hbar\Omega_1(t)(|-1\rangle\langle 0| + |0\rangle\langle -1|) + \hbar\Omega_2(t)(|+1\rangle\langle 0| + |0\rangle\langle +1|)$ with $\Omega_1(t)$ being the Rabi frequency of the $|-1\rangle \rightarrow |0\rangle$ transition, $\Delta_1(x,t)$ being the corresponding detuning, $\Omega_2(t)$ being the Rabi frequency of the $|0\rangle \rightarrow |+1\rangle$ transition and $\Delta_2(y,t)$ being the corresponding detuning. Here, $\Delta_1(x,t)(\Delta_2(y,t))$ is determined by the X(Y) gradient magnetic field, and $\Omega_1(t)(\Omega_2(t))$ is determined by the microwave pulse A(B). The last term

$\mathcal{L}_\rho = (\rho_{0,\pm1} + \rho_{\pm1,0})/T_2$ is the dephasing of the system. The whole process needs to be done before the population decay time T_1 which can be of the order of milliseconds. Here, we neglect population decay but keep the dephasing process. In the numerical simulation, we apply the Fourth-Order Runge-Kutta method to solve the time evolution of Equation (3) with Rabi frequencies and detunings vary as the pulse sequence shown in Figure 1c.

For the 1D pattern example, we write a chirped grating within a 150 nm field of view using numerical simulation. Here, only an X-gradient and the $|-1\rangle \rightarrow |0\rangle$ transition are involved. The spins are initially prepared in $|-1\rangle$ state. We assume the wire diameter is 1 μm and the distance between the centers of the two wires is 2 μm. The currents in both wires are 50 mA and in the same direction. The currents generate a magnetic gradient of about 200 Gauss/μm in the X-direction. The magnetic field envelope of the microwave excitation pulse for the $|-1\rangle \rightarrow |0\rangle$ transition in the time domain is shown in Figure 2a. The total time duration is about 20 μs. This desired pulse could be generated with a commercial arbitrary waveform generator (AWG). Using an AWG also allows any non-uniformities in magnetic bias or gradient fields to be compensated by applying the appropriate inverse non-linearity to the AWG instruction set. In the numerical simulation, we assume the Rabi frequency has 10% amplitude fluctuation. The paramagnetic material is assumed to be 5 nm deep. The probability that the spins are in the $m = 0$ state is shown in Figure 2b where we can see that the desired pattern is encoded in the $m = 0$ bright state. The pattern of the internal state can then be transfered to the photoresist by the FRET. The FRET efficiency on the substrate surface is shown in Figure 2c where R_0 is assumed to be $5nm$. We can see that a chirped grating can be generated with very high contrast (It is noted that the efficiency presented in Figure 2c can be greater than 1 because it is the summation of the FRET efficiencies of multiple emitters).

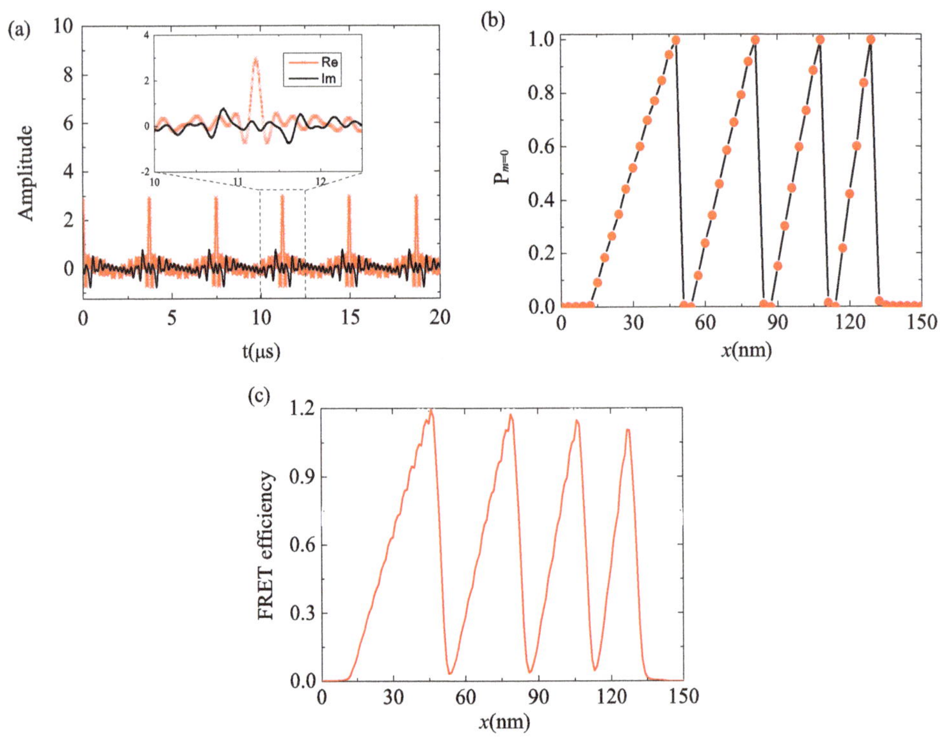

Figure 2. (Color online) (**a**) Complex RF pulse used to write a chirped grating. Red line with cross symbols is the real part while blue line without symbols is the imaginary part; (**b**) The spatial distribution of the probability that the spins are prepared in the bright state; (**c**) The Föster resonance energy transfer (FRET) efficiency on the substrate surface when the readout contrast is 100%.

For a 2D pattern numerical example, we write the Texas A&M University logo ("ATM") into a 120 nm × 300 nm region. This area is divided into 24 slices in the x direction and 60 slices in the y direction. The separation of the spins is chosen to be 5 nm on a square grid. The implantation of square grid of spin defects may be generated by electron holography lithography [34] or nano-sphere lithography [35]. The pulse sequence is shown in Figure 1d where both X-gradient and Y-gradient magnetic fields are involved. In the numerical simulation, the gradients in both directions are generated by 100 mA currents using the same wire separation as in the 1D example. The narrow band microwave π pulse used to drive transitions $|-1\rangle \rightleftharpoons |0\rangle$ is assumed to have a Gaussian time envelope. The time duration of the entire pulse A which selects a line is about 2.7 μs where the Gaussian envelope portion has a width of about 0.75 μs. The time duration of the complex pulse B is also about 2.7 μs which is discretized into 5000 time steps. The total time duration to write one line is about 10.8 μs and it is about 250 μs for the whole process. The dephasing time is assumed to be about 20 μs [36]. Again, the amplitudes of the pulses have a fluctuation of about 10%. The probability that the spins are in the bright state ($m = 0$) is shown in Figure 3a where we can see that the desired pattern is successfully encoded into the internal state of the system. Finally, an optical readout can transfer the pattern of the internal states to the photoresist. Assuming that the paramagnetic spins are about 5 nm deep with 1 nm uncertainty and the FRET radius R_0 is 5 nm, the relative FRET efficiency at the surface is shown in Figure 3b where we see that the desired pattern can be formed. The full width at half maximum of the line in the middle is about 8 nm.

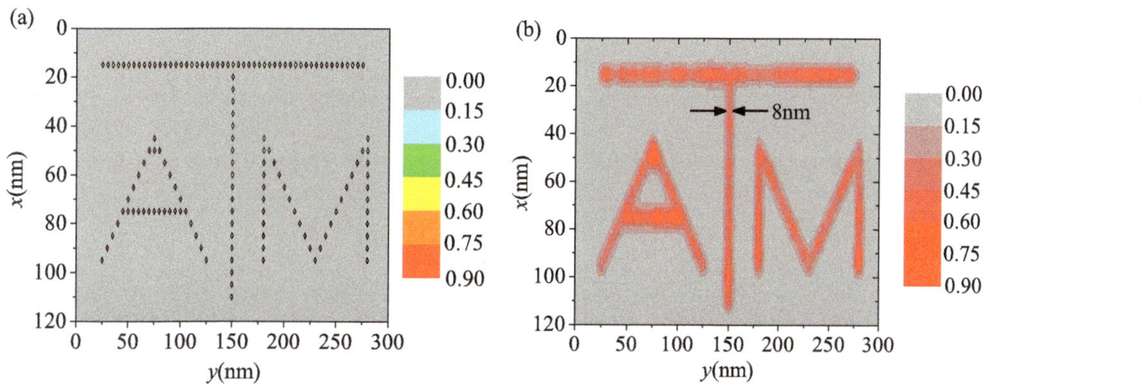

Figure 3. (Color online) Numerical simulation to write the Texas A&M University logo into a two-dimensional area using MRL. (**a**) The spatial distribution of the probability that the spins are prepared in the bright state; (**b**) The FRET efficiency on the substrate surface when the readout contrast is 100%.

4. Possible materials for MRL

As is mentioned in Sec. II, MRL requires a solid system with high spin transition quality factors and high contrast optical readout of the spin state. There are a number of possible systems which can satisfy these requirements. One promising candidate is the negativley charged NV color center in diamond. The NV exhibits an optically detected magnetic resonance (ODMR) spin system with long coherence times, even at room temperature [37–40]. Coherence times larger than a millisecond have been experimentally achieved for NVs leading to their frequent use as nanoscale magnetic field sensors [40]. Such properties could give MRL resolutions as high as a few nanometers. Unfortunately at room temperature the optical readout of the NV spin state has less than 30% contrast. Nonetheless, the inherent nonlinearity of resist exposure may allow us to produce high contrast patterns, even with this relatively low optical contrast.

One alternative candidate is the silicon vacancy (SiV) center in diamond. The SiV has an optical readout contrast near 100% at low temperature since its ground state sublevel transitions can be optically resolved [41]. Unlike NV and many other materials, this high resolution persists even for ensembles, due to the low inhomogeneous broadening and low spectral diffusion of SiV. Granted that working at low temperature might produce some technical difficulties; nonetheless, photoresist exposure has been demonstrated in cryogenic temperatures [42]. Another important advantage of the SiV is that it can have optical Raman transitions connecting the ground state sublevels. Here, light shift gradients could also replace magnetic field gradients [43] so that no wires would be needed on the substrate at all.

Finally, we briefly consider other possible materials systems for performing MRL at room temperature. One of these is the ST1 color center exhibiting a high optical readout contrast up to 50%, albeit with shorter spin lifetimes [44]. Ruby is also of interest as it has shown room temperature optical Raman transitions with readout contrast comparable to NV [45]. Other alternatives are the rare-earth doped materials. For instance, super-resolution microscopy has been demonstrated using praseodymium-doped yttrium aluminum garnet (Pr:YAG) at room temperature [46]. The upconversion UV emission of (Pr:YAG) is optimal for photoresist exposure. Hence, the prospects are good that a material suitable for room temperature MRL will eventually be identified.

5. Summary

In summary, we have proposed an approach to achieve super-resolution in the optical lithography. This technique is based on the inverse of magnetic resonance imaging. The resolution is not limited by the wavelength of the microwave or the light source but depends on the spin transition linewidth and the maximum gradient of the magnetic field. We design a pulse sequence to write arbitrary 1D and 2D patterns into the spin states with very high resolution. The spin patterns may then be optically detected and transferred to the photoresist pattern by the FRET. We presented a numerical simulation to prove the fundamental physics of MRL which shows that our method is capable of producing arbitrary 1D and 2D high-resolution patterns in one of the spin sublevels with high contrast. We may also further improve the performance of our scheme by designing an echo sequence to eliminate the pure dephasing effect and narrow the spin transition linewidth from $1/T_2^*$ to $1/T_2$. Finally, we propose several possible paramagnetic materials for MRL which may be experimentally implemented in the future. The realization of our technique could find important applications in the semiconductor industry.

Acknowledgments: This work is supported by grants from the Qatar National Research Fund (QNRF) under the NPRP project 8-352-1-074 and the King Abdulaziz City for Science and Technology (KACST), DARPA QuASAR, and NSF 1202258. We acknowledge helpful discussions with Charles Santori of HP Laboratories.

Author Contributions: Philip Hemmer and M. Suhail Zubairy conceived the study. Fahad AlGhannam, Philip Hemmer, and Zeyang Liao designed the pulse sequences and performed the numerical simulations. All authors contributed to the basic theory and the manuscript preparation. All authors have read and approved the final manuscript.

Conflicts of Interest: The authors declare no conflict of interest.

Abbreviations

The following abbreviations are used in this manuscript:

MRI: Magnetic resonance imaging
MRL: Magnetic resonance lithography
GSD: Ground state depletion
FRET: Föster resonance energy transfer
AWG: Arbitray waveform generator
NV: Nitrogen Vacancy

ODMR: Optically detected magnetic resonance
SiV: Silicon Vacancy
Pr:YAG: Praseodymium-doped yttrium aluminum garnet

References

1. Rayleigh, L. Investigations in Optics, with Special Reference to the Spectroscope. *Philos. Mag. Ser.* **1880**, *9*, 40–55.
2. Abbe, E. Beitrge zur Theorie des Mikroskops und der Mikroskopischen. *Arch. Mikr. Anat.* **1873**, *9*, 413–418.
3. Al-Amri, M.; Liao, Z.; Zubairy, M.S. Beyond the Rayleigh Limit in optical Lithography. In *Advances in Atomic, Molecular, and Optical Physics*; Arimondo, E., Berman, P.R., Lin, C.C., Eds.; Academic Press: Cambridge, MA, USA, 2012; Volume 61, pp. 409–466.
4. Hemmer, P.R.; Zapata, T. The universal scaling laws that determine the achievable resolution in diffraction schemes for super-resolution imaging. *J. Opt.* **2012**, *14*, 083002.
5. Gwyn, C.W.; Stulen, R.; Sweeney, D.; Attwood, D. Extreme ultraviolet lithography. *J. Vac. Sci. Technol. B* **1998**, *16*, 3142–3149.
6. Heuberger, A. X-ray lithography. *J. Vac. Sci. Technol. B* **1988**, *6*, 107–121.
7. Williams, C.; Kok, P.; Lee, H.; Dowling, J.P. Quantum lithography: A non-computing application of quantum information. *Inform. Forsch. Entw.* **2006**, *21*, 73–82.
8. Yablonovitch, E.; Vrijen, R.B. Optical Projection Lithography at Half the Rayleigh Resolution Limit by Two-photon Exposure. *Opt. Eng.* **1999**, *38*, 334–338.
9. Pe'er, A.; Dayan, B.; Vucelja, M.; Silberberg, Y.; Friesem, A.A. Quantum lithography by coherent control of classical light pulses. *Opt. Express* **2014**, *12*, 6600–6605.
10. Bentley, S.J.; Boyd, R.W. Nonlinear optical lithography with ultra-high sub-Rayleigh resolution. *Opt. Express* **2004**, *12*, 5735–5740.
11. Boto, A.N.; Kok, P.; Abrams, D.S.; Braunstein, S.L.; Williams, C.P.; Dowling, J.P. Quantum interferometric optical lithography: Exploiting entanglement to beat the diffraction limit. *Phys. Rev. Lett.* **2000**, *85*, doi:10.1103/PhysRevLett.85.2733.
12. D'Angelo, M.; Chekhova, M.V.; Shih, Y. Two-Photon Diffraction and Quantum Lithography. *Phys. Rev. Lett.* **2001**, *87*, 013602.
13. Agarwal, G.S.; Boyd, R.W.; Nagasako, E.M.; Bentley, S.J. Quantum interferometric optical lithography: Exploiting entanglement to beat the diffraction limit. *Phys. Rev. Lett.* **2001**, *86*, 1389.
14. Hemmer, P.R.; Muthukrishnan, A.; Scully, M.O.; Zubairy, M.S. Quantum lithography with classical light. *Phys. Rev. Lett.* **2006** , *96*, 163603.
15. Sun, Q.; Hemmer, P.R.; Zubairy, M.S. Quantum lithography with classical light. Generation of arbitrary patterns. *Phys. Rev. A* **2007**, *75*, 065803.
16. Ge, W.; Hemmer, P.R.; Zubairy, M.S. Quantum lithography with classical light. *Phys. Rev. A* **2013** , *87*, 023818.
17. Kiffner, M.; Evers, J.; Zubairy, M.S. Resonant Interferometric Lithography Beyond the Diffraction Limit. *Phys. Rev. Lett.* **2008**, *100*, 073602.
18. Liao, Z.; Al-Amri, M.; Zubairy, M.S. Quantum Lithography beyond the Diffraction Limit via Rabi Oscillations. *Phys. Rev. Lett.* **2010**, *105*, 183601.
19. Liao, Z.; Al-Amri, M.; Becker, T.; Schleich, W.P.; Scully, M.O.; Zubairy, M.S. Atomic Lithography with Subwavelength Resolution via Rabi Oscillations. *Phys. Rev. A* **2013**, *87*, 023405.
20. Liao, Z.; Al-Amri, M.; Zubairy, M.S. Coherent Atom Lithography with Nanometer Spacing. *Phys. Rev. A.* **2013**, *88*, 053809.
21. Brimhall, N.; Andrew, T.L.; Manthena, R.V.; Menon, R. Breaking the Far-Field Diffraction Limit in Optical Nanopatterning via Repeated Photochemical and Electrochemical Transitions in Photochromic Molecules. *Phys. Rev. Lett.* **2011**, *107*, 205501.
22. Gan, Z.; Cao, Y.; Evans, R.A.; Gu, M. Three-dimensional deep sub-diffraction optical beam lithography with 9 nm feature size. *Nat. Commun.* **2013**, *4*, doi:10.1038/ncomms3061.
23. Hell, S.W.; Wichmann, J. Breaking the diffraction resolution limit by stimulated emission: Stimulated-emission-depletion fluorescence microscopy. *Opt. Lett.* **1994**, *19*, 780–782.

24. Hell, S.W.; Kroug, M. Ground-state-depletion fluorscence microscopy: A concept for breaking the diffraction resolution limit. *Appl. Phys. B Lasers Optics* **1995**, *60*, 495–497.
25. Thywissen, J.H.; Prentiss, M. Demonstration of frequency encoding in neutral atom lithography. *New J. Phys.* **2005**, *7*, doi:10.1088/1367-2630/7/1/047.
26. Haacke, E.M.; Brown, R.W.; Thompson, M.R.; Venkatesan, R. *Magnetic Resonance Imaging: Physical Principles and Sequence Design*; John Wiley and Sons: New York, NY, USA, 1999.
27. Hemmer, P.; AlGhannam, F.; Zubairy, M.S. Toward sub-wavelength lithography with atomic coherence. *Proc. SPIE* **2013**, *8875*, doi:10.1117/12.2025180.
28. McFarland, E.W. Time-independent point-spread function for NMR microscopy. *Magn. Reson. Imaging* **1992**, *10*, 269–278.
29. Ofori-Okai, B.K.; Pezzagna, S.; Chang, K.; Loretz, M.; Schirhagl, R.; Tao, Y.; Moores, B.A.; Groot-Berning, K.; Meijer, J.; Degen, C.L. Spin properties of very shallow nitrogen vacancy defects in diamond. *Phys. Rev. B* **2012** , *86*, 081406.
30. Scully, M.O.; Zubairy, M.S. *Quantum Optics*; Cambridge University Press: Cambridge, UK, 1997.
31. Clegg, R.M. *Laboratory Techniques in Biochemistry and Molecular Biology*; Gadella, T.W.J., Ed.; Elsevier: Amsterdam, The Netherlands, 2009; Volume 33, pp. 1–57.
32. Tisler, J.; Reuter, R.; Lämmle, A.; Jelezko, F.; Balasubramanian, G.; Hemmer, P.R.; Reinhard, F.; Wrachtrup, J. Highly Efficient FRET from a Single Nitrogen-Vacancy Center in Nanodiamonds to a Single Organic Molecule. *ACS Nano* **2011**, *5*, 7893–7898.
33. Spectra Group Limited Inc. H-Nu Blue 640, 660 Visible Light Photoinitiators. Available online: http://www.sglinc.com/Portals/0/Images/limited/PDFS/photoinitiators/H-Nu20Blue-IR20TDS.pdf (accessed on 20 April 2012).
34. Dunin-Borkowski, R.E.; Kasama, T.; Wei, A.; Tripp, S.L.; H′ytch, M.J.; Snoeck, E.; Harrison, R.J.; Putnis, A. Off-axis electron holography of magnetic nanowires and chains, rings, and planar arrays of magnetic nanoparticles. *Microsc. Res. Tech.* **2004**, *64*, 390–402.
35. Cheung, C.L.; Nikolić, R.J.; Reinhardt, C.E.; Wang, T.F. Fabrication of nanopillars by nanosphere lithography. *Nanotechnology* **2006**, *17*, doi:10.1088/0957-4484/17/5/028.
36. Ohno, K.; Heremans, F.J.; Bassett, L.C.; Myers, B.A.; Toyli, D.M.; Bleszynski Jayich, A.C.; Palmstrøm, C.J.; Awschalom, D.D. Engineering shallow spins in diamond with nitrogen delta-doping. *Appl. Phys. Lett.* **2012**, *101*, 082413.
37. Gaebel, T.; Domhan, M.; Popa, I.; Wittmann, C.; Neumann, P.; Jelezko, F.; Rabeau, J.R.; Stavrias, N.; Greentree, A.D.; Prawer, S.; *et al.* Room-temperature coherent coupling of single spins in diamond. *Nat. Phys.* **2006**, *2*, 408–413.
38. Childress, L.; Gurudev Dutt, M.V.; Taylor, J.M.; Zibrov, A.S.; Jelezko, F.; Wrachtrup, J.; Hemmer, P.R.; Lukin, M.D. Coherent Dynamics of Coupled Electron and Nuclear Spin Qubits in Diamond. *Science* **2006**, *314*, 281–285.
39. Wrachtrup, J.; Jelezko, F. Processing quantum information in diamond. *J. Phys. Condens. Matter* **2006**, *18*, doi:10.1088/0953-8984/18/21/S08.
40. Balasubramanian, G.; Neumann, P.; Twitchen, D.; Markham, M.; Kolesov, R.; Mizuochi, N.; Isoya, J.; Achard, J.; Beck, J.; Tissler, J.; *et al.* Ultralong spin coherence time in isotopically engineered diamond. *Nat. Matter* **2009**, *8*, 383–387.
41. Müller, T.; Hepp, C.; Pingault, B.; Neu, E.; Gsell, S.; Schreck, M.; Sternschulte, H.; Steinmller-Nethl, D.; Becher, C.; Atatüre, M. Optical signatures of silicon-vacancy spins in diamond. *Nat. Commun.* **2014**, *5*, doi:10.1038/ncomms4328.
42. Lee, K.H.; Green, A.M.; Taylor, R.A.; Sharp, D.N.; Turberfield, A.J.; Brossard, F.S.F.; Williams, D.A.; Briggs, G.A.D. Cryogenic two-photon laser photolithography with SU-8. *Appl. Phys. Lett.* **2006**, *88*, 143123.
43. Gardner, J.R.; Marable, M.L.; Welch, G.R.; Thomas, J.E. Suboptical wavelength position measurement of moving atoms using optical fields. *Phys. Rev. Lett.* **1993**, *70*, doi:10.1103/PhysRevLett.70.3404.
44. Lee, S.-Y.; Widmann, M.; Rendler, T.; Doherty, M.W.; Babinec, T.M.; Yang, S.; Eyer, M.; Siyushev, P.; Hausmann, B.J.M.; Loncar, M.; *et al.* Readout and control of a single nuclear spin with a metastable electron spin ancilla. *Nat. Nanotechnol.* **2013**, *8*, 487–492.

45. Kolesov, R. Coherent population trapping in a crystalline solid at room temperature. *Phys. Rev. A* **2005**, *72*, 051801.
46. Kolesov, R.; Reuter, R.; Xia, K.; Stöhr, R.; Zappe, A.; Wrachtrup, J. Super-resolution upconversion microscopy of praseodymium-doped yttrium aluminum garnet nanoparticles. *Phys. Rev. B* **2011**, *84*, 153413.

5

Membrane-Cryogenic Post-Combustion Carbon Capture of Flue Gases from NGCC

Colin A. Scholes [1,*], Minh T. Ho [2] and Dianne E. Wiley [2]

1 Department of Chemical and Biomolecular Engineering, The University of Melbourne, Melbourne 3010, Australia
2 School of Chemical Engineering, University of New South Wales, Kensington 2052, Australia; minh.ho@unsw.edu.au (M.T.H.); d.wiley@unsw.edu.au (D.E.W.)
* Correspondence: cascho@unimelb.edu.au

Academic Editor: Gustavo A. Fimbres Weihs

Abstract: Membrane gas separation for carbon capture has traditionally been focused on high pressure applications, such as pre-combustion capture and natural gas sweetening. Recently a membrane-cryogenic combined process has been shown to be cost competitive for post-combustion capture from coal fired power stations. Here, the membrane-cryogenic combined process is investigated for application to post-combustion carbon capture from the flue gas of a Natural Gas Combined Cycle (NGCC) process. This process involves a three-membrane process, where the combustion air is used as the sweep gas on the second membrane stage to recycle CO_2 through the turbine. This ensures high CO_2 recovery and also increases the CO_2 partial pressure in the flue gas. The three-CO_2-selective membrane process with liquefaction and O_2-enrichment was found to have a cost of capture higher than the corresponding process for coal post-combustion capture. This was attributed to the large size and energy duty of the gas handling equipment, especially the feed blower, because of the high gas throughput in the system caused by significant CO_2 recycling. In addition, the economics were uncompetitive compared to a modelled solvent absorption processes for NGCC.

Keywords: membrane gas separation; cryogenic; NGCC; post-combustion

---

1. Introduction

Recent design developments in membrane gas separation are moving this technology towards being economically competitive for post-combustion carbon capture from coal-fired power stations [1,2]. Here these competitive membrane process designs are extended to post-combustion capture from a Natural Gas Combined Cycle (NGCC) process, to evaluate the viability of membrane technology.

Traditionally, membrane-based carbon capture has only been considered for natural gas sweetening before NGCC [3] and for pre-combustion capture before the IGCC process [4], while post-combustion capture has only been studied in depth using solvent absorption [5–9], or as a novel Integrated Gasification Combined Cycle (IGCC) process with flue gas recycling [10]. This is because for NGCC the flue gas has a low CO_2 partial pressure which has previously been seen as a limitation for membrane gas separation. Belaissaoui *et al.* [11] recently proposed overcoming this for membrane systems through flue gas recycle, where some of the CO_2 produced is recycled through the gas turbine resulting in a higher CO_2 partial pressure. Similar flue gas recycle has been proposed for solvent absorption technology for NGCC [12]. For coal post-combustion, flue gas recycling has been shown by Merkel *et al.* [2] to make membrane separation viable for post-combustion capture from a coal-fired power station. In that case, the process consists of three CO_2-selective membrane stages with a cryogenic liquefaction process for CO_2 separation. Sweep gas on the second membrane stage is designed to recycle CO_2 through the coal burner, and importantly enable the entire membrane process to operate at low pressure. This differs from Belaissaoui *et al.* process, which requires compression before the membrane stage. The three stages membrane process is able to recover over 90% of the CO_2 and achieves a product purity of over 95% CO_2 at a cost of capture comparable with current state-of-art solvent absorption technology. Hence, the three membrane stage combination presents itself as the current state of the art membrane process for post-combustion capture. As such, the three membrane stages design is applied here to post-combustion carbon capture from a NGCC process. This differs from Belaissaoui *et al.* design, in that the process presented here operates at low pressures.

The capture design utilises two CO_2-selective membrane stages in series to concentrate CO_2 in the permeate stream of the first membrane stage, and uses the second membrane stage to recycle CO_2 back through the turbine (Figure 1). The CO_2 rich permeate from the first membrane stage then undergoes water removal followed by compression and liquefaction to produce CO_2 at the necessary purity. The membrane stages minimise the energy requirements of the liquefaction process by the exclusion of N_2. The second membrane stage processes the retentate from the first membrane stage, and recycles CO_2 back to the process feed stream through the turbine. This is possible because the combustion air for the turbine is used as a sweep gas on the second membrane stage. Combustion air as the sweep gas is a critical feature of the design to generate the pressure driving force across the membrane in order to maximise CO_2 recovery while minimising membrane area and energy duty. However, recycling CO_2 through the turbine (known as exhaust gas recirculation (EGR)) can result in limitations to the overall efficiency because of flame stability issues, incomplete oxidation of CO and changes to convection thermal transfer [13]. The third CO_2-selective membrane stage recovers CO_2 from the liquefaction column off-gas and recycles the permeate stream back to the first membrane stage permeate stream. The retentate from the third membrane stage is recycled back to the first membrane stage feed stream. This ensures all the CO_2 around the liquefaction process is recovered while minimising the cooling duty.

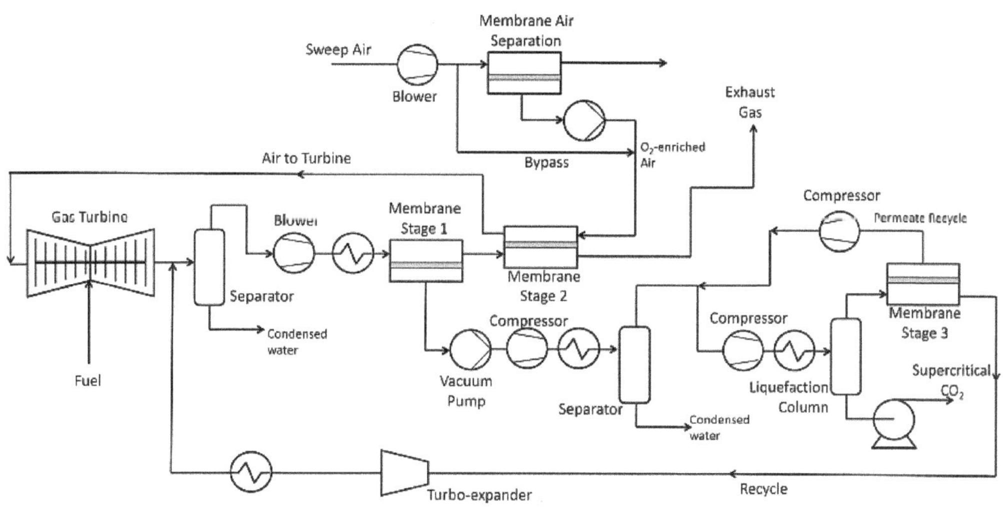

Figure 1. Three CO_2-membrane stages with liquefaction and O_2-enrichment for post-combustion capture from a gas turbine.

The recycling of CO_2 through the turbine also has the problem of reducing the O_2 partial pressure in the combustion zone, and hence reduces the efficiency of the turbine because of incomplete combustion. Therefore, in this paper we have introduced a fourth O_2-enrichment membrane stage on the combustion air upstream of the second membrane stage to counter the problem of reduced O_2 partial pressure. This increases the O_2 concentration of combustion air/sweep gas and ensures the correct O_2 partial pressure is present in the gas turbine, as well as lowering N_2 concentration to ensure the volumetric flowrate is similar to that of the non-EGR NGCC process. Substituting CO_2 for N_2 in the process will impact blade cooling, convection transfer and Heat Recovery Steam Generator (HRSG) operations [14]. However, the concept of O_2 enrichment of the combustion air has previously been considered for oxy-fuel NGCC capture and is likely to be applicable in this configuration [10].

To enable comparison with other carbon capture technologies, economic analysis of the carbon capture options is also presented [15].

2. Methodology

2.1. Experimental

Aspen HYSYS software package (AspenTech, Bedford, MA, USA), version 7.5, was used for all simulations. The fluid package was Peng-Robinson, which is reliable over a wide range of temperatures and pressures, and has previously been applied to CO_2 capture [16,17]. The gas separation membrane process was based on an in-house programmed module, using standard mass transfer equations for cross-flow and counter-flow configurations [18].

In the Aspen HYSYS simulations all units where operated with their default parameters, with pressure drop in piping negligible, while pumps, compressors and turbo-expanders operating at 75% efficiency. The membrane's CO_2 permeance was set at 1000 GPU, with a CO_2/N_2 selectivity of 50. This permeance and selectivity matches MTR's Polaris membrane and is within the membrane performance range suggested for post-combustion capture by Merkel *et al.* [2] The O_2 permeance was assumed equal to N_2, while the H_2O permeance was held constant at 10,000 GPU, consistent with literature [19]. A pressure drop of 10 kPa was assumed across both sides of the membrane stages. To ensure flow through the first and second membrane stages, the feed blower compressed the gas to 150 kPa. The first and third membrane stages were operated in a cross-flow configuration, while the second membrane stage with air sweep was operated in counter-current flow configuration. The O_2-enrichment membrane was based on a cross-flow Air Products PRISM membrane process, with a blower on the feed to ensure flow through the membrane as well as the second membrane stage. A vacuum (22 kPa) on the air separation membrane permeate stream generates the pressure driving force. The O_2-enrichment was to 50%, and the O_2 partial pressure of the combustion air controlled through a bypass stream around the O_2-enrichment membrane.

The permeate stream from the first CO_2-selective membrane was compressed to 25 bar and cooled to 15 °C to remove the bulk of the water. The dry gas was then cooled to −20 °C and sent to a separator to recover liquefied CO_2 at high purity. The off-gas from the separator was sent through a third CO_2-selective membrane to ensure high CO_2 recovery. Plasticization of the membrane at this pressure condition is possible, but not considered in these simulations as this third membrane stage has a minor role in the overall process ensuring recovery. Cooling duty was supplied by an ammonia based refrigeration cycle. All designs were set to capture over 90% of the CO_2 from the flue gas and achieve at least 95% purity of CO_2 in the product stream.

2.2. Economics

The capital expenditure (CAPEX) estimates for individual pieces of equipment were determined by Aspen's Economic Evaluation package, 2010 1st Quarter, with default parameters. The cost of the CO_2 selective membrane was taken from Merkel *et al.* [2] at US\$50/m^2, with a membrane installation factor of 1.6. The cost of the O_2-enrichment membrane was US\$20/m^2, with a membrane installation factor of 1.6, the cost difference recognises this is mature technology. The CAPEX of the total process was the sum of the individual equipment costs, plus US\$10 million for flue gas pretreatment. An additional loading of 25% was added to account for instrumentation, piping and electrical costs, plus an additional 13% for engineering and start-up costs. A final additional 10% was added as a project contingency to the CAPEX, in keeping with literature [20]. For operating expense (OPEX) estimates, the cost of electricity was \$0.0576/kWh for NGCC which corresponds to a natural gas price of about US\$5.81 per GJ [21]. Also included was a 5 year replacement period for all membranes and a 6% maintenance cost. To simplify the process simulations all membrane separation performance was assumed constant over this time period (*i.e.*, no aging). The project life was 25 years, with 2 years construction period and 85% load factor. The real discount rate was 7%. The economics presented here do not take into account the transportation or injection of the CO_2 product stream for geo-sequestration. Full details of the economic methodology can be found in Ho *et al.* [20].

The post-combustion flue gas conditions and composition are from a Natural Gas Combined Cycle power plant, based on a net output of 300 MW [5], provided in Table 1.

Table 1. Flue gas conditions and composition from a Natural Gas Combined Cycle (NGCC) process, which is the basis of the process simulations.

Flue gas conditions	NGCC
Flowrate (tonne/hr)	2268
Temperature (°C)	100
Composition (mol %)	-
CO_2	4.97
N_2	74.28
O_2	9.73
H_2O	11.02

3. Results and Discussion

The combination of two membrane stages with CO_2 recycle through the turbine overcomes the limitation experienced with only one membrane stage; that is choosing between recovery or purity in the permeate stream. The two CO_2-selective membrane stages in series on the flue gas feed enables both high CO_2 recovery and high CO_2 purity to be achieved in the permeate stream of the first membrane stage. This ensures that N_2 is excluded from the liquefaction process and reduces the subsequent compression and cooling duty. In a NGCC power plant with capture, an important parameter in the design is the amount of CO_2 being recycled through the turbine, via the combustion air sweep gas. This is varied by alternating the relative CO_2 recovery in the first and second membrane stages, while always ensuring that the combined CO_2 recovery is greater than 90%. The amount of CO_2 being recycled through the turbine in this analysis is varied between 2.5% and 30%, with literature suggesting the impact from the higher recycle level, while significant, is within operational conditions of gas turbines [22]. Where a small recovery in the first membrane stage results in a significant amount of CO_2 being recycled via the second membrane stage and vice versa. Importantly, the two stage membrane process ensures that the CO_2 purity in the first stage permeate stream is always above 75%, irrespective of the individual membrane stage recoveries.

Altering the amount of CO_2 being recycled through the process impacts the equipment size and energy duty. A large CO_2 recycle reduces the membrane area for both the first and second membrane stages, because of the higher partial pressure. However, a large recycle rate also increases the energy duty of the process as the feed blower must handle increased throughput. This represents a trade-off between equipment cost through reduced membrane area and increased energy duty. The impact CO_2 recycle has on cost of capture is shown in Figure 2. The NGCC process achieves a minimum in the cost of capture when the CO_2 recycle going through the turbine is 0.25 mole fraction for NGCC. At lower CO_2 recycle rates, the membrane area of the first and second stages are substantially large because the CO_2 partial pressure is low, producing only a small driving force for separation. This resulted in a substantial CAPEX due to significant increases in membrane cost as well as OPEX increase because of higher membrane replacement cost. In contrast, at higher CO_2 recycle of 0.3, there is a smaller membrane area, but the increase in energy demand of the feed blower, and a significant increase in OPEX offsets any reduction in membrane CAPEX. In terms of the gas turbine, operating the NGCC process with the combustion air having a CO_2 mole fraction of 0.24 is theoretically possible. However, industrial testing will be required to demonstrate its feasibility.

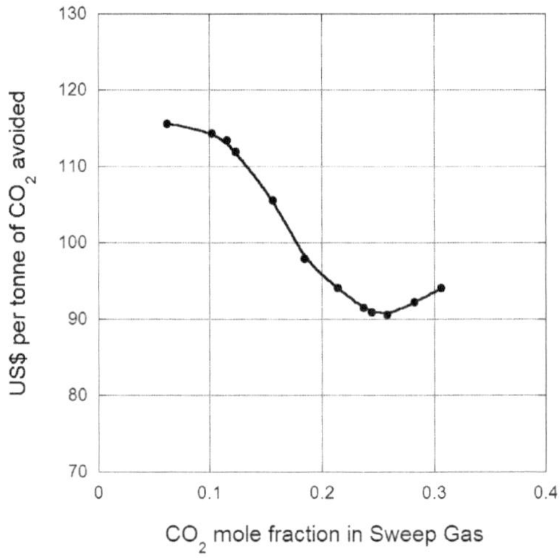

Figure 2. Cost of capture (US$ per tonne of CO_2 avoided) for NGCC as a function of recycle CO_2 mole fraction.

For the NGCC scenario where the recycle rate is 0.24, the combined CO_2-selective membrane area is 3.3×10^5 m^2, which is comparable with existing natural gas sweetening membrane areas [3]. Of the CO_2-selective membranes, the total area is mostly associated with the second membrane stage, given the high CO_2 recycle through the turbine, while the third membrane stage area is less than 100 m^2 to recycle the small amount of CO_2 in the off-gas from the liquefaction process and the first membrane requires an area of about 1.6×10^4 m^2. In comparison the O_2-enrichment process membrane area is 7.9×10^5 m^2. This area is more than double that of the CO_2-selective membranes, and indicates the quantity of air for combustion needed to be processed to ensure the correct O_2 partial pressure in the turbines.

As shown in Figure 2, the cost of post-combustion capture for a NGCC power plant with a combined three-CO_2 selective membrane process and O_2-enrichment membrane has a minimum capture cost of US$91 per tonne of CO_2 avoided. This cost is greater than values reported for post-combustion NGCC capture using solvent technology. For a Fluor solvent the cost of capture is US$58 per tonne of CO_2 avoided, and with a MHI solvent the cost of capture is US$48 per tonne of CO_2 avoided [23]. The corresponding energy penalties for these processes are about 15% and 11% respectively. It should be noted that these costs are for a 776 MW NGCC process with the same operating lifetime and load factor, but a higher COE at $0.067/kWh. As such, the results suggest that the membrane process does not compete with solvent technology for NGCC. Furthermore, the cost for the proposed post-combustion NGCC membrane process has been shown to be lower if used at a supercritical coal-fired power station, with reported values of US$28–US$32 per tonne of CO_2 avoided [24]. One of the main reasons for the difference in cost is the lower CO_2 purity in the liquefaction process, 0.75 for the post-combustion NGCC compared to 0.9 CO_2 mole fraction in the coal-fired power station process. This increases the energy duty for the vacuum pump, compressor and cooling. For the NGCC process the energy demand is 35% (117 MW). Hence, over a third of the power output from the power plant is consumed by the capture plant, whilst for the coal power plant it was less than 30%.

The breakdown of the energy duty for the capture plant is provided in Figure 3. The feed blower accounts for almost half of the energy duty of the capture plant. This unit operation is a blower, not a compressor, as it provides the necessary pressure (150 kPa) for the flue gas to pass through membrane stages 1 and 2. The significant energy duty is because of the large amount of CO_2 recycle and considerable gas throughput in the process. The next major component, accounting for over a quarter of the energy demand is the vacuum pump on the first membrane stage to ensure sufficient partial pressure driving force across the membrane module. Hence, gas handling accounts for three quarters or more of the energy demand of this capture process, and highlights the major limitation of achieving significant energy savings for membrane processes with low CO_2 partial pressure systems. The refrigeration cooling duty in the liquefaction process accounts for 19%.

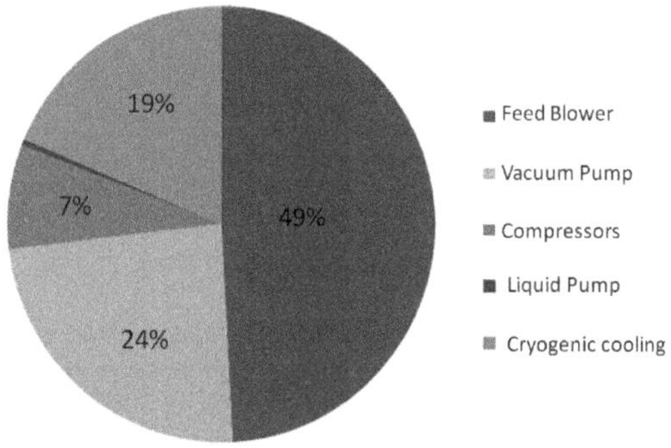

Figure 3. Energy duty breakdown of the membrane-liquefaction process.

The CO_2/N_2 selectivity is a major membrane variable as it controls the purity of CO_2 entering the liquefaction process. Improving the selectivity reduces both the compression and cooling duty of the process, because of exclusion of N_2 from the liquefaction process. This represents a reduction in cost of capture, as shown in Figure 4 for the NGCC process. The cost of capture falls as selectivity increases and reaches a plateau at $CO_2/N_2 = 150$. Above this selectivity there is no savings made in the cost of capture because essentially enough of the N_2 has been excluded from the permeate stream in the first membrane stage to meet the required CO_2 concentration for storage, once water has been removed. At this selectivity, the cost of capture reduces to about US$83 per tonne of CO_2 avoided, though it is still higher than conventional solvent technology. However, while a number of polymeric membrane materials have reported selectivities this high [25], to the authors' knowledge none have been commercialized. One of the current best performing membrane for post-combustion capture is MTR's Polaris, which has a CO_2/N_2 selectivity of 50 [2]. Hence, to obtain these potential decreases in cost of capture, further development of membrane materials and improvement in selectivity will be required.

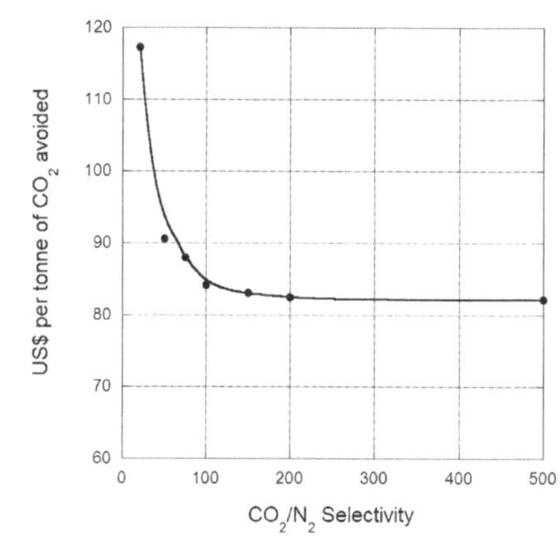

Figure 4. Cost of capture (US$ per tonne of CO_2 avoided) for NGCC as a function of membrane CO_2/N_2 selectivity.

The CAPEX breakdown of the capture plant is provided in Figure 5. The biggest capital expense is the blower/compressor/vacuum pump for gas handling. The next biggest expense is the general cost, of piping, instrumentation, *etc.* The CO_2-selective membranes contribute 18% to the CAPEX. Hence, to achieve a CAPEX reduction, savings must be made to the gas handling equipment, in particular the feed blower which accounts for 26% of the CAPEX.

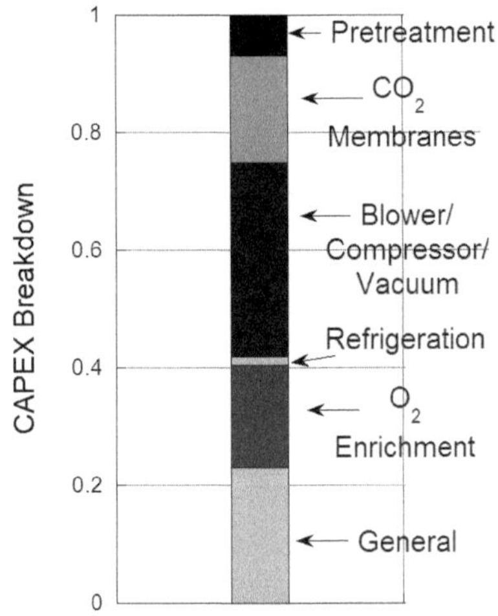

Figure 5. CAPEX breakdown of NGCC post-combustion membrane capture plant.

The OPEX breakdown is provided in Figure 6. Again, the biggest contribution is the energy cost for gas handling, account for 50%. The next largest contributor to the OPEX is the general cost, including cooling water and plant maintenance. The refrigeration duty is 3% of the OPEX, O_2 enrichment process is 7%, because of the high air throughput, while membrane replacement is less than 7%.

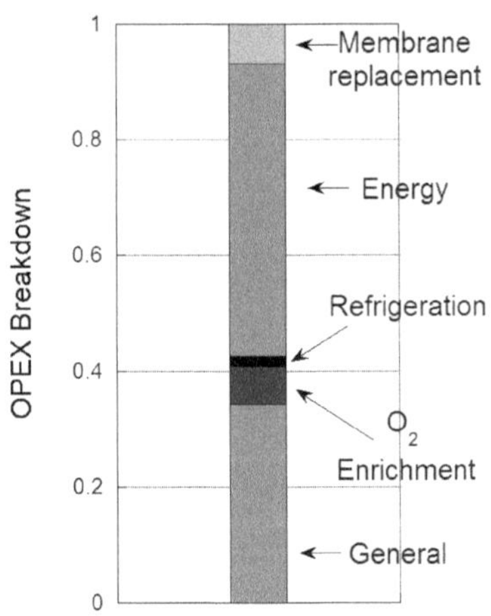

Figure 6. OPEX breakdown of NGCC post-combustion membrane capture plant.

One option to reduce the cost of capture could be to eliminate the feed blower from the process, given that it accounts for almost half the energy duty of the process because of the high gas throughput. This will require the exhaust gas from the Heat Recovery Steam Generator (HRSG) to be at sufficient pressure to ensure flow through the associated duct work. However, reducing the HRSG back pressure results in a reduction to the power output and efficiency of the combined cycle, ~0.25% loss per

50 mbar [26]. Removing the blower and incorporating the power loss because of increased flue gas pressure, indicates that the energy demand for the membrane capture plant becomes 60 MW, and reduces the parasitic load on the NGCC plant to 20%. The cost of capture for the process without the feed blower is provided in Figure 7, as a function of membrane CO_2/N_2 selectivity. At a CO_2/N_2 selectivity of 50, the cost of capture for NGCC reduces to US$67 per tonne of CO_2 avoided if the blower is excluded, which represents a saving of 26% compared to the baseline case. Again, the cost of capture reduces with increased selectivity until it reaches a constant value at CO_2/N_2 = 200, because of the high CO_2 purity in the permeate stream eliminates the need for liquefaction.

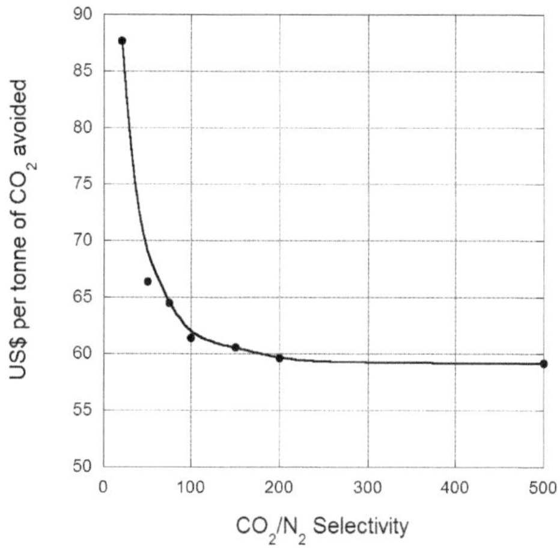

Figure 7. Cost of capture (US$ per tonne of CO_2 avoided) for NGCC as a function of membrane CO_2/N_2 selectivity, with no blower.

4. Conclusions

A three-CO_2-selective membrane with liquefaction and O_2-enrichment process was shown not to be competitive for post-combustion carbon capture compared to conventional chemical absorption technology when applied to the flue gas of NGCC. The process achieves a minimum in cost of capture when a significant amount of CO_2 is recycled through the turbine in the combustion air. The feed blower accounts for almost half of the energy duty of the plant, due to the high gas throughput. This highlights the strong dependence the membrane-liquefaction process design has on CO_2 partial pressure in the flue gas to be competitive with other CO_2 separation technologies.

Acknowledgments: The authors would like to thank the Cooperative Research Centre for Greenhouse Gas Technologies (CO2CRC) for funding.

Author Contributions: Colin A. Scholes and Minh T. Ho conceived and designed the experiments; Colin A. Scholes performed the process simulations; Minh T. Ho undertook the economic analysis; Colin A. Scholes, Minh T. Ho and Dianne E. Wiley analyzed the data and wrote the paper. All authors have read and approved the final manuscript.

Conflicts of Interest: The authors declare no conflict of interest.

Reference

1. Favre, E.; Bounaceur, R.; Roizard, D. A hybrid process combining oxygen enriched air combustion and membrane separation for post-combustion carbon dioxide capture. *Sep. Purif. Technol.* **2009**, *68*, 30–36. [CrossRef]
2. Merkel, T.C.; Lin, H.; Wei, X.; Baker, R. Power plant post-combustion carbon dioxide capture: An opportunity for membranes. *J. Membr. Sci.* **2010**, *359*, 126–139. [CrossRef]

3. Scholes, C.A.; Stevens, G.W.; Kentish, S.E. Membrane gas separation applications in natural gas processing. *Fuel* **2012**, *96*, 15–28. [CrossRef]
4. Scholes, C.A.; Smith, K.H.; Kentish, S.E.; Stevens, G.W. CO_2 capture from pre-combustion processes—Strategies for membrane gas separation. *Int. J. Greenh. Gas Control* **2010**, *4*, 739–755. [CrossRef]
5. Amann, J.-M.G.; Bouallou, C. CO_2 capture from power stations running with natural gas (NGCC) and pulverized coal (PC): Assessment of a new chemical solvent based on aqueous solutions of n-methyldiethanol amine + triehtylene tetramine. *Energy Procedia* **2009**, *1*, 909–916. [CrossRef]
6. Falk-Pedersen, O.; Bjerve, Y.; Glittum, G.; Ronning, S. Separation of carbon dioxide from offshore gas turbine exhaust. *Energy Convers. Manag.* **1995**, *6–9*, 393–396. [CrossRef]
7. Li, H.; Haugen, G.; Ditaranto, M.; Berstad, D.; Jordal, K. Impacts of exhaust gas recirculation (egr) on the natural gas combined cycle integrated with chemical absorption CO_2 capture technology. *Energy Procedia* **2011**, *4*, 1411–1418. [CrossRef]
8. Mofarahi, M.; Khojasteh, Y.; Khaledi, H.; Farahnak, A. Design of CO_2 absorption plant for recovery of CO_2 from flue gases of gas turbine. *Energy* **2008**, *33*, 1311–1319. [CrossRef]
9. Peeters, A.N.M.; Faaij, A.P.C.; Turkenburg, W.C. Techno-economic analysis of natural gas combined cycles with post-combustion CO_2 absorption, including a detailed evaluation of the development potential. *Int. J. Greenh. Gas Control* **2007**, *1*, 396–417. [CrossRef]
10. Shao, Y.; Golomb, D. Power plants with CO_2 capture using integrated air separation and flue gas recycling. *Energy Convers. Manag.* **1996**, *37*, 903–908. [CrossRef]
11. Belaissaoui, B.; Cabot, G.; Cabot, M.-S.; Willson, D.; Favre, E. An energetic analysis of CO_2 capture on a gas turbine combining flue gas recirculation and membrane separation. *Energy* **2012**, *38*, 167–175. [CrossRef]
12. Shao, Y.; Golomb, D.; Brown, G. Natural gas fired combined cycle power plant with CO_2 capture. *Energy Convers. Manag.* **1995**, *36*, 1115–1128. [CrossRef]
13. Sipocz, N.; Tobiesen, A.; Assadi, M. Integrated modelling and simulation of a 400 MW NGCC power plant with CO_2 capture. *Energy Procedia* **2011**, *4*, 1941–1948. [CrossRef]
14. Bolland, O.; Mathieu, P. Comparison of two CO_2 removal options in combined cycle power plants. *Energy Convers. Manag.* **1998**, *39*, 1653–1663. [CrossRef]
15. Allinson, G.; Ho, M.T.; Neal, P.N.; Wiley, D.E. The methodology used for estimating the costs of CCS. In Proceedings of the 8th International Conference on Greenhouse Gas Technologies, Tronheim, Norway, 19–22 June 2006.
16. Desideri, U.; Paolucci, A. Performance modelling of a carbon dioxide removal system for power plants. *Energy Convers. Manag.* **1999**, *40*, 1899–1915. [CrossRef]
17. Scholes, C.A.; Anderson, C.J.; Cuthbertson, R.; Stevens, G.W.; Kentish, S.E. Simulations of membrane gas separation: Chemical solvent absorption hybrid plants simulations for pre- and post-combustion carbon capture. *Sep. Sci. Technol.* **2013**, *48*, 1954–1962. [CrossRef]
18. Zolandz, R.R.; Fleming, G.K. Design of gas separation systems. In *Membrane Handbook*; Ho, W.S.W., Sirkar, K.K., Eds.; Kluwer Academic: Norwell, MA, USA, 1992; pp. 54–77.
19. Scholes, C.A.; Kentish, S.E.; Stevens, G.W. Effects of minor components in carbon dioxide capture using polymeric gas separation membranes. *Sep. Purif. Rev.* **2009**, *38*, 1–44. [CrossRef]
20. Ho, M.T.; Allinson, G.W.; Wiley, D.E. Reducing the cost of CO_2 capture from flue gases using membrane technology. *Ind. Eng. Chem. Res.* **2008**, *47*, 1562–1568. [CrossRef]
21. Fout, T.; Zoelle, A.; Keairns, D.; Turner, M.; Woods, M.; Kuehn, N.; Shah, V.; Chou, V.; Pinkerton, L. *Cost and Performance Baseline for Fossil Energy Plants. Volume 1a: Bituminous Coal (pc) and Natural Gas to Electricity*; National Energy Technology Laboratory, U.S. Department of Energy: Washington, DC, USA, 2015.
22. Jonshagen, K.; Sipocz, N.; Genrup, M. Optimal combined cycle for CO_2 capture with EGR. In Proceedings of the ASME Turbo Expo, Glasgow, UK, 14–18 June 2010.
23. Davison, J. Performance and costs of power plants with capture and storage of CO_2. *Energy* **2007**, *32*, 1163–1176. [CrossRef]
24. Scholes, C.A.; Ho, M.T.; Stevens, G.W.; Kentish, S.E. Cost competitive membrane-Cryogenic post-combustion carbon capture. *Int. J. Greenh. Gas Control* **2013**, *17*, 341–348. [CrossRef]

25. Powell, C.E.; Qiao, G.G. Polymeric CO_2/N_2 gas separation membranes for the capture of carbon dioxide from power plant flue gases. *J. Membr. Sci.* **2006**, *279*, 1–49. [CrossRef]
26. Kehlhofer, R.; Rukes, B.; Hannemann, F.; Stirnimann, F. *Combined-Cycle Gas & Steam Turbine Power Plants*; PennWell: Tulsa, OK, USA, 2009.

6

Single Atoms Preparation Using Light-Assisted Collisions

Yin Hsien Fung, Pimonpan Sompet and Mikkel F. Andersen*

Dodd-Walls Centre for Photonics and Quantum Technologies, Department of Physics, University of Otago, 9016 Dunedin, New Zealand; funyi014@student.otago.ac.nz (Y.H.F.); mim.sompet@otago.ac.nz (P.S.)
* Correspondence: mikkel.andersen@otago.ac.nz

Academic Editor: Alexander V. Sergienko

Abstract: The detailed control achieved over single optically trapped neutral atoms makes them candidates for applications in quantum metrology and quantum information processing. The last few decades have seen different methods developed to optimize the preparation efficiency of single atoms in optical traps. Here we review the near-deterministic preparation of single atoms based on light-assisted collisions and describe how this method can be implemented in different trap regimes. The simplicity and versatility of the method makes it feasible to be employed in future quantum technologies such as a quantum logic device.

Keywords: light-assisted collisions; quantum information processing; qubits; single atom; dipole trap; few-body physics; collisional blockade; atomic physics

---

1. Introduction

Optically trapped neutral atoms have proven to be an excellent experimental platform in the pursuit of understanding the physical reality of quantum mechanics. Fundamental research of atom-light interaction at the individual atoms level has consolidated our knowledge of microscopic systems and improved the level of control of such systems tremendously. This has led to the idea of using the knowledge and control tools to develop new technologies based on quantum physics as part of the second quantum revolution.

One of these important future quantum based technologies is quantum information processing. According to DiVincenzo criteria, a physical realization of quantum information processing requires a string of qubits that can interact with each other but have long relevant coherence times [1,2]. Individual neutral atoms have presented themselves as candidates for qubits [3–6]. An array of optical microtraps with each site storing a single atom [7–10] is a promising working arrangement to fulfill the DiVicenzo criteria. The probability of having an array of N sites all filled with an atom simultaneously scales with the single atom preparation efficiency to the power of N. Therefore high efficiency loading of single neutral atoms is an important step to realize a scalable quantum information processing device [7,11–13]. Additionally, trapped single cold atoms in separate traps has minimal perturbation due to interaction with the environment, which is utilized in quantum metrology in such apparatus as a 3D optical lattice clock [14,15].

Over the past few decades, several important techniques have been developed to spatially confine neutral atoms by using optical traps and to control the motion of atoms [16–19]. Of these, Far-Off Resonance Traps (FORT) [19] provide conservative potentials for the atoms and they can be dynamically reconfigured [12,20]. This makes FORTs an attractive tool to trap single neutral atoms. Furthermore, atoms in FORTs have long coherence times and both the long and short range atomic interactions can be controlled [3,21–23]. A few different approaches have been developed to achieve efficient preparation of single neutral atoms in FORTs. These include the redistribution of individual atoms into arrays of microtraps by atom-sorting [24], isolation of single atoms in microtraps by using the superfluid to Mott-insulator transition from a Bose-Einstein condensate [11,25,26] or using Pauli's exclusion principle [23], and using the Rydberg blockade mechanism [27], Penning ionization [28], or repulsive light-assisted collisions [29–31] to ensure that the trap is not occupied by more than one atom.

Using light-assisted collisions to prepare single neutral atoms in FORTs have gained considerable interests in recent years due to its relative simplicity and speed. Sub-poissonian loading was first observed in optical lattices [32] and in tight optical tweezers [33,34] where the collisional blockade takes effect. The preparation efficiencies were initially reported to be about 50%. Near-deterministic single atoms preparation in a FORT using repulsive light-assisted collisions were reported in [29,30], with efficiency of 91%. With similar methods, the preparation efficiency in collisional blockade loading can be enhanced to about 80% [35]. Recently, repulsive light-assisted collisions have been employed to load single atoms in multiple traps with efficiency of up to 90% [31].

The main subject of this review is on the advances of experimental methods to near-deterministically prepare a single neutral atom in an optical microtrap by using light-assisted collisions. Investigation of the dynamics of two atoms under the influence of near resonant light inside an optical microtrap reveals the physics behind the method [36]. This kind of light-assisted two-atom collision experiments allow us to discriminate and quantify different two-atom loss channels, which could be of fundamental interest [37,38].

In Section 2, we will review the concepts of light-assisted collisions between two atoms. Section 3 describes the experimental apparatus used in the works of this review. Section 4 describes the dynamics between two atoms in a FORT under the presence of near resonant light. Section 5 presents the method to achieve high preparation efficiency of single atom in a FORT using repulsive light-assisted collisions. Section 6 presents how the method can be extended to efficient collisional blockade loading of single atom into a tight microtrap. Section 7 provides some discussion, future directions and summary.

2. Light-Assisted Collisions Between Two Atoms

The subject of collisions between two atoms assisted by near resonant light occupies an important position in a wide range of research in atomic, molecular and optical physics. The mechanisms and effects of binary collisions between cold atoms have been widely studied since the advent of laser cooling and trapping of atomic gasses. The model of light-assisted collisions that we will review in this section is based on the semiclassical approach described in [39–41].

The basic features of collisions between two Rubidium-85 (^{85}Rb) atoms can be described schematically with Figure 1. $U_g(R)$ is the ground state interaction potential as a function of internuclear separation R and its long range asymptote is $U_g(R)_{R\to\infty} \approx -\frac{C_6}{R^6}$, where C_6 is a constant [42]. The atom-atom interaction is negligible when R is large and here the ground state is just two atoms each in their ground state, represented by $|S+S\rangle$. The first excited molecular state, at large R, is denoted by $|S+P\rangle$ where one of the atoms is in its ground state and the other atom is in its excited state. This first excited molecular potential has a long range asymptote of $U_e(R)_{R\to\infty} \approx \hbar\omega_0 \pm \frac{C_3}{R^3}$, where $\hbar\omega_0$ is the atomic excitation energy and C_3 is a constant arising from the resonant dipole interaction [42,43]. At this excited state the atoms approach along the $\frac{C_3}{R^3}$ potential, that can be either attractive or repulsive depending on the relative phase of the interacting dipoles.

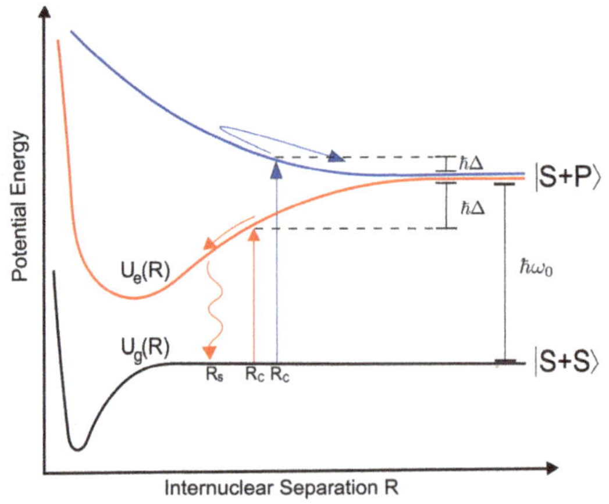

Figure 1. Light assisted collisions model: interaction potentials as a function of internuclear separation R. Red arrows represent the process of excitation of two ground state atoms to an attractive potential (red curve). Blue arrows represent the process of excitation of the atoms to a repulsive potential (blue curve). Figure adapted from [44].

As two atoms approach each other in a collision, the transition frequency changes. The Condon point R_c is the internuclear separation where the atom pair is on resonance with the light field of frequency ω_L, *i.e.*, $U_g(R_c) - U_e(R_c) = \hbar\omega_L$. For near resonant light fields (relative to the atomic transition), the excited state potential can be approximated by the $\frac{C_3}{R^3}$ asymptote. In addition, the ground state potential is approximated to be independent of R. This gives a Condon point of $R_c = \left(\frac{C_3}{\hbar\Delta}\right)^{1/3}$, where $\Delta = \omega_L - \omega_0$ is the detuning of the light field from the free atom resonance. At R_c, two ground state atoms may absorb a photon and get transferred to the excited state. If the light is red-detuned, then the atoms will be excited to the attractive potential. If the light is blue-detuned, then the atoms will be excited to the repulsive potential.

To understand the outcomes of the collisions, we use the Landau-Zener (LZ) formalism for avoided crossings in the dressed state picture [41] to quantify the probability that the atoms get transferred to the excited states when the internuclear separation crosses R_c. The dressed state picture for the collisional processes are shown in Figures 2 and 3. The asymptotic states are the atom-field states, denoted by $|S+S,n\rangle$ (n is the photon number in the light field) and $|S+P,n-1\rangle$. If there is no optical coupling, the dressed states will cross each other at R_c. The coupling by the light field forms a Landau-Zener avoided crossing at R_c. As the internuclear separation crosses R_c in a collision, the atom pair may either undergo adiabatic following, where they move through the coupling region adiabatically, or undergo a LZ transition. The probability that the two colliding atoms undergo a LZ transition is given by $P_{LZ} = \exp\left(\frac{-2\pi\Omega^2}{3v\Delta}\left(\frac{C_3}{\hbar\Delta}\right)^{1/3}\right)$ with v being the relative radial speed of the atoms and Ω being the on-resonance (with molecular transitions) Rabi frequency proportional to the light intensity [41]. The probability of adiabatic following is then $P_A = 1 - P_{LZ}$.

In a collisional process involving the repulsive excited state potential, which is induced by a blue-detuned light field, the atoms cross R_c twice: first when they are approaching each other and again when they are separating from each other. The collision may be an elastic process where the atoms enter and exit the collision in the $|S+S,n\rangle$ state and there is no net gain in kinetic energy. This can happen when the colliding atoms undergo adiabatic followings or LZ transitions both times at R_c (shown as paths ER 1 and ER 2 in Figure 2A). Alternatively, the collision may be an inelastic process where the atoms enter the collision in the $|S+S,n\rangle$ state but exit in the $|S+P,n-1\rangle$ state. This happens when the atoms first undergo an LZ transition and then an adiabatic following, or vice versa (shown as paths IR 1 and IR 2 in Figure 2B). The probability for this inelastic collision to occur is $P_{IR} = 2P_{LZ}(1 - P_{LZ})$. If no spontaneous emission takes place during collision, then the energy gained by the atoms in an inelastic collision is given by $\hbar\Delta$, since $\hbar\Delta$ is the energy difference between the states at large R [41,45].

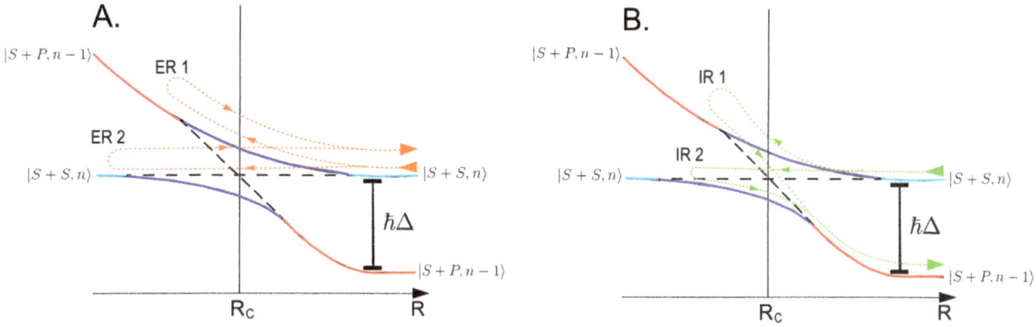

Figure 2. Dressed state picture for a repulsive collisional process showing an avoided crossing at R_c: (**A**) Possible paths for elastic collisions. Path ER 1 is the case where the atoms undergo adiabatic following both times at R_c. Path ER 2 is the case where the atoms undergo LZ transition both times at R_c. (**B**) Possible paths for inelastic collisions (IR 1 and IR 2).

In a collisional process involving the attractive excited state potential, which is induced by a red-detuned light field, elastic (paths EA 1 in Figure 3A) and inelastic collisions can happen. There are two types of inelastic collisions. The first one is shown as path IA 1 in Figure 3B: the atoms undergo an adiabatic following through the coupling region at the first instance and exit the collision via the $|S + P, n-1\rangle$ state. The second is shown as path IA 2 in Figure 3B: the atoms initially undergo an LZ transition and then adiabatically follow their state that becomes attractive. The atoms will therefore return to R_C for a third time where they can undergo an LZ transition to the other dressed state. If an LZ transition does not occur at this point, then the process may repeat. The probability for inelastic collision to occur can be approximated by [41]: $P_{IA} \simeq 1 - \frac{P_{LZ}}{2-P_{LZ}}$. In an inelastic collision, the atoms accelerate towards each other and roll down the excited state potential curve before spontaneous emission to the ground state happens at a position R_s. This results in the atoms gaining kinetic energy E_r given by the difference between the attractive excited state interaction energy at R_c and at R_s, the internuclear separation where the spontaneous emission takes place (see Figure 1).

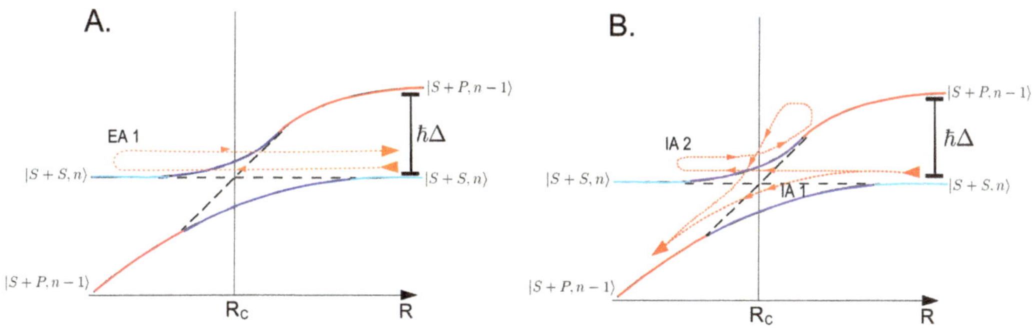

Figure 3. Dressed state picture for an attractive collisional process showing an avoided crossing at R_c: (**A**) Possible paths for elastic collisions (EA 1). (**B**) Possible paths for inelastic collisions (IA 1 and IA 2).

From an experimental point of view, we can tune the laser frequency. By using blue-detuned light to induce collisions, we have a certain level of control that limits the energy released in each inelastic collision. This is in stark contrast to the case of collisions induced by red-detuned light, where the amount of energy released in an inelastic collision is bounded by the excited molecular potential, which is beyond our control.

3. Experimental Apparatus

FORTs have become a popular method to optically confine neutral atoms by using their electric dipole interaction with far-detuned light [19]. A FORT is usually formed by tightly focussing a single Gaussian-mode beam tuned far off any atomic resonance frequencies. As such, spontanenous scattering of the trap light is negligible while maintaining trap potentials of few millikelvins, which is sufficient to hold laser cooled atoms. Therefore, FORTs are an excellent platform to study few-body dynamics such as light-assisted collisions between atoms since they provide a conservative non-dissipative potential and near perfect isolation from the environment.

The apparatus for the experiments that are presented in this review are shown schematically in Figure 4. We trap a ^{85}Rb atomic cloud by using a MOT, which consists of three pairs of counter-propagating MOT cooling beams (wavelength $\lambda = 780$ nm) that are near the D2 $F = 3$ to $F' = 4$ cyclic transition and a repump beam that is on resonance with the D2 $F = 2$ to $F' = 3$ transition. The FORT is formed by focussing a far-off resonant laser beam ($\lambda = 828$ nm or 1064 nm) through a 0.55 high numerical aperture objective lens. The atoms are loaded into the FORT via a Compressed-MOT (CMOT) stage that overlaps the MOT cloud with the FORT.

The imaging beam and collision beam are mode-matched $\lambda = 795$ nm beams that operate close to the D1 $F = 2$ to $F' = 3$ transition. These beams have Gaussian intensity profiles and are focused to a $\frac{1}{e^2}$ radius of about 150μm at the position of the atoms. The collision beam induces light-assisted collisions between atoms. The imaging beam is used to induce the trapped atoms to fluoresce. The imaging/collision beam is retro-reflected to form a standing wave. A mechanical shutter in the returning beam path toggles between running and standing wave configurations for the beam. More details on the experimental apparatus and imaging method can be found in [29,46,47].

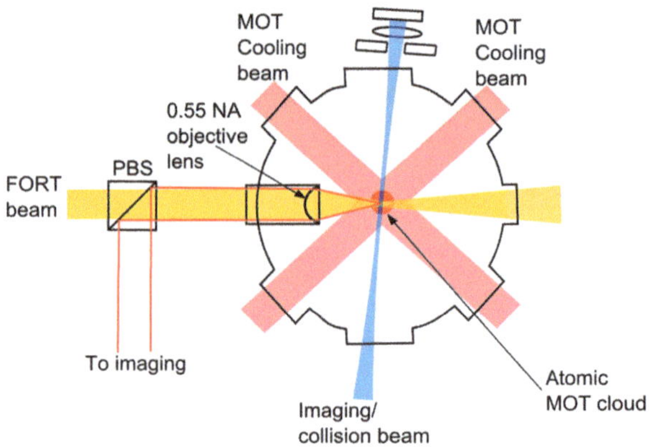

Figure 4. Schematic of the experimental apparatus. Figure adapted from [35].

4. Dynamics of Two Atoms Undergoing Light-Assisted Collisions in a FORT

In the field of optical trapping of atoms, it has been known that light-assisted collisions between atoms can lead to trap loss. In an inelastic collision, the kinetic energy gained by an atom can exceed the potential of the trap, causing the atom to escape from the trap. Depending on the dynamics of the atoms inside the trap, the collisional trap loss events can be either the loss of both colliding atoms ($2-0$ loss) or the loss of only one of the two colliding atoms ($2-1$ loss). Another trap loss is the one-body loss that is due to the finite lifetime of an atom in the trap, which for example, is dependent on the atomic equilibrium temperature and the vacuum condition.

For efficient preparation of single atom in a FORT using light-assisted collisions, it is crucial that 2 − 1 loss dominates over 2 − 0 loss [30]. In this section, we study the atoms dynamics leading to the different loss channels and how it can be influenced by experimental parameters. We do this by observing the outcomes of the ideal collision experiments that starts with exactly two atoms. This allows us to discriminate between 2 − 0 loss and 2 − 1 loss events, and thereby observe how their relative rates depend on the relevant parameters. This information is usually not accessible in experiments conducted using large samples, where the trap losses typically are described by ensemble average parameters.

Through the two-atom experiments, we can better understand the mechanisms of the 2 − 1 and 2 − 0 loss channels due to light-assisted collisions in the presence of laser cooling. In particular, we can identify the roles played by red-detuned collisional light, blue-detuned collisional light and the effect of laser cooling on the rates of the loss channels. Understanding how the loss channel rates depend on various light parameters forms the grounds that will allow us to implement an efficient method to prepare single atoms in a FORT in the subsequent sections.

4.1. Light-Assisted Collisions Between Two Atoms in a FORT

Figure 5 shows the various outcomes of light-assisted collision between two atoms in a FORT, combined with the effect of in-trap laser cooling. Path (a) shows two colliding atoms escaping (2 − 0 loss), ending with no atoms left inside the trap. For this channel to be present, the collision must release sufficient energy such that the total energy of the atoms after the collision, E_p, is enough for both of the two atoms to escape the trap potential. This can arise when high energy is released upon collision or/and when the atoms have high thermal energies prior to the collision.

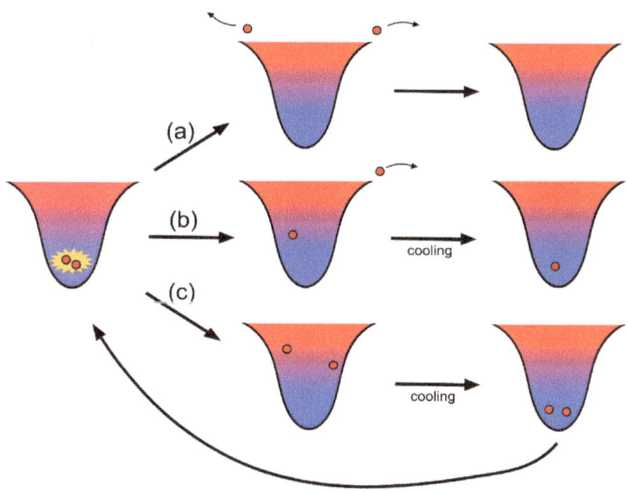

Figure 5. Light-assisted collisions between two atoms with laser cooling. Path (a) shows two colliding atoms escaping (2 − 0 loss), leaving no atoms inside the trap. Path (b) shows one of the two colliding atoms escaping (2 − 1 loss), leaving one atom inside the trap. Path (c) shows neither of the two colliding atoms escaping. Laser cooling removes the energy of the atoms.

Path (b) shows one of the two colliding atoms escaping (2 − 1 loss), leaving one atom inside the trap. This happens when the two colliding atoms have unequal energies after the collision such that one of the two atoms has enough energy to escape the trap potential, while the other atom stays in the trap. Therefore it does not require as high E_p as path (a). However, the atoms should share the energy released in the collision unevenly, which may happen if the thermal energy of the two atoms give them sufficient centre of mass momentum prior to the collision.

Path (c) shows neither of the two colliding atoms escaping. This happens when low energy is released in a collision such that neither of the atoms has sufficient energy to escape the trap potential. There are two scenarios that may lead to this. Trivially if E_p is insufficient for any atoms to escape then no atom will be lost. But in the situation where E_p is sufficient for one atom to escape but insufficient for both to escape, the atoms may share the released energy more or less evenly, leading to no atom escaping. The laser cooling taking place inside the trap can then remove the energy released in the collision, returning the atoms to the condition prior to the collision. This can prevent the atoms from having high energies before the next collision, which can increase the chance of $2-0$ loss. We will discuss the role of laser cooling later in this section.

4.2. Two-Atom Light-Assisted Collisions Induced by Red-Detuned Light

In this subsection, we want to investigate the dynamics of just two atoms in a FORT, undergoing light-assisted collisions induced by red-detuned light. Typically, cold atoms are loaded into a FORT from a MOT, which utilizes red-detuned light beams. Studies of individual collisional loss events in a MOT with high gradient magnetic field has shown that about 10 percent of collisional loss events are $2-1$ loss [48].

As mentioned in Section 2, inelastic collisions induced by red-detuned light can release a kinetic energy E_r given by the difference between the attractive excited state potential energy at R_c and R_s. The probability densities of the released energy $D(E_r)$ for different red detunings calculated with methods described in [36] are shown in Figure 6. For small detunings, the gentle slope of the attractive excited potential at R_c makes it more likely that spontaneous emission has occurred before the atom pair managed to gather any significant relative acceleration. The energy released in these collisions might not be large enough to always cause $2-0$ loss. By observing individual loss events, depending on the light parameters used, we see that the $2-1$ loss channel is indeed also present in red-detuned collisions.

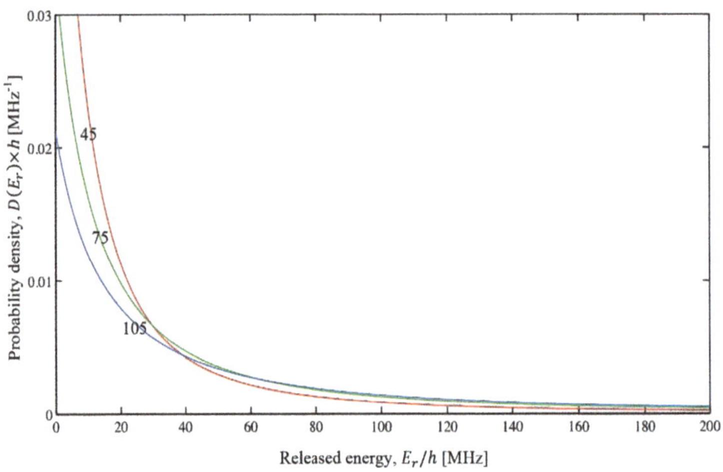

Figure 6. The probability densities of the released energy $D(E_r)$ for head-on collisions induced with light with red detunings of 45, 75 and 105 MHz. The probability of a high energy release are generally small. Figure adapted from [36].

We experimentally prepare two atoms inside the FORT and observe the evolution of the trap population under the influence of near resonant lights. The experimental procedure is illustrated in Figure 7: We prepare a low number of atoms inside the FORT and determine the number of atoms using methods described in [46,47]. We select those realizations with only two atoms inside the trap.

We then expose the atoms with near resonant light beams for a variable period and again determine the number of atoms remaining in the trap.

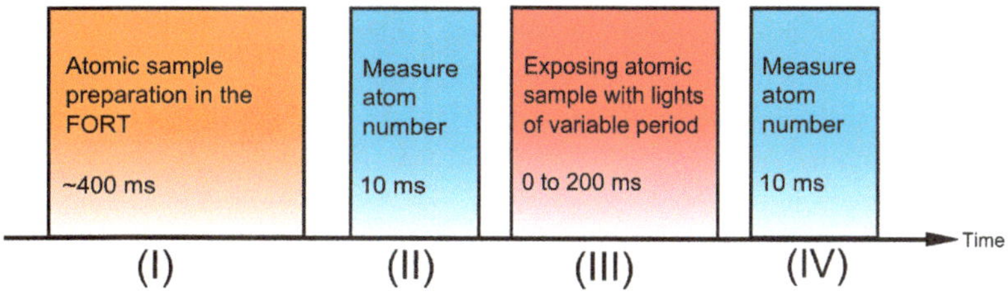

Figure 7. Experimental procedure to observe the dynamics of two atoms inside a FORT under the influence of near resonant lights: (II) Verify that there are two atoms in the trap initially by fluorescence imaging. (III) Expose the atoms with near resonant light beams (Six D2 cooling beams and a D1 collision beam) with variable duration Δt. (IV) Measure the remaining number of atoms.

We use a $U_0 = h \times 85$ MHz FORT formed by focussing a $\lambda = 828$ nm laser to a spotsize of $w_0 = 1.8\ \mu$m. The two atoms are exposed with a collision beam that is red-detuned (negative Δ) from the $D1\ F = 2$ to $F' = 3$ transition at the centre of the FORT to induce the collisions. At the same time, we apply the six cooling beams (the MOT beams) that are 4 MHz red-detuned from the $D2\ F = 3$ to $F' = 3$ transition at the centre of the FORT. The experimental results for each exposure time Δt are presented as the probabilities of two, one or zero atoms remaining in the trap after averaging over 180 repetitions of the experimental sequence (with two atoms in the trap initially). This gives us the decay rate of 2 atoms and growth rates of 0 and 1 atom, as shown in Figure 8, with cooling beam intensities of 11 W/m^2 for (A) and 19 W/m^2 for (B).

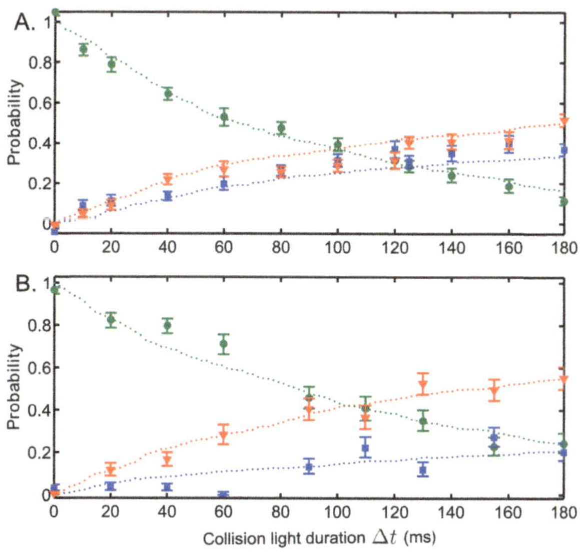

Figure 8. Two-atom evolution when exposed to red-detuned light as a function of Δt. Light parameters: The collision beam is 45 MHz red-detuned from the $D1\ F = 2$ to $F' = 3$ transition at the centre of the FORT. Figures (**A**) and (**B**) have cooling beams intensities of 11 W/m^2 and 19 W/m^2 respectively. The probabilities of two, one and zero atoms remaining in the trap are indicated by green circles, blue squares and red triangles respectively. The dotted lines are the simulated plots. Error bars represent a statistical confidence of 68.3%. Figure adapted from [36].

The results show that there is a non-zero probability of $2-1$ loss events when the atoms undergo light-assisted collisions induced by the red-detuned cooling light. There is a broad range of intermediate E_rs that can lead to a $2-1$ loss. A high E_r collision, which has low probability (see Figure 6), leads to a $2-0$ loss. However, most of the collisions lead to low energy releases where none of the atoms escape the trap. The atoms may therefore go through a number of collisions until there is one that can lead to $2-0$ loss or $2-1$ loss.

With only two atoms, we can simulate the dynamics of the atom pair inside the FORT numerically and compare it with our experimental results. The simulated results are shown as the dotted lines in Figure 8. In this simulation, which is described in detail in [36], the main ingredients are: (1) selection of two atoms according to a Maxwell-Boltzmann distribution with measured initial temperature of $280\ \mu K$; (2) calculation of classical trajectories of two atoms inside the Gaussian potential FORT with the effect of laser cooling simulated by a Doppler cooling model; and (3) calculation of the probability of two atoms undergoing an inelastic collision P_{IA} when their internuclear separation is R_c using the LZ formalism (described in section 1), which is used to decide if a collision has happened. To determine whether any atoms are lost from the trap, when a collision happens, we generate the energy release probability distribution (similar to Figure 6), which is dependent on the collisional impact parameters. We then pick the released energy E_r randomly based on this probability distribution. The simulated results shows good agreement with the experiments, confirming our physical description of the process. Figure 9 shows the simulated evolution of energies of the atoms in the trap, where we can see the difference between $2-0$ loss and $2-1$ loss.

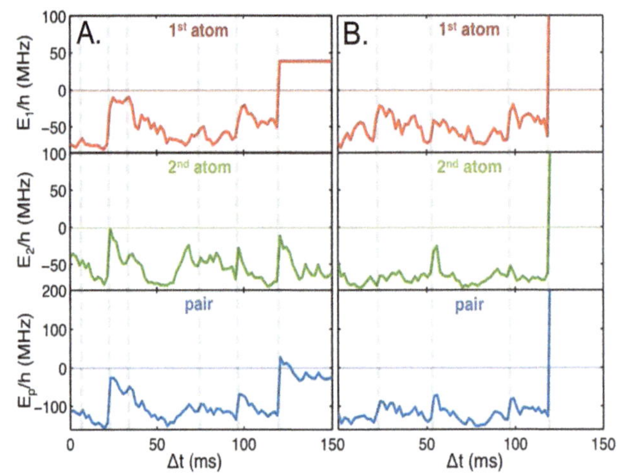

Figure 9. Simulated evolution of individual and the combined energies of the atom pair: (**A**) is the case of $2-1$ loss and (**B**) is the case of $2-0$ loss. The dashed lines indicate when inelastic collisions occurred. In between the collisions, the energies of the atoms are lowered due to the laser cooling effect. Figure adapted from [36].

We found empirically that the cooling beams provide some laser cooling on the atoms in the deep FORT. Higher cooling beam intensity yielded faster cooling rate and generally lower equilibrium temperatures. Thus laser cooling plays an important role in the dynamics of the atoms, with different cooling beam intensities giving different probabilities for collisions leading to $2-0$ loss or $2-1$ loss (as shown in Figure 8A,B.) Higher cooling beam intensities generally provide more efficient cooling to lower the E_p. This increases the required E_r and the atoms are more likely to have less centre of mass momentum before the next collision, hence reducing the probability of $2-1$ loss.

Scanning the collision beam detuning Δ in the range of -105 MHz$\leq \delta \leq -30$ MHz shows that $2-1$ loss events are more likely for smaller detunings while $2-0$ loss events dominate at larger detunings. For larger Δ, the atoms get transferred to the excited molecular state at smaller R_c where the potential gradient is steeper, this leads to large E_r being released in the collision.

The dominance of the $2-0$ loss channel explains why past experiments such as [32,33] have about 50% chance in ending with one atom in the trap, depending on whether the number of atoms loaded is even or odd. In order to significantly improve the single atom loading efficiencies, the $2-1$ loss channel must dominate over the $2-0$ loss channel.

4.3. Two-Atom Light-Assisted Collisions Induced by Blue-Detuned Light

We now study the dynamics of two atoms that undergo light-assisted collisions induced by blue-detuned light. In particular, we want to demonstrate that the $2-1$ loss can be implemented effectively while suppressing the $2-0$ loss by using a combination of a blue-detuned beam and in-trap laser cooling. Recall that the maximal energy gained by the atoms in a repulsive inelastic collision induced by blue-detuned light is given by $\hbar\Delta$, so by using blue-detuned light to induce the collisions, we can limit the energy released by the atoms, such that we give finite probability to $2-1$ loss (path (b) in Figure 5) while eliminating $2-0$ loss (path (a) in Figure 5). For cases where the collision did not cause any trap losses (path (c) in Figure 5), the energy released is removed by the laser cooling and the atoms are returned to the initial condition. The process can thus repeat until $2-1$ loss happens and one atom remains in the trap.

The process described can be achieved by the following: firstly, we want the trap loss to happen only from light-assisted collisions induced by a blue-detuned light with Δ in the right range. Trap loss caused by other light sources (e.g., red-detuned light) and trap loss mechanisms that are not due to collisions between the trapped atoms (e.g., one-body loss) have to be eliminated. Before a collision, the atoms must be in an internal ground state that has a transition such that Δ is in the desired range.

Secondly, we want the two colliding atoms to have unequal energies after the collision such that one of the two atoms has enough energy to escape the trap potential, while the other atom stays in the trap. For this to happen, the thermal energy of the two atoms must be large enough such that they can have sufficient centre of mass momentum prior to the collision. However, the thermal energy of the atoms cannot be too high prior to the collision to avoid $2-0$ loss and one-body loss.

We conduct the experiment with a sequence similar to that is described in the last subsection (shown as Figure 7), except that in stage (III), the collision beam is now blue-detuned (positive Δ) from the $D1$ $F=2$ to $F'=3$ transition at the centre of the FORT and has 11 μW in power. The cooling beams have power of 0.64 mW per beam and are 4 MHz red-detuned from the $D2$ $F=3$ to $F'=3$ transition at the centre of the FORT. The probabilities of two, one or zero atoms remaining in the trap are shown in Figure 10, with $\Delta = 85$ MHz for A and $\Delta = 185$ MHz for B.

The two-atom decay time is 70 ms for Figure 10A and we can see that the $2-1$ loss is the dominant trap loss channel, where its probability is 93% of all the collisional trap loss events. Here the $2-1$ loss channel is favoured by choosing the proper blue-detuned Δ to suppress $2-0$ loss by limiting the kinetic energies that can be gained by the atoms in an inelastic collision. Almost all of the collisions happen in the $F=2$ ground state such that $\Delta = 85$ MHz is in the right range to induce the blue-detuned light-assisted collisions. Large Δ can cause significant $2-0$ loss due to the high energy released upon collisions. This can be seen from Figure 10B, where $\Delta = 185$ MHz leads to significant $2-0$ loss.

Any build up of atomic population in the $F=3$ ground state due to off-resonant spontaneous Raman scattering will cause Δ to be off from the right range by the ground state hyperfine splitting. This is prevented by effective optical pumping of the atoms back into the $F=2$ ground state, provided by the six cooling beams. As mentioned before, the equilibrium temperature has to be such that the thermal energy of the two atoms is sufficient for them to share the released energy unevenly but insufficient to cause one-body and $2-0$ loss. Recall that the cooling rate and equilibrium temperature depend on the cooling beam intensity, so this provides a parameter that can be used to optimize $2-1$ loss. Pumping the atoms into the $F=2$ ground state renders the cooling beams to be about 3 GHz off-resonant, thereby suppresing the rate of light-assisted collisions induced by the cooling beams. The two-atom decay timescale due the cooling beams is 4 s, so while $2-0$ loss can be induced, it is negligible in short timescales.

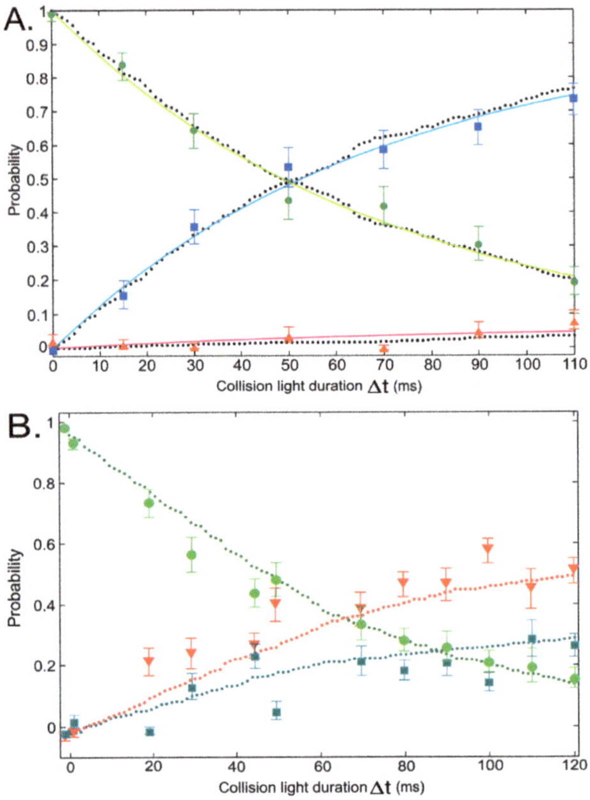

Figure 10. Two-atom evolution when exposed to blue-detuned light as a function of Δt: (**A**) The collision beam is 85 MHz blue-detuned from the $D1$ $F=2$ to $F'=3$ transition at the centre of the FORT. (**B**) Other parameters remained the same but the collision beam is 185 MHz blue-detuned. Green circles, blue squares and red triangles are the probabilities of the experiment ending with two, one and zero atoms respectively. The dotted lines are the simulated plots. Error bars represent a statistical confidence of 68.3%. Figure adapted from [44].

The simulated results in Figure 10 is obtained with simulation similar to that of Figure 8. However, for the blue-detuned collisions case here, the calculation of the probability of two atoms undergoing an inelastic collision at R_c is based on P_{IA}, with collisional energy released given by $\hbar\Delta$. This calculation is to determine whether any collisional trap losses can happen. Note here that in the simulation we adjusted the Ω in P_{LZ} so that the two-atom decay time of the simulation agrees with the experiment. The Ω in the simulation is within a factor of two of the expected Ω based on the beam parameters. The Ω of the simulated results in Figure 10A,B matched each other within 7.5%. The simulated results are consistent with the experimental data.

5. High Efficiency Preparation of Single Atoms Using Repulsive Light-Assisted Collisions

The combination of optimized values of the blue-detuned collision beam and cooling beams to favour the 2 − 1 loss forms the basis of a near-deterministic scheme in the preparation of single atoms in a FORT [29,30]. The experimental sequence for the scheme is illustrated in Figure 11. The FORT is initially loaded with about 20 atoms on average from a CMOT. This average sample size of 20 atoms is large enough such that the probability of loading zero atoms into the FORT is effectively eliminated. Then a single atom isolation stage is initiated. This stage involves light-assisted collisions between the trapped atoms until only one atom is left in the trap and the loss process ceases. After a 157 ms sample preparation stage inside the FORT and a 385 ms single atom isolation stage, 91% of 3200 experimental realizations ended with one atom being prepared in the trap. This is represented as the large peak in the histogram in Figure 12. The isolation stage has a collision beam and six cooling beams with same parameters as those used in Figure 10A.

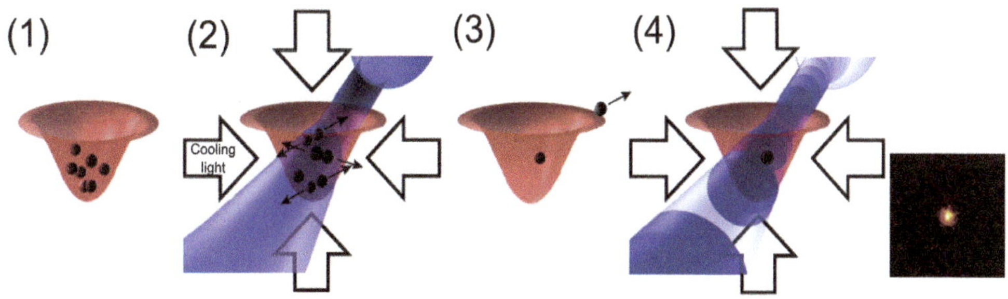

Figure 11. Single atom preparation sequence using blue-detuned collision beam and cooling light: (**1**) Sample preparation stage: about 20 atoms are loaded into the FORT from a compressed Magneto-Optical Trap (CMOT); (**2**) Isolation stage: a combination of in-trap laser cooling and light-assisted collisions induced by blue-detuned beam are used to make the atoms escape the trap one by one via 2 − 1 loss channel. (**3**) Collisions stop when only one atom is left inside the trap. (**4**) Imaging stage: measurement of the number of atoms inside the trap by fluorescent imaging. The inset shows an image of one atom. Figure adapted from [30].

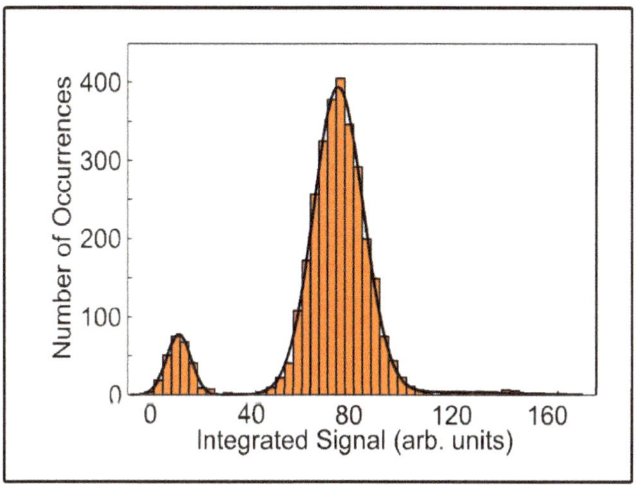

Figure 12. Histogram of the integrated fluorescence for 3200 realizations, representing 91% single atom loading efficiency (statistical error less than 0.01%) [30].

There are two scenarios than can result in no atoms being left in the trap: a single atom is prepared before the end of the isolation stage but is lost before the imaging stage; or the last pair of atoms in the trap are lost via the $2-0$ channel. Therefore, to achieve a high single atom loading efficiency, these two effects have to be mitigated by prolonging the single atom lifetime in the trap and maximizing the probability of $2-1$ loss. In the subsequent subsections, we investigate how these two quantities depend on the parameters of collision beam and cooling beams.

5.1. Dependence on Collision Beam Parameters

Figure 13 shows the single atom loading efficiency as a function of collision beam detuning Δ. The beam detuning is investigated for 15 MHz $\leq \Delta \leq$ 185 MHz, with the collision beam power at each Δ adjusted to obtain the highest probability of loading one atom in the FORT. At smaller Δ, there is stronger radiation pressure exerted on the atom by the collision beam, causing increased heating and higher atom equilibrium temperature. This means that there is higher chance for a prepared single atom to be lost from the trap as it visits the high energy tail of the thermal distribution. At larger Δ, the light-assisted collision between atoms will release higher energies such that both colliding atoms have sufficient energy to escape the trap potential, thereby increasing the probability of $2-0$ loss. This can be seen from Figure 10 where there is a significant $2-0$ loss when $\Delta = 185$ MHz. $\Delta = 85$ MHz gives the optimum single atom loading efficiency, which occurs as a compromise between the short single atom lifetime at small Δ and $2-0$ loss due to large collisional energy released at large Δ.

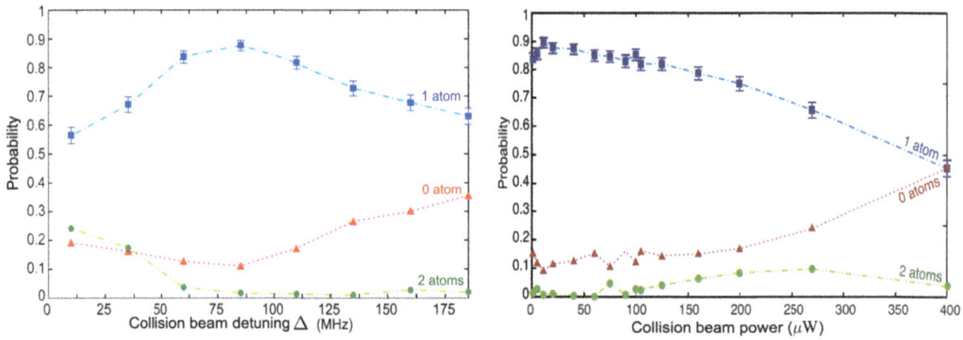

Figure 13. Probabilities of loading one (blue squares), two (green circles) and zero (red triangles) atom as a function of collision beam detuning Δ (**left figure**) and collision beam power (**right figure**). Figure adapted from [30].

We also investigate the single atom loading efficiency as a function of collision beam power at $\Delta = 85$ MHz. The beam pulse duration is adjusted at each beam power to maximize the probability of loading one atom [49]. Figure 13 shows that the peak probability for preparing one atom is obtained with collision beam power of 11 μW. The optimized collision pulse duration at this point is 385 ms. For high collision beam powers, the single atom lifetime is short due to heating from radiation pressure. Whereas, at low collision beam powers the single atom lifetime is long. For example, at 1 μW of collision beam power the single atom lifetime is 63 s. However, the optimized pulse duration at this power is 2.75 s. This duration is comparable to the two-atom decay timescale due to light-assisted collisions induced by the red-detuned cooling light, which is 4 s. For this decay, the $2-0$ loss channel dominates since the cooling light is about 3 GHz red-detuned for atoms colliding in the $F = 2$ ground state. As mentioned in the previous sections, red-detuned collisions at large detuning often release high enough energies for both atoms to escape from the trap. Therefore, lower collision beam power does not guarantee a higher single atom loading probability, since for low powers, the rate of blue-detuned light induced collisions may not dominate over those induced by the cooling beams.

5.2. Dependence on Cooling Light Parameters

Cooling light during the isolation stage plays an important role in the probability of loading one atom since it provides both the cooling and optical pumping of the atoms. With the collision beam parameters being fixed (11 μW, 85 MHz blue-detuned, with 385 ms pulse duration), the single atom loading efficiency as a function of cooling light power (per beam) is shown in Figure 14. The peak loading efficiency is obtained at cooling beam power of 0.64 mW, which occurs as a compromise between short single atom lifetime at low powers and increased rate of light-assisted collisions induced by the red-detuned cooling light at high powers. At low powers, the cooling effect is not efficient enough to overcome the heating caused by the collision beam. At high powers, the $2-0$ loss caused by the cooling light is becoming significant and this impedes higher single atom loading probability.

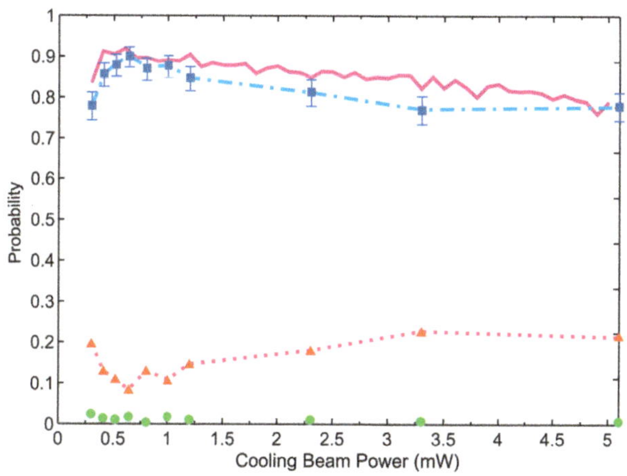

Figure 14. Loading probabilities as a function of cooling beams power (per beam) during the isolation stage. Blue squares, red triangles and green circles are probabilities of loading one, zero and two atoms respectively. The thick pink continuous line is a result of the loading simulation, described in [44]. Figure adapted from [30].

5.3. Limiting Factors

The 9% of unsuccessful runs that limited the single atom loading efficiency to 91% can be accounted for by the following reasons: (1) 3.6% of it is contributed from the $2-0$ loss events induced by the collision beam due to the collisions between atoms that already have high total energies prior to the collision. This is obtained from the simulation in Figure 10A. (2) 1.7% is from the $2-0$ loss events caused by the cooling/optical pumping light, which has a timescale of 4 s at the optimum parameters. (3) 1.5% comes from the finite single atom lifetime of about 22 s, which is dependent on the condition of the vacuum system and the equilibrium temperature of the atom. (4) 0.5% is due to the detection efficiency of the imaging system. (5) The remaining contribution could come from an inelastic collision where the atoms exit through states other than those mentioned in Figure 2, which may release high energy and causing $2-0$ loss.

5.4. Loading Single Atom without Blue-Detuned Induced Collision Beam

Without the presence of blue-detuned light to induce the repulsive collision during the isolation stage, the $2-1$ loss channel can still present, along with the dominant $2-0$ loss (see Figure 8). This can enhance the the single atom loading efficiency beyond 50% in the case of loading with red-detuned light only. This has been observed in [36], where the efficiency is 63% and similar effects have been observed in [20,50,51].

6. Repulsive Light-Assisted Collisions in Collisional Blockade Preparation of Single Atoms

We now extend the methods of using blue-detuned collision beam to induce the $2-1$ loss into the collisional blockade regime. The basic ideas and the physics behind the preparation methods here are similar to those in the last section (not in collisional blockade regime). However, as we shall see in this section, there are additional challenges and restrictions related to the implemention of the methods in the collisional blockade regime.

The high efficiency of single atom preparation outlined in the last section builds on the fact that an initial sample of about 20 atoms on average can be loaded from the MOT into the FORT during the sample preparation stage. This average initial sample size is large enough to ensure that the probability of starting the isolation stage without any atoms in the FORT is effectively eliminated. A finite probability of having no atoms prior to the isolation stage will naturally translate to lower efficiency of the method. As the trap volume of a FORT gets smaller, it becomes increasingly prohibitive to load high number atomic samples directly from a MOT into a FORT since the tightness of the trap along with the light-assisted collisions induced by the red-detuned MOT light can cause rapid trap loss. This is because of the high atomic density increases the probability of the atoms reaching the Condon point and thus increasing the rate of light-assisted collisions. We denote the regime where the efficiency of the method in the last section is restricted by the volume of the FORT as the "tight microtrap regime" [35].

The collisional blockade is the extreme manifestation of the tight microtrap regime [34]. The collisional blockade mechanism prevents the tight microtrap from being occupied by more than one atom at a time since rapid trap loss happens as soon as a second atom enters the trap. The rapid trap loss is a result of the small trap volume and light-assisted collisions induced by the the red-detuned MOT light, often causing $2-0$ loss. Due to this, previous experiments that have used the collisional blockade to prepare single atoms in a tight FORT from a MOT have reported efficiencies of about 50% [33,34]. Apart from the trap volume, the loading rate of the atoms also affects whether or not the collisional blockade occurs since it kicks in when the $2-0$ loss rate is much larger than the loading rate.

To enhance the preparation efficiency of a single atom in the collisional blockade regime, we can introduce the $2-1$ loss channel by including a near-resonant blue-detuned collision beam during the loading process of the tight FORT from a MOT. The idea is that we want to make the $2-1$ loss rate much higher than the $2-0$ loss rate such that there is a high chance for only one of the two atoms to get lost from the trap when the second atom enters and collides with the atom originally in the trap. With careful implementation of the $2-1$ loss channel (see Section 4.3) in the collisional blockade loading regime, we can improve the preparation efficiency to about 80% [35].

6.1. Collisional Blockade Model

In this subsection, we model the loading dynamics in the collisional blockade regime. During the loading process of the microtrap, there can be three different loss processes: $2-0$ loss, $2-1$ loss and one-body loss. In the collisional blockade regime, we can assume that the $2-0$ loss and $2-1$ loss are much faster than the loading rate R such that the probability of more than one atom occupying the microtrap at a given time is effectively zero. This means that the number of atom occupying the microtrap fluctuates only between zero and one. We also assume the one-body loss coefficient γ is smaller than or close to R.

There are two discrete events that can change the atom number inside the microtrap: (1) An atom loading event: when an atom enters the microtrap with no atoms before the loading event, the number of atom inside the trap will change from 0 to 1. When an atom enters the microtrap with already one atom before the loading event, light-assisted collisions will either lead to $2-1$ loss or

$2-0$ loss almost instantaneously, leaving 1 or 0 atom inside the microtrap with probabilities $p_{2\rightarrow 1}$ and $p_{2\rightarrow 0}$. (2) One-body loss event: this can only happen when there is an atom in the microtrap.

We assume the loading events and one-body loss events are uncorrelated and follow Poisson statistics. For cases where there is one atom in the microtrap, the total rate of events is $R+\gamma$ and the probability that the atom number changes following an event is $\frac{R}{R+\gamma}p_{2\rightarrow 0}+\frac{\gamma}{R+\gamma}$, where $\frac{R}{R+\gamma}$ is the probability for a loading event and $\frac{\gamma}{R+\gamma}$ the probability for a one-atom loss event. The probability that the atom number does not change is $\frac{R}{R+\gamma}p_{2\rightarrow 1}$. For cases where there is no atom in the microtrap, the actual rate of events is R. However, we can have the same total rate of events as $R+\gamma$ by introducing a series of "fictitious" events that has no physical role (i.e. do not change the atom number) with rate γ. Having the same rate of events independent of whether there are one or zero atom in the microtrap simplifies the expressions. Only a loading event will change the atom number to one and these have a probability $\frac{R}{R+\gamma}$ out of all events.

The atom number between events therefore form a Markov chain with transition matrix:

$$P=\begin{bmatrix} \left(\frac{R}{R+\gamma}p_{2\rightarrow 1}\right) & \frac{R}{R+\gamma} \\ \left(\frac{R}{R+\gamma}p_{2\rightarrow 0}+\frac{\gamma}{R+\gamma}\right) & \frac{\gamma}{R+\gamma} \end{bmatrix} \tag{1}$$

and the probabilities for obtaining zero or one atom as a function of time is

$$\begin{aligned} \mathbf{p}(t) &= \begin{pmatrix} p_1(t) \\ p_0(t) \end{pmatrix} \\ &= \sum_{n=0}^{\infty}\frac{[(R+\gamma)t]^n e^{-(R+\gamma)t}}{n!}P^n\mathbf{p}(0) \\ &= e^{-(R+\gamma)t}e^{-(R+\gamma)Pt}\mathbf{p}(0) \end{aligned} \tag{2}$$

with $p_1(t)$ and $p_0(t)$ being the probabilities that there is one or zero atoms in the microtrap, respectively, as a function of time. The Perron-Frobenius eigenvector of P yields the long time steady state probabilities. For detecting one atom, this is $p_1(t\rightarrow\infty)=\frac{1}{1+\frac{\gamma}{R}+p_{2\rightarrow 0}}$ and for zero atom, $p_0(t\rightarrow\infty)=\frac{\frac{\gamma}{R}+p_{2\rightarrow 0}}{1+\frac{\gamma}{R}+p_{2\rightarrow 0}}$ [35]. We observe that if $R>>\gamma$ and $p_{2\rightarrow 0}$ is a small number, then the probability for the microtrap to be occupied by only one atom is large.

6.2. *Efficient Preparation of Single Atom in Collisional Blockade Regime*

Efficient preparation of a single atom in a tight microtrap where the collisional blockade mechanism takes effect is a non-trivial subject since the rate of the $2-0$ loss channel during loading is large due to the presence of the red-detuned MOT cooling beams and repump beam. These beams are present to provide a reservoir of atoms to load the microtrap from and to effect the actual loading. For example, due to light beams other than the blue-detuned collision beam, the loss rate timescale with two atoms in the microtrap in this section is at least four order magnitudes faster than that during the isolation stage of the method in the previous section. The experimental setup is similar to Figure 4, except that the FORT is now a laser beam with wavelength of 1064 nm focussed to a spot size of about 1 μm. The trap depth is $U_0=h\times 47$ MHz, unless otherwise stated. Figure 15 shows the experimental sequence. Similar to the previous section, the collision beam is blue-detuned from the $F=2$ to $F'=3$ transition on the $D1$ line to induce the desired $2-1$ loss channel. The scheme can only be successful if the rate of $2-1$ loss exceeds the rate of $2-0$ loss induced by the red-detuned beams. The probability of light-assisted collisions induced by the red-detuned beams are small for low intensity and/or a large detuning of the light as mentioned in Section 2. However, the MOT and repump beam detunings and intensities are restricted by the need to sustain a reservoir of atoms to load from and to conduct loading. Another parameter that we can tune is the trap depth. Different trap depth causes different light shifts on the atomic transitions [19]. The trap light can shift the atomic transitions such that the repump beam frequency is away from any transitions and the MOT beam frequency is close to a transition that can pump atoms into the $F = 2$ ground state. These effects, when used correctly, will suppress unwanted light-assisted collisions that are most likely to cause $2 - 0$ loss.

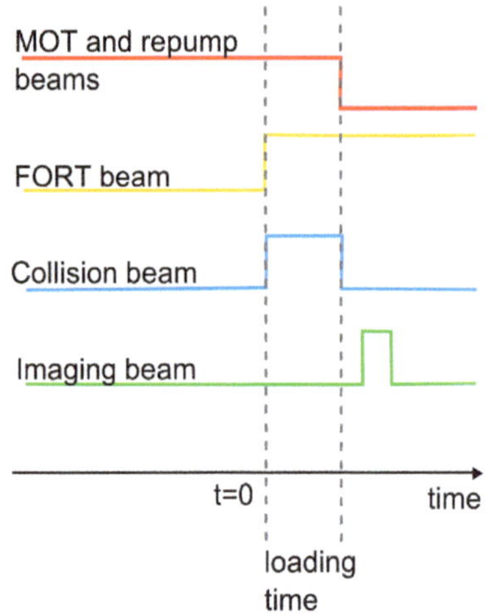

Figure 15. Experimental sequence for efficient preparation of single atom in a tight microtrap where the collisional blockade takes place. Figure adapted from [35].

Figure 16 shows the probability of loading one atom (p_1) as a function of loading time. Without the collision beam during loading, the probability of loading one atom in the microtrap (p_1) reaches the steady-state at about 56% (red crosses). The atom occupancy of the microtrap does not exceed one, which means that the light-assisted collisions induced by the red-detuned MOT and repump beams have put the loading into the collisional blockade regime. When we repeat the experiment with the presence of a collision beam (power of 12 μW and detuning of 68 MHz), we observe that the loading efficiency increases to about 80% at the steady-state (blue circles).

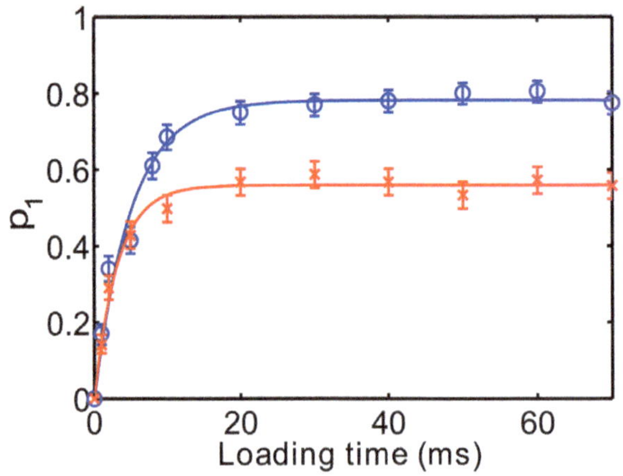

Figure 16. Probability of loading one atom (p_1) as a function of loading time with a trap depth of $h \times 47$ MHz. The solid lines are fitted functions with Equation (2) in Section 6.1. Figure adapted from [35].

6.3. Dependence on Trap Depth

Figure 17A shows the single atom loading efficiencies dependence on the trap depth used during loading. Without the collision beam, for the intermediate trap depths ($h \times 35$ MHz to $h \times 55$ MHz), the trap loss is mostly $2-0$ loss and hence the single atom loading efficiency stays around 50% . For trap depths outside of this range, there are increasingly more one-body losses (γ) due to the atom equilibrium temperature being comparable to the trap depth. This leads to lower loading efficiency.

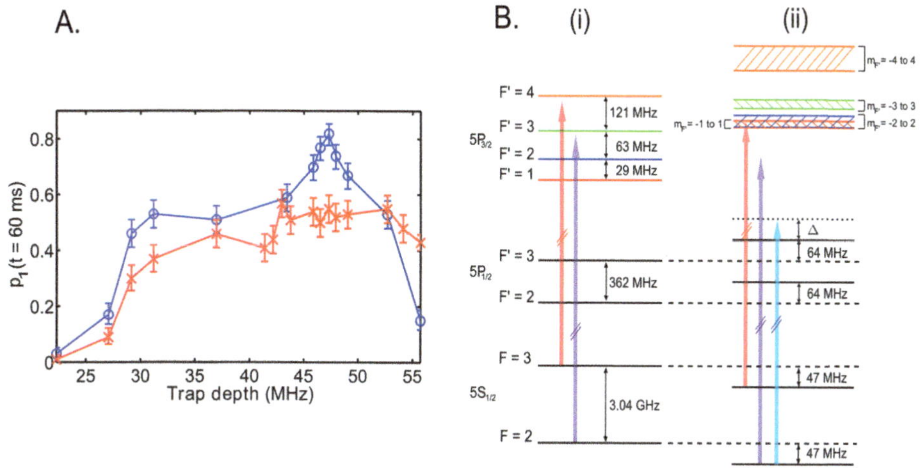

Figure 17. (**A**) Probability of loading one atom with 60 ms loading time p_1 ($t = 60$ ms) as a function of trap depth. Blue circles represent experimental data taken with the presence of collision beam, while red crosses are those taken without collision beam. The single atom loading efficiency peaks at trap depth around $h \times 47$ MHz. (**B**) (**i**) Level diagram for ^{85}Rb atom in free space. (**ii**) Light-shifted level diagram for ^{85}Rb atom in the centre of the FORT with $h \times 47$ MHz trap depth. Red, purple and blue arrows represent the frequencies of MOT beams, repump beam and the collision beam respectively. Figure adapted from [35].

As mentioned earlier, the in-trap atomic transitions experience light shifts from the trap light and this can aid the suppression of light-assisted collisions induced by the MOT beams (red detuned from the D2 $F = 3$ to $F' = 4$ transition in free space) and repump beam (close to D2 $F = 2$ to $F' = 3$ transition in free space). At a trap depth around $h \times 47$ MHz, the D2 $F = 3$ to $F' = 2$ transition is shifted close to resonance with the MOT beams frequency. This leads to the effective depumping of atoms into the $F = 2$ ground state, which is about 3 GHz away from the MOT beam frequency. At the same time, the repump transition is also shifted by 156 MHz by the trap light (see Figure 17B), suppressing the pumping of atoms back into $F = 3$ ground state. These effects reduce the unwanted $2-0$ loss, making the rate of the $2-1$ loss channel introduced by the collision beam to prevail over the rate of the $2-0$ loss channel. We therefore observe the maximal single atom loading efficiency of about 80% at trap depth around $h \times 47$ MHz. If other trap depths are required for subsequent experiments, then after loading a single atom in the microtrap, we can adiabatically ramp the trap depths.

When the method from Section 5 was applied at different trap depths, we managed to achieve high efficiencies at all trap depths tested. However, to achieve this we needed to adjust the collision and cooling beams parameters at each trap depths. Since trap loading and the blue-detuned collisions happened at different stages of the experiment (see. Figure 11), then adjusting the cooling light parameters during the collision stage to match the trap light shift did not affect the initial sample preparation. When doing the collisional blockade loading (see. Figure 15), we cannot freely change the MOT light frequency to match the trap light shifts. Without a separate optical pumping beam [31], we restrict ourselves to trap depths that have a light shift that can allow the MOT light to serve its multiple roles without changing its frequencies. Due to the large excited states light shifts caused by the $\lambda = 1064$ nm trap laser, small variations in trap depths will affect the loading efficiencies. Please note that in Figure 17A, we did not change the collision and cooling beam parameters as we scanned the trap depth.

6.4. Dependence on Blue-Detuned Collision Beam Parameters

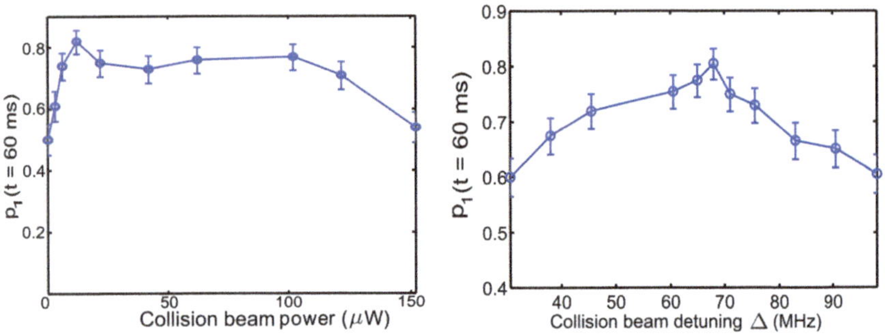

Figure 18. Probability of loading one atom as a function of power of the collision beam (**left figure**) and detuning of the collision beam, Δ (**right figure**). Figure adapted from [35].

In order to maximize the single atom loading efficiency, the right parameters of the collision beam have to be chosen. The rate of $2-1$ loss increases with the intensity of the collision beam and we want the rate of $2-1$ loss to exceed the rate of $2-0$ loss. However at high powers (beyond 12μW, see Figure 18), the loading effiency decreases since the rate of $2-1$ loss reaches saturation while the power of the collision beam increases the atom equilibrium temperature, leading to one-body loss. Next we look at the detuning of the collision beam Δ. Δ determines the energy released in the inelastic repulsive collisions caused by the collision beam. As shown in Figure 18, the single atom loading efficiency peaks at $\Delta = 68$ MHz, which is 1.4 times the trap potential, agreeing with the expectation that for a $2-1$ loss, the energy released should be sufficient for only one of the two atoms to escape but not sufficient for both.

7. Summary and Future Directions

We have studied light-assisted collisions at the individual events level for both red-detuned light and blue-detuned light, which allows us to discriminate between $2-0$ loss and $2-1$ loss directly. We found that a combination of in-trap laser cooling and blue-detuned repulsive light-assisted collisions can lead to the dominance of the $2-1$ loss channel. On the other hand, red-detuned light induced collisions lead to both the $2-0$ loss and the $2-1$ loss [36], although the $2-1$ loss probability will generally be lower than that with optimized blue-detuned induced collisions. Our studies provide us with insights on the effective implementation of $2-1$ loss, which is the essential ingredient for the method of near-deterministic preparation of single atoms in a FORT by using repulsive light-assisted collisions. By optimizing the collision and in-trap cooling/pumping parameters, an efficiency up to 91% can be achieved [29,30].

Some scenarios require direct loading of single atoms from a MOT into tight FORTs where the collisional blockade takes effect. We showed that in the collisional blockade regime, we can also implement the $2-1$ loss channel during loading by introducing the repulsive light-assisted collisions while suppressing the $2-0$ loss channel caused by the red-detuned MOT lights. When we choose the right trap depth and beam parameters to achieve this, the probability of $2-1$ loss of all two-atom decays in this regime was 0.72, which yielded a preparation efficiency of 80% in the collisional blockade regime [35]. The efficiency is mainly limited by the fact that the MOT lights parameters cannot be freely adjusted to reduce the $2-0$ loss since the MOT lights are present to sustain an atom reservoir and load the microtrap.

Our work was conducted with ^{85}Rb, but efficient loading of single atoms using repulsive light-assisted collisions is not limited to this atomic species. Reference [31] has recently demonstrated that using a similar method, preparation of single ^{87}Rb atoms into an array of tight FORTs can be achieved with efficiencies of up to 90% in less than 170 ms for trap separation of more than 1.7μm. Reference [52] mentioned that repulsive light-assisted collisions played a role in their loading of single Cs atoms into an array of blue-detuned FORTs with an efficiency of about 70%. Figure 19 shows some examples of works that used light-assisted collisions as an approach to prepare a single atom [20,30,31,33,35,36,50–53]. Repulsive light-assisted collisions that favours the $2-1$ loss channel generally results in higher preparation efficiency across different trap sizes and atomic species.

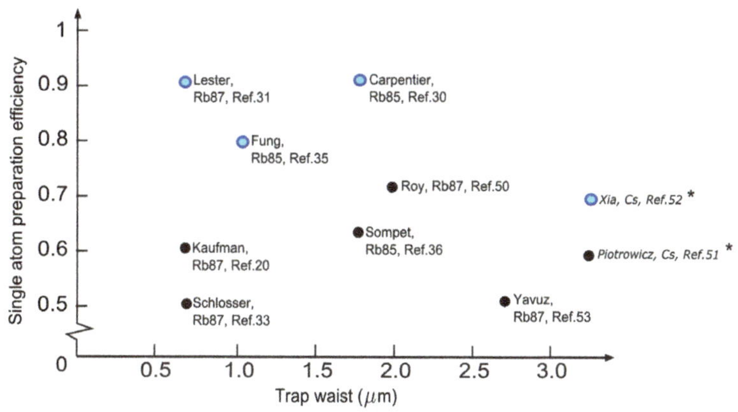

Figure 19. Some examples of works with different atomic species using light-assisted collisions to prepare a single atom in a FORT with known trap waist. The single atom preparation efficiency is quoted as the single trap efficiency. Blue dots represent works that employs repulsive light-assisted collisions. Note: * The works using Cs are conducted in a blue-detuned FORT, so the individual trap size are not directly comparable to the other works that use red-detuned FORT.

Employing light-assisted collisions for the preparation of single atoms may be useful for future applications in quantum information processing with neutral atoms, where high filling efficiencies have to be achieved when loading atoms into multiple trap arrays with arbitary geometries, thus improving the scalability [54]. The construction of hybrid quantum devices that interface microfabricated solid state structures and individual atoms in tight optical microtraps to attain better control in quantum metrology and sensing [55,56] may also implement these methods. With the high single atom prepartion efficiency in a microtrap, construction of a low entropy systems using a bottom up approach may be realized by combining efficient loading and ground state cooling of single neutral atom [57,58].

Acknowledgments: We wish to acknowledge Marsden Fund Council from Government funding, administered by the Royal Society of New Zealand (Contract number UOO1320) for the support in these work. We would like to thank our colleagues and students who have participated in various stages of this research: In particular Alicia Carpentier and Thad Walker, and also Tzahi Grünzweig, Andrew Hilliard, Matt McGovern and Eyal Schwartz. We also thank Carolyn Cowdell for proofreading the manuscript.

Author Contributions: Yin Hsien Fung wrote the manuscript, contributed to the implementation of the experiments, data acquisition and analysis. Pimonpan Sompet editted the manuscript, performed the numerical simulations and contributed to the implementation of the experiments, data acquisition and analysis.

Mikkel F. Andersen editted the manuscript, conceptualized and supervised the research. All authors have read and approved the final manuscript.

Conflicts of Interest: The authors declare no conflict of interest.

References and Notes

1. Bennett, C.H. Quantum information and computation. *Phys. Today* **1995**, *48*, 24–30.
2. Divincenzo, D.P. The Physical Implementation of Quantum Computation. *Fortschritte der Physik* **2000**, *48*, 771–783.
3. Schrader, D.; Dotsenko, I.; Khudaverdyan, M.; Miroshnychenko, Y.; Rauschenbeutel, A.; Meschede, D. Neutral atom quantum register. *Phys. Rev. Lett.* **2004**, *93*, 150501.
4. Monroe, C. Quantum information processing with atoms and photons. *Nature* **2002**, *416*, 238–246.
5. Jessen, P.S.; Deutsch, I.H.; Stock, R. Quantum information processing with trapped neutral atoms. *Quantum Inf. Process.* **2004**, *3*, 91–103.
6. Hijlkema, M.; Weber, B.; Specht, H.P.; Webster, S.C.; Kuhn, A.; Rempe, G. A single-photon server with just one atom. *Nat. Phys.* **2007**, *3*, 253–255.
7. Greiner, M.; Mandel, O.; Esslinger, T.; Hänsch, T.W.; Bloch, I. Quantum phase transition from a superfluid to a Mott insulator in a gas of ultracold atoms. *Nature* **2002**, *415*, 39–44.
8. Lukin, M.D.; Fleischhauer, M.; Cote, R.; Duan, L.M.; Jaksch, D.; Cirac, J.I.; Zoller, P. Dipole blockade and quantum information processing in mesoscopic atomic ensembles. *Phys. Rev. Lett.* **2001**, *87*, 037901.
9. Nelson, K.D.; Li, X.; Weiss, D.S. Imaging single atoms in a three-dimensional array. *Nat. Phys.* **2007**, *3*, 556–560.
10. Brennen, G.K.; Caves, C.M.; Jessen, P.S.; Deutsch, I.H. Quantum logic gates in optical lattices. *Phys. Rev. Lett.* **1999**, *82*, 1060–1063.
11. Itah, A.; Veksler, H.; Lahav, O.; Blumkin, A.; Moreno, C.; Gordon, C.; Steinhauer, J. Direct observation of a sub-poissonian number distribution of atoms in an optical lattice. *Phys. Rev. Lett.* **2010**, *104*, 113001.
12. Dumke, R.; Volk, M.; Müther, T.; Buchkremer, F.B.J.; Birkl, G.; Ertmer, W. Micro-optical realization of arrays of selectively addressable dipole traps: A scalable configuration for quantum computation with atomic qubits. *Phys. Rev. Lett.* **2002**, *89*, 097903.
13. Anderlini, M.; Lee, P.J.; Brown, B.L.; Sebby-Strabley, J.; Phillips, W.D.; Porto, J.V. Controlled exchange interaction between pairs of neutral atoms in an optical lattice. *Nature* **2007**, *448*, 452–456.
14. Derevianko, A.; Katori, H. Colloquium: Physics of optical lattice clocks. *Rev. Mod. Phys.* **2011**, *83*, doi:10.1103/RevModPhys.83.331.
15. Katori, H. Optical lattice clocks and quantum metrology. *Nat. Phot.* **2011**, *5*, 203–210.
16. Phillips, W.D. Laser cooling and trapping of neutral atoms. *Rev. Mod. Phys.* **1998**, *70*, 721–742.
17. Cohen-Tannoudji, C. Manipulating atoms with photons. *Physica Scripta* **1998**, *T76*, doi:10.1103/RevModPhys.70.707.
18. Wieman, C.E.; Pritchard, D.E.; Wineland, D.J. Atom cooling, trapping, and quantum manipulation. *Rev. Mod. Phys.* **1999**, *71*, doi:10.1103/RevModPhys.71.S253.
19. Grimm, R.; Weidemüller, M. Optical dipole traps for neutral atoms. *Adv. At. Mol. Opt. Phys.* **2000**, *42*, 95–170.
20. Kaufman, A.M.; Lester, B.J.; Reynolds, C.M.; Wall, M.L.; Foss-Feig, M.; Hazzard, K.R.A.; Rey, A.M.; Regal, C.A. Two-particle quantum interference in tunnel-coupled optical tweezers. *Science* **2014**, *345*, 306–309.
21. Urban, E.; Johnson, T.A.; Henage, T.; Isenhower, L.; Yavuz, D.D.; Walker, T.G.; Saffman, M. Observation of Rydberg blockade between two atoms. *Nat. Phys.* **2009**, *5*, 110–114.
22. Gaëtan, A.; Miroshnychenko, Y.; Wilk, T.; Chotia, A.; Viteau, M.; Comparat, D.; Pillet, P.; Browaeys, A.; Grangier, P. Observation of collective excitation of two individual atoms in the Rydberg blockade regime. *Nat. Phys.* **2009**, *5*, 115–118.
23. Serwane, F.; Zürn, G.; Lompe, T.; Ottenstein, T.B.; Wenz, A.N.; Jochim, S. Deterministic preparation of a tunable few-fermion system. *Science* **2011**, *332*, 336–338.

24. Miroshnychenko, Y.; Alt, W.; Dotsenko, I.; Förster, L.; Khudaverdyan, M.; Meschede, D.; Schrader, D.; Rauschenbeutel, A. Quantum engineering: An atom-sorting machine. *Nature* **2006**, *442*, doi:10.1038/442151a.
25. Sherson, J.F.; Weitenberg, C.; Endres, M.; Cheneau, M.; Bloch, I.; Kuhr, S. Single-atom-resolved fluorescence imaging of an atomic Mott insulator. *Nature* **2010**, *467*, 68–72.
26. Bakr, W.S.; Peng, A.; Tai, M.E.; Ma, R.; Simon, J.; Gillen, J.I.; F"olling, S.; Pollet, L.; Greiner, M. Probing the superfluid-to-mott insulator transition at the single-atom level. *Science* **2010**, *329*, 547–550.
27. Ebert, M.; Gill, A.; Gibbons, M.; Zhang, X.; Saffman, M.; Walker, T.G. Atomic Fock state preparation using Rydberg blockade. *Phys. Rev. Lett.* **2014**, *112*, 043602.
28. Manning, A.G.; Khakimov, R.; Dall, R.G.; Truscott, A.G. Single-atom source in the picokelvin regime. *Phys. Rev. Lett.* **2014**, *113*, 130403.
29. Grünzweig, T.; Hilliard, A.; McGovern, M.; Andersen, M.F. Near-deterministic preparation of a single atom in an optical microtrap. *Nat. Phys.* **2010**, *6*, 951–954.
30. Carpentier, A.V.; Fung, Y.H.; Sompet, P.; Hilliard, A.J.; Walker, T.G.; Andersen, M.F. Preparation of single atom in an optical microtrap. *Laser Phys. Lett.* **2013**, *10*, 125501.
31. Lester, B.J.; Luick, N.; Kaufman, A.M.; Reynolds, C.M.; Regal, C.A. Rapid production of uniformly filled arrays of neutral atoms. *Phys. Rev. Lett.* **2015**, *115*, 073003.
32. DePue, M.T.; McCormick, C.; Winoto, S.L.; Oliver, S.; Weiss, D.S. Unity occupation of sites in a 3D optical lattice. *Phys. Rev. Lett.* **1999**, *82*, 2262–2265.
33. Schlosser, N.; Reymond, G.; Protsenko, I.; Grangier, P. Sub-poissonian loading of single atoms in a microscopic dipole trap. *Nature* **2001**, *411*, 1024–1027.
34. Schlosser, N.; Reymond, G.; Grangier, P. Collisional blockade in microscopic optical dipole traps. *Phys. Rev. Lett.* **2002**, *89*, 023005.
35. Fung, Y.H.; Andersen, M.F. Efficient collisional blockade loading of a single atom into a tight microtrap. *New J. Phys.* **2015**, *17*, 073011.
36. Sompet, P.; Carpentier, A.V.; Fung, Y.H.; McGovern, M.; Andersen, M.F. Dynamics of two atoms undergoing light-assisted collisions in an optical microtrap. *Phys. Rev. A* **2013**, *88*, 051401.
37. Fuhrmanek, A.; Bourgain, R.; Sortais, Y.R.; Browaeys, A. Light-assisted collisions between a few cold atoms in a microscopic dipole trap. *Phys. Rev. A* **2012**, *85*, 062708.
38. Yanyshev, D.N.; Balykin, V.I.; Vladimirova, Y.V.; Zadkov, V.N. Dynamics of atoms in a femtosecond optical dipole trap. *Phys. Rev. A* **2013**, *87*, 033411.
39. Gallagher, A.; Pritchard, D.E. Exoergic collisions of cold Na*-Na. *Phys. Rev. Lett.* **1989**, *63*, 957–960.
40. Julienne, P.S.; Vigue, J. Cold collisions of ground-and excited-state alkali-metal atoms. *Phys. Rev. A* **1991**, *44*, doi:10.1103/PhysRevA.44.4464.
41. Weiner, J. *Cold and Ultracold Collisions in Quantum Microscopic and Mesoscopic Systems*; Cambridge University Press: Cambridge, UK, 2003.
42. Fontana, P.R. Theory of long-range interatomic forces. i. dispersion energies between unexcited atoms. *Phys. Rev.* **1961**, *123*, doi:10.1103/PhysRev.123.1865.
43. Bussery, B.; Aubert-Frecon, M. Multipolar long-range electrostatic, dispersion, and induction energy terms for the interactions between two identical alkali atoms Li, Na, K, Rb, and Cs in various electronic states. *J. Chem. Phys.* **1985**, *82*, doi:10.1063/1.448222.
44. Fung, Y.H.; Carpentier, A.V.; Sompet, P.; Andersen, M.F. Two-atom collisions and loading of atoms in microtraps. *Entropy* **2013**, *16*, 582–606.
45. Hoffmann, D.S.; Bali, S.; Walker, T.G. Trap-depth measurements using ultracold collisions. *Phys. Rev. A* **1996**, *54*, doi:10.1103/PhysRevA.54.R1030.
46. McGovern, M.; Hilliard, A.; Grünzweig, T.; Andersen, M.F. Counting atoms in a deep optical microtrap. *Opt. Lett.* **2011**, *36*, 1041–1043.
47. Hilliard, A.J.; Fung, Y.H.; Sompet, P.; Carpentier, A.V.; Andersen, M.F. In-trap fluorescence detection of atoms in a microscopic dipole trap. *Phys. Rev. A* **2015**, *91*, 053414.
48. Ueberholz, B.; Kuhr, S.; Frese, D.; Meschede, D.; Gomer, V. Counting cold collisions. *J. Phys. B At. Mol. Opt. Phys.* **2000**, *33*, doi:10.1088/0953-4075/33/4/105.
49. This is because the inelastic collision rate and single atom lifetime change significantly with collision beam power.

50. Roy, A.; Jing, A.B.S.; Barrett, M.D. The trapping and detection of single atoms using a spherical mirror. *New J. Phys.* **2012**, *14*, 093007.
51. Piotrowicz, M.J.; Lichtman, M.; Maller, K.; Li, G.; Zhang, S.; Isenhower, L.; Saffman, M. Two-dimensional lattice of blue-detuned atom traps using a projected Gaussian beam array. *Phys. Rev. A* **2013**, *88*, 013420.
52. Xia, T.; Lichtman, M.; Maller, K.; Carr, A.W.; Piotrowicz, M.J.; Isenhower, L.; Saffman, M. Randomized benchmarking of single-qubit gates in a 2D array of neutral-atom qubits. *Phys. Rev. Lett.* **2015**, *114*, 100503.
53. Yavuz, D.D.; Kulatunga, P.B.; Urban, E.; Johnson, T.A.; Proite, N.; Henage, T.; Walker, T.G.; Saffman, M. Fast ground state manipulation of neutral atoms in microscopic optical traps. *Phys. Rev. Lett.* **2006**, *96*, 063001.
54. Nogrette, F.; Labuhn, H.; Ravets, S.; Barredo, D.; Béguin, L.; Vernier, A.; Lahaye, T.; Browaeys, A. Single-atom trapping in holographic 2D arrays of microtraps with arbitrary geometries. *Phys. Rev. X* **2014**, *4*, 021034.
55. Thompson, J.D.; Tiecke, T.G.; de Leon, N.P.; Feist, J.; Akimov, A.V.; Gullans, M.; Zibrov, A.S.; Vuletic, V.; Lukin, M.D. Coupling a single trapped atom to a nanoscale optical cavity. *Science* **2013**, *340*, 1202–1205.
56. Reitz, D.; Sayrin, C.; Mitsch, R.; Schneeweiss, P.; Rauschenbeutel, A. Coherence properties of nanofiber-trapped cesium atoms. *Phys. Rev. Lett.* **2013**, *110*, 243603.
57. Kaufman, A.M.; Lester, B.J.; Regal, C.A. Cooling a single atom in an optical tweezer to its quantum ground state. *Phys. Rev. X* **2012**, *2*, 041014.
58. Thompson, J.D.; Tiecke, T.G.; Zibrov, A.S.; Vuletic, V.; Lukin, M.D. Coherence and Raman sideband cooling of a single atom in an optical tweezer. *Phys. Rev. Lett.* **2013**, *110*, 133001

7

Scale-Up Effects of CO_2 Capture by Methyldiethanolamine (MDEA) Solutions in Terms of Loading Capacity

Samuel P. Santos [1], João F. Gomes [1,2,*] and João C. Bordado [1]

1 Centro de Recursos Naturais e Ambiente (CERENA), Instituto Superior Técnico, Universidade de Lisboa, Av. Rovisco Pais, s/n, Lisboa 1049-001, Portugal; samuelpsantos@tecnico.ulisboa.pt (S.P.S.); jcbordado@tecnico.ulisboa.pt (J.C.B.)

2 Área Departamental de Engenharia Química (ADEQ), Instituto Superior de Engenharia de Lisboa, Instituto Politécnico de Lisboa, R. Conselheiro Emídio Navarro 1, Lisboa 1959-007, Portugal

* Correspondence: jgomes@deq.isel.ipl.pt

Academic Editor: Manoj Gupta

Abstract: In the present study, results from three different CO_2 capture experimental scales (laboratory, pilot unit, and a larger pilot unit), using aqueous amine solutions of methyldiethanolamine (MDEA) 20 wt %, are compared in terms of loading capacity. All three tested scales produced results regarding CO_2 absorption using MDEA aqueous solutions, which were largely in accordance with the theoretical loading capacity of the used amine. Nevertheless, the observed differences between the theoretical and actual absorption behaviors of MDEA solutions for the different scales can be justified with the relative weight that process variables exhibit when the process is scaled up. Therefore, in order to achieve a correct scale-up of the process, simulations should be performed in order to define the best set of operational parameters in order to achieve high production yields and therefore more process profitability.

Keywords: carbon dioxide; chemical absorption; aqueous amine solutions; scale-up; methyldiethanolamine (MDEA)

1. Introduction

Environmental issues, due to the emissions of pollutants from combustion of solid, liquid, and gaseous fuels on various stationary and mobile energy systems, as well as the emissions from manufacturing plants, have contributed to major global problems involving not only pollutants such as sulphur, nitrogen oxides (SO_x, NO_x), and particulate matter, but also greenhouse gases (GHG) such as carbon dioxide (CO_2) and methane (CH_4) [1]. Among GHG, CO_2 is the largest contributor, accounting for more than 60 percent of global warming effects, which is due to its high amount in the atmosphere [2].

The concept of separating CO_2 from flue gas streams started to be applied in the 1970s, not only due to concerns about the greenhouse effect, but also as a potentially economic source of CO_2, mainly for enhanced oil recovery (EOR) operations. Several commercial CO_2 captures were constructed in the U.S. in the late 1970s and early 1980s. CO_2 was also produced for other industrial applications such as the carbonation of brine and production of products such as dry ice, urea, and beverages [3,4].

From the existing capture technologies, the only proven and mature technology is, currently, chemical absorption using aqueous amine solutions [5]. Chemical absorption is a well-known technology, and it has been widely deployed on a large scale across several industries [6].

In terms of solvent selection, amines have traditionally been considered as reagents of choice, whereas a primary alkanolamine, monoethanolamine (MEA), is typically considered the benchmark to which alternative solvents are compared. Other compounds that are often considered are piperazine (PZ), diethanolamine (DEA), and methyldiethanolamine (MDEA) [6].

One major setback in this process is that MEA has the limitation of a maximum CO_2 loading capacity, based on stoichiometry of approximately 0.5 mol CO_2/mol amine—unlike tertiary amines such as MDEA, which has an equilibrium CO_2 loading capacity nearly of 1.0 mol CO_2/mol amine [7].

The mechanism involved in the absorption of CO_2 by aqueous solutions of tertiary amines such as MDEA is somewhat different to those of primary and secondary amines, as they do not react directly with CO_2. In fact, they act as a base, catalyzing the hydration of CO_2. Thus, the reaction of interest in aqueous solutions of tertiary amines is

$$CO_2 + H_2O + R_1R_2R_3N \rightleftharpoons R_1R_2R_3NH^+ + HCO_3^-.$$

Technologically, a typical chemical absorption process consists of an absorber and a stripper, in which the absorbent is thermally regenerated. In a chemical absorption process, as shown in Figure 1 [8], the flue gas containing CO_2 enters a packed bed absorber from the bottom and contacts in counter-current with a CO_2-lean absorbent. After absorption, the CO_2-rich absorbent flows into a stripper for thermal regeneration. After regeneration, the CO_2-lean absorbent is pumped back to the absorber as a cyclic mode. The pure CO_2 released from the stripper is then compressed for the subsequent transportation and storage [3].

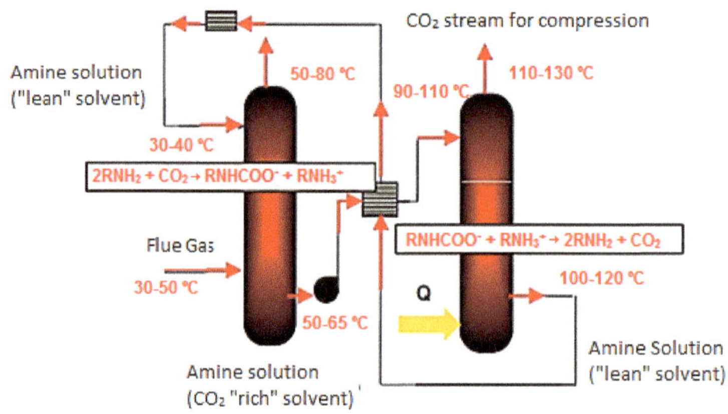

Figure 1. Process diagram for CO_2 absorption using amines with the respective temperatures and reactions [8].

As stated by Arias [9], the main challenge for scaling-up is to progress from the experimental demonstration of the concept at an increasing scale and under realistic conditions, simultaneously validating the expected benefits and overcoming the obstacles that may appear in the path towards large-scale demonstration. In particular, the actual need to perform scale-up for CO_2 absorption plants is considered critical [2], as the existing ones are not enough to deal with the current huge amount of CO_2 emissions to the atmosphere. This requirement is related to both the CO_2 flow rate and the loading capacity of solutions being used to absorb it [3].

In the present study, results from three different CO_2 capture experimental scales (laboratory, pilot unit, and a larger pilot unit) in terms of loading capacity, using aqueous amine solutions of MDEA 20 wt %, are compared, following previous studies from the authors [10].

2. Experimental

For this study, on the behavior of MDEA solutions regarding CO_2 absorption, three different experimental scales were used in order to collect consistent data, a laboratory scale, where a 500-mL three-necked flask was used, and two pilot-units, whose dimensions are shown in Table 1. All tests were performed at 1 atm and 20 °C.

Table 1. Dimensions of the absorption column used for the different scales.

Scale	Column Height (dm)	Column Inner Diameter (dm)	Column Volume (L)	CO_2 Flowrate (L/h)	Aqueous Amine Solution Flow (L/h)
Pilot-unit	15	0.505	3	1.2	24
Large pilot-unit	110	0.750	62	1200	40

2.1. *Technique for CO_2 Dosing*

For CO_2 dosing, samples were collected from the absorption column through a sampling valve, in order to ascertain whether the solution was or was not saturated. The sample was then analyzed by the $BaCO_3$ precipitation method. To use this method, the procedures described by Li et al. [11] and Santos [12] were followed: After the amine solution had been saturated with CO_2, a sample of 20 mL was collected from the pilot unit. Then, a solution of sodium hydroxide (NaOH) 1.0 M, previously prepared from commercial sodium hydroxide (Solvay), was added in excess to the sample so that the dissolved CO_2 was converted to non-volatile ionic species. Then, a solution of barium chloride dihydrate ($BaCl_2 \cdot 2H_2O$) 1.0 M, prepared previously from barium chloride dehydrate (Panreac Quimica Slu, Barcelona, Spain), was added in excess. The solution was well stirred to ensure that all the CO_2 was absorbed, and was precipitated as barium carbonate ($BaCO_3$). Afterwards, the solution containing the precipitate was filtrated, dried, and weighted. The amount of precipitate was used to calculate the CO_2 loading capacity as CO_2 moles per mol of amine. The equations used in the calculations are as follows:

$$w_{amine} = w_{sample} \times \%_w, \quad (1)$$

where

w_{amine} is the amine weight (g);
w_{sample} is the weight of the sample collected from the pilot unit (g);
$\%_w$ is the concentration of aqueous amine solution.

$$n_{amine} = \frac{w_{amine}}{MW_{amine}}, \quad (2)$$

where

n_{amine} is the number of moles of amine (mol amine);
w_{amine} is the amine weight calculated before in Equation (1);
MW_{amine} is the molecular weight of the amine (g/mol).

$$n_{CO_2} = \frac{m_{precipitate}}{MW_{BaCO_3}}, \quad (3)$$

where

n_{CO2} is the number of moles of the obtained CO_2 (mol CO_2);
$m_{precipitate}$ is the weight of the obtained precipitate (g);
MW_{BaCO3} is the molecular weight of $BaCO_3$ (g/mol).

$$\alpha = \frac{n_{CO_2}}{n_{amine}}, \quad (4)$$

where

α is the CO_2 loading capacity of the aqueous amine solution (mol CO_2/mol amine);
n_{CO2} is the number of moles of the obtained CO_2 calculated before as in Equation (3) (mol CO_2);
n_{amine} is the number of moles of the amine calculated before as in Equation (2) (mol amine).

2.2. Laboratory Scale

In a first stage, a laboratory-scale trial was performed to assess the maximum loading capacity of the prepared MDEA solution. A MDEA (Merck & Co., Kenilworth, NJ, USA) aqueous solution (20 wt %, with distilled water), was introduced in a flask. Initially, the system was purged with N_2 for several minutes in order to remove any gaseous contaminants which could be possibly present. Then, CO_2 (99.99% Air Liquide, Paris, France) was carefully introduced into the flask to promote the saturation of the solution, as shown in Figure 2 [13]. Saturation was considered complete throughout the observation of the evolution of CO_2 bubbles into the aqueous amine solution.

Figure 2. Illustration of experimental tests carried out to study the absorption of CO_2 from aqueous amines solutions [13].

In order to determine the CO_2 loading capacity of each aqueous amine solution under test until saturation, the method of barium chloride precipitation described in the previous section was used.

2.3. Pilot Unit Scale

In the following stage, in order to study the absorbing behaviour of MDEA in a carbon capture and separation system (CCS), two pilot units were used. Both units include an absorption column, as well as a stripping column, a heat exchanger between the two columns, a reboiler for the stripping column, pumping systems, surge tanks, and all necessary instrumentation and control systems.

The second stage of this study was performed using a smaller pilot unit in order to assess whether the results obtained in the trials of the first stage were still valid at a larger scale. The flowsheet of this unit is shown in Figure 3, and an actual picture is shown in Figure 4 [13].

To simulate the expected CO_2 concentration in the gaseous streams, CO_2 from a cylinder (99.99% Air Liquide, Paris, France) (3) was mixed with air from a compressor (4). Then, the gaseous solution entered the absorption column (2) and contacted with the aqueous amine solution coming from the feed tank (1), and the chemical absorption of CO_2 was able to take place. The gaseous stream leaving the top of the absorption column was connected to a CO_2 analyzer (Witt Gastechnik GmbH & Co KG, Witten, Germany) (8) so that the amount of CO_2 existing in that stream could be measured. The cold liquid stream leaving the bottom of the absorption column flowed through a dosing pump (5) and then went through a heat exchanger (7) where it was heated. The liquid stream reached the top of the stripping column (6) where the CO_2 release occurs. The gaseous stream, leaving the top of the stripping column, was connected to the stream leaving the top of the absorption column to the CO_2 meter (8) so that the amount of CO_2 in that stream could be also measured. The liquid stream leaving the stripping column at high temperature flowed through a dosing pump (5) and went through the heat exchanger (7), where it heated the stream leaving the bottom of the absorption column. This stream was cooled down in the heat exchanger and then went back into the feeding tank (1), where amine make-up was completed. Then, it entered the top of the absorption column (2), thus closing the cycle. The absorption column was filled with Berl saddles packing. Samples were collected just after the absorption column.

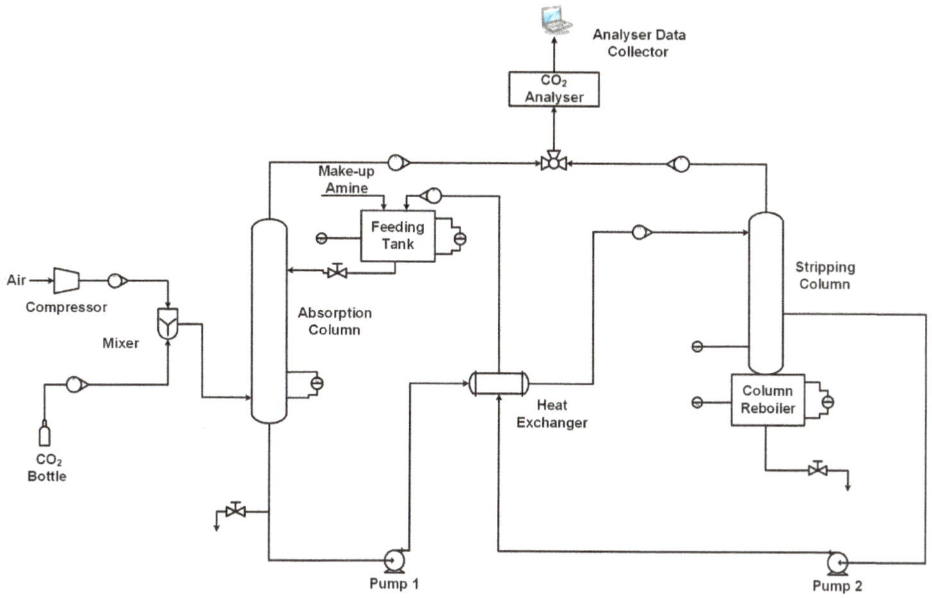

Figure 3. Flowsheet of the pilot units [13].

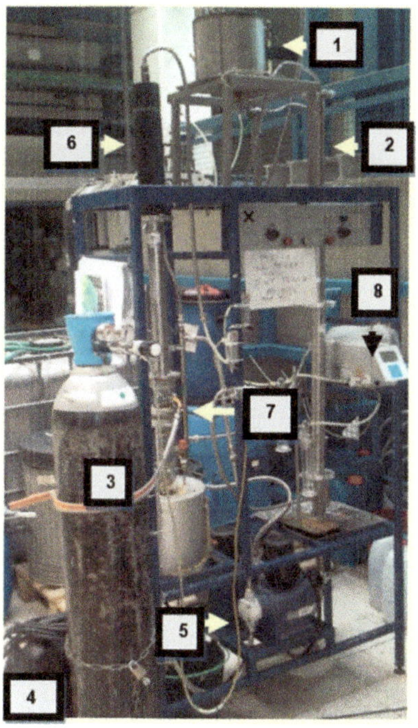

Figure 4. Small pilot unit [13] and its components: **1**—Feeding tank; **2**—Absorption column; **3**—CO_2 cylinder; **4**—Air compressor; **5**—Dosing pump; **6**—Stripping column; **7**—Heat exchanger; **8**—CO_2 analyzer.

In order to study the loading capacity of the amine under test, aqueous solutions 20 wt % were prepared and placed in the feeding tank. CO_2 was then circulated jointly with the aqueous amine solution, and samples were collected every 30 min until the amine saturation was achieved. The amount of formed precipitate from the addition of $BaCl_2 \cdot 2H_2O$ was used to calculate the CO_2 loading capacity in terms of moles of CO_2 per moles of amine, following the procedure previously described.

In the third stage, tests were performed in a larger pilot unit constructed in accordance with the same flowsheet, but using larger columns and higher flowrates. Moreover, due to the size of pilot unit, a closed circuit television (CCTV) system was installed so that, with the help of the monitors placed throughtout the pilot unit, it was possible to observe what was happening during the tests. A Berl saddles packing was placed inside the absorption column in order to enable better contact between the aqueous MDEA solution and the CO_2 stream. CO_2 flows jointly with the aqueous amine solution, and samples were collected every 30 min until the amine saturation was achieved, again after the absorption column. In order to determine the CO_2 loading capacity by the aqueous MDEA solutions, the precipitation method was used again. A photo of this pilot unit was shown in Figure 5. Both units were operated so that the system achieves nearly a steady state.

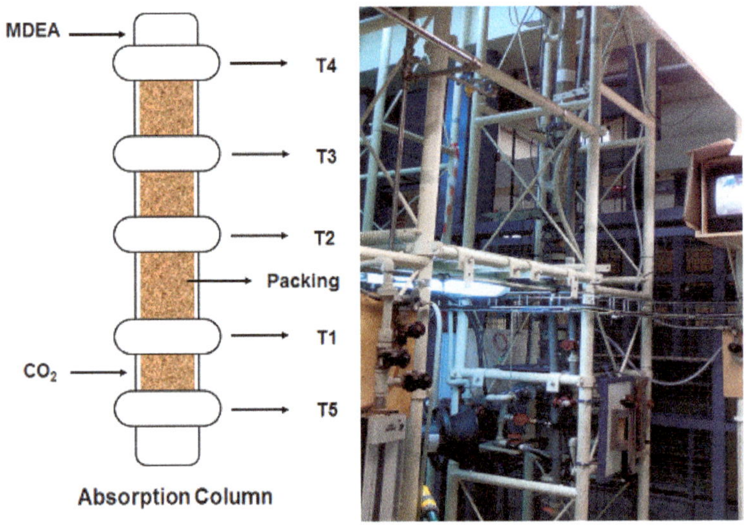

Figure 5. Large pilot unit and absorption column detail.

3. Results

3.1. Results from Laboratory Scale

In the laboratory scale trial performed using a 500-mL three-necked flask, the maximum loading capacity of the prepared MDEA solution was assessed. CO_2 was bubbled inside the aqueous MDEA solution until it was saturated. This process took around four hours to be achieved.

Using the precipitation method, as described before, the loading capacity of the saturated MDEA solution, after four hours, was determined to be 0.979 mol CO_2/mol amine. This value can be used as a benchmark to compare the results obtained in both pilot units. The results presented here, as well as the ones for the other scales described thereafter in Section 3.2, are the averaged values from a series of three tests each.

3.2. Results from Pilot Unit Scale

For the smaller pilot unit, using the same dosing method, the obtained results are shown in Table 2 and represented graphically in Figure 6.

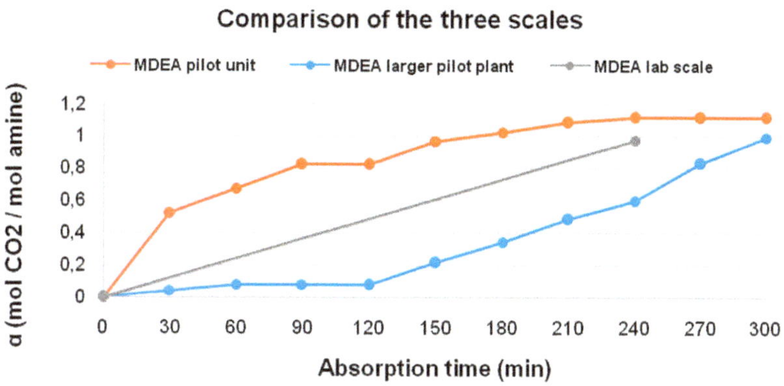

Figure 6. Comparison of the obtained results for CO_2 loading capacity (α) over the absorption time for the tested aqueous MDEA solution (20 wt %) in the three different testing scales.

Table 2. Results from CO_2 absorption by the aqueous methyldiethanolamine (MDEA) solution at 20% (w/w) in the small pilot unit.

Absorption Time (min)	Loading Capacity (α) (mol CO_2/mol amine)
30	0.529
60	0.676
90	0.824
120	0.824
150	0.971
180	1.029
210	1.089
240	1.118
270	1.118
300	1.118

In the tests performed in the larger pilot plant, the precipitation method was also used to determine the CO_2 loading capacity by the aqueous MDEA solutions in order to assess whether the results obtained in this pilot unit would be in accordance with the ones obtained in the small pilot unit. The obtained results are shown in the Table 3 and graphically in Figure 6.

Table 3. Results from CO_2 absorption by the aqueous MDEA solution at 20% (w/w) by the precipitation method in the larger pilot unit.

Absorption Time (min)	Loading Capacity (α) (mol CO_2/mol amine)
30	0.046
60	0.084
90	0.082
120	0.079
150	0.218
180	0.347
210	0.486
240	0.602
270	0.831
300	0.998

4. Discussion

4.1. *Laboratory Scale*

It should be noted that MDEA is a tertiary amine and does not react directly with CO_2, but acts as a base, catalyzing the hydration of CO_2, which means that its equilibrium CO_2 loading capacity is approximately 1.0 mol CO_2/mol amine. Therefore, the results obtained at laboratory scale (α = 0.979 mol CO_2/mol amine) are consistent with what was to be expected [7,14].

4.2. *Pilot Unit Scale*

When analyzing the obtained results, it can be concluded that, by using the smaller pilot unit, it was possible to achieve a CO_2 absorption profile nearly in accordance with the theoretical basis, whereas the MDEA aqueous solution reached CO_2 saturation around the 240th min; although the actual obtained value was slightly higher than it was expected (1.0 mol CO_2/mol amine), this could be attributed to the experimental inaccuracy of the precipitation method used to determine CO_2. Comparing the obtained results for the three scales, which are graphically shown in Figure 6, marked differences appear between the absorption patterns regarding the tests performed in each pilot unit. It should be noted that, although the loading capacity is still increasing, in the larger pilot unit, from the 270th to the 300th min, it seems to be enough, for comparison purposes, to use the loading capacity at 300 min of operation.

These differences may result from process scale-up. In fact, scaling up any process can result in several difficulties for chemical engineers due to the fact that the relative importance of process variables affecting performance considerably increases. In fact, process variables, such as CO_2 injection point in the absorption column, reactant flow, contact time between the reactants throughout the absorption column, temperature and pressure control, and maintenance, have a marked influence on the performance of the units. These variables are somewhat easier to control at laboratory scale and at a small pilot unit, but its significance becomes considerably relevant for larger scales, as noticed by Zlokarnik [15]. This can be easily concluded by comparing the results from both pilot plant units presented in this paper. Although theoretically at the same pressure and temperature (two variables which were kept constant for all performed tests), the saturated CO_2 loading should be the same, regardless of the scale used. However, this does not happen in practice, as demonstrated by this study. In this study, other variables such as column internal packing, pressure drops, residence time, and flooding conditions were kept approximately constant, in spite of scale differences. Nevertheless, it should be noted that, for this preliminary analysis, mass transfer coefficients have not been calculated.

As stated in Table 1, the small pilot unit has a column height of 15 dm, while the larger pilot unit has a column height of 110 dm, representing an increase of 7.3 times, and the inner column diameter increased from 0.505 dm to 0.750 dm in the larger pilot unit, representing an increase of 1.5 times. This results in quite different volumes for the columns of 3 L in the small pilot unit and 62 L in the larger pilot unit, which represents a scale increase of 20.7. In such a larger volume, it is much more difficult to obtain a good mixing contact between the gas phase (CO_2) and the liquid phase (aqueous amine solution); therefore, the system CO_2 loading capacity (α) will be much slower to reach equilibrium, as observed in this study, as shown in Figure 6. Furthermore, the maximum obtained for the latter situation is even lower than the one obtained in the small pilot plant.

These scale-up effects are, in fact, quite difficult to quantify but tend to considerably affect the performance of the absorption process, particularly when progressing for even larger scales, such as industrial plants, as pointed out by previous authors [3,4].

However, a way to decrease the significance of the inconvenient scale-up effects will be to complement the scaling-up process by additionally including the use of simulation tools in order to help in the selection of the best operating conditions for the process, contributing to high production yields and thus a more profitable operation, as required for the integration of CCS within current production processes [16]. In fact, the use of steady-state or dynamic models for simulation, taking into account precise geometry factors of the columns, as well as operating factors such as the flow rate of each phase, could help to reduce the inaccuracy of the estimation process, as suggested by Panahi and Skogestad [17]. Particular care should be taken in the scale-up of chemical processes involving unit operations such as absorption and stripping, which are, in fact, affected by a multitude of operational factors, apart from dimensional ones.

Acknowledgments: The authors gratefully acknowledge the support from the KIC Inno Energy Project: ACoPP—Advanced Near-Zero Emissions Coal-Fired Plant—and thank Jorge Rodrigues de Carvalho (IST) for laboratory and equipment utilization. No funds were received to cover publication costs.

Author Contributions: João C. Bordado and João F. Gomes conceived and designed the experiments. Samuel P. Santos performed the experiments. All authors were involved in the analysis of data, the discussion of results, and the writing of the paper. All authors have read and approved the final manuscript.

Conflicts of Interest: The authors declare no conflict of interest.

References

1. Song, C. Global challenges and strategies for control, conversion and utilization of CO_2 for sustainable development involving energy, catalysis, adsorption and chemical processing. *Catal. Today* **2006**, *115*, 2–32. [CrossRef]
2. Yang, H.; Xu, Z.; Fan, M.; Gupta, R.; Slimane, R.; Bland, A.; Wright, I. Progress in carbon dioxide separation and capture: A review. *J. Environ. Sci.* **2008**, *20*, 14–27. [CrossRef]
3. Yu, C.; Huang, C.; Tan, C. A review of CO_2 capture by absorption and adsorption. *Aerosol Air Qual. Res.* **2012**, *12*, 745–769. [CrossRef]
4. Rao, A.; Rubin, E. A technical, economic and environmental assessment of amine-based CO_2 capture technology for power plant greenhouse gas control. *Environ. Sci. Technol.* **2002**, *36*, 4467–4475. [CrossRef] [PubMed]
5. Figueroa, J.; Fout, T.; Plasynski, S.; McIlvried, H.; Srivastava, R. Advances in CO_2 capture technology—The U.S. Department of Energy's Carbon Sequestration Program. *Int. J. Greenh. Gas Control* **2008**, *2*, 9–20. [CrossRef]
6. MacDowwell, N.; Florin, N.; Buchard, A.; Hallet, J.; Galindo, A.; Adjiman, C.; Williams, C.; Shah, N.; Fennell, P. An overview of CO_2 capture technologies. *Energy Environ. Sci.* **2010**, *3*, 1645–1669. [CrossRef]
7. Supap, T.; Idem, R.; Tontiwachwuthikul, P.; Saiwan, C. Kinetics of sulfur dioxide and oxygen-induced degradation of aqueous monoethanolamine solution during CO_2 absorption from power plant flue gas streams. *Int. J. Greenh. Gas Control* **2009**, *3*, 133–142. [CrossRef]
8. Gomes, J. Sequestração de CO_2—Uma Visão Integrada das Tecnologias Químicas Disponíveis. Habilitation Thesis, Universidade Técnica de Lisboa, Lisboa, Portugal, 2010. (In Portuguese).
9. Arias, B.; Diego, M.; Abanades, J.; Lorenzo, M.; Dias, L.; Martínez, D.; Alvarez, A.; Sánchez-Biezma, A. Demonstration of steady state CO_2 capture in a 1.7 MW_{th} calcium looping pilot. *Int. J. Greenh. Gas Control* **2013**, *18*, 237–245. [CrossRef]
10. Gomes, J.; Santos, S.; Bordado, J. Choosing amine-based absorbents for CO_2 capture. *Environ. Technol.* **2015**, *36*, 19–25. [CrossRef] [PubMed]
11. Li, M.; Chang, B. Solubilities of carbon dioxide in Water + Monoethanolamine + 2-Amino-2-methyl-1-propanol. *J. Chem. Eng. Data* **1994**, *39*, 448–452. [CrossRef]
12. Santos, A. Estudos de Absorção de CO_2 em Soluções Aminadas. Master's Thesis, Instituto Superior de Engenharia de Lisboa, Lisboa, Portugal, 2012. (In Portuguese).
13. Santos, S.; Gomes, J.; Bordado, J. *Comparative Study of Amine Solutions Used in Absorption/Desorption Cycles of CO_2*; LAP Lambert Academic Publishing: Saarbrucken, Germany, 2014.
14. Budzianowski, W. Single solvents, solvent blends and advanced solvent systems in CO_2 capture by absorption: A review. *Int. J. Glob. Warm.* **2015**, *7*, 184–225. [CrossRef]
15. Zlokarnik, M. *Scale up in Chemical Engineering*, 2nd ed.; Wiley-VCH: Weinheim, Germany, 2006.

16. Budzianowski, W. Low-carbon power generation cycles: The feasibility of CO_2 capture and opportunities for integration. *J. Power Technol.* **2011**, *91*, 6–13.
17. Panahi, M.; Skogestad, S. Economically efficient operation of CO_2 capturing process part I: Self-optimizing procedure for selecting the best controlled variables. *Chem. Eng. Process.* **2011**, *50*, 247–253. [CrossRef]

8

Dry Separation of Palm Kernel and Palm Shell Using a Novel Five-Stage Winnowing Column System

Rohaya Mohamed Halim [1], Ridzuan Ramli [2], Che Rahmat Che Mat [1], Choo Yuen May [3], Nasrin Abu Bakar [4] and Nu'man Abdul Hadi [1,*]

1 Milling & Processing Unit, Engineering & Processing Research Division, Malaysian Palm Oil Board, Ministry of Plantation Industries & Commodities, 6, Persiaran Institusi, Bandar Baru Bangi, 43000 Kajang, Selangor, Malaysia; rohaya@mpob.gov.my (R.M.H.); cherahmat@mpob.gov.my (C.R.C.M.)
2 Biomass Technology Unit, Engineering & Processing Research Division, Malaysian Palm Oil Board, Ministry of Plantation Industries & Commodities, 6, Persiaran Institusi, Bandar Baru Bangi, 43000 Kajang, Selangor, Malaysia; ridzco@gmail.com
3 Director General's Office, Malaysian Palm Oil Board, Ministry of Plantation Industries & Commodities, 6, Persiaran Institusi, Bandar Baru Bangi, 43000 Kajang, Selangor, Malaysia; choo@mpob.gov.my
4 Energy & Environment Unit, Engineering & Processing Research Division, Malaysian Palm Oil Board, Ministry of Plantation Industries & Commodities, 6, Persiaran Institusi, Bandar Baru Bangi, 43000 Kajang, Selangor, Malaysia; nasrin@mpob.gov.my
* Correspondence: numan@mpob.gov.my

Academic Editor: Manoj Gupta

Abstract: The conventional separation system for the recovery of palm kernel from its palm shell–kernel mixture using water as process media generates a considerable amount of waste effluent that harms the environment. The aim of this study is to develop a dry separation process for the recovery of palm kernel by using winnowing columns. A commercial system consisting of a series of five winnowing columns was developed and installed at a local palm oil mill. The system parameters, including column height, blower capacity, airflow rate and mesh screen size for shell removal, were studied and optimized to ensure good separation of kernel and shell in the column to enable collection of different sizes of kernel and shell at each column outlet. The performance of the separation process was evaluated in terms of its kernel losses, dirt content and kernel recovery rate. The average kernel losses based on oil palm fresh fruit bunches processed were found to vary from 0.11 to 0.30 wt %, with most of the values obtained being below the targeted limit of 0.30 wt %. The dirt content was in the range 4.56–6.03 wt %, which was mostly below the targeted limit of 5.5 wt %. The kernel recovery rate was in the range 5.69–6.89 wt %, with most of the values achieving the minimum targeted limit of 6.00 wt %. The system operates under completely dry conditions and, therefore, produces zero waste effluent.

Keywords: dry separation system; winnowing column; kernel recovery; palm kernel; palm shell; palm oil mill

---

1. Introduction

Palm fruit of oil palm (*Elaeis guineensis*) contains mesocarp and palm kernel, the latter contributing 45–48 wt % of the total fruit [1]. The kernel contains *ca.* 45%–50% kernel oil, which is rich in lauric acid C12 (45%–52%), myristic acid C14 (15%–17%) and oleic acid C18:1 (13%–19%) [2]. Palm kernel oil is widely used in food applications, including the production of margarine, confectionary fat, shortening and vegetable ghee [3–5]. Furthermore, research and development have made it possible for the palm kernel oil to be used in nonfood lipid sources for the production of bioresin, lubricants, chelating agents for metal extraction and other important chemicals [6,7].

In palm oil mill (POM), oil palm fresh fruit bunches (FFB) are processed to produce crude palm oil (CPO). Palm fiber, empty fruit bunches, sludge, decanter cake and palm nuts are produced as byproducts. The nut is usually collected after the removal of the mesocarp fiber, and the kernel inside the nut is separated from the nut's shell for subsequent oil extraction [8,9]. The shell can be utilized as fuel, used in the production of activated carbon for waste water treatment, and used in analytical applications [10–12]. The production of palm kernel starts with the cracking of palm nuts using a palm nutcracker to produce a cracked mixture of palm kernel and palm shell [13]. Subsequently, the separation of shell and kernel from the cracked mixture is performed using a combination of dry and wet separation systems.

The existing system commonly adopted at a POM in Malaysia separates *ca.* 20 wt % light shell fragments using a dry separation winnowing column, while *ca.* 80 wt % heavy shell and kernel are separated using wet separation systems either by claybath or hydrocyclone [14]. The wet separation process of a cracked mixture is based on the differential specific gravities of shell and kernel [15]. Both the wet separation systems use water for performing the desired separation and are not considered environmentally friendly due to the generation of a large volume of waste effluent [16]. These processes produce wet kernel that requires a drying process prior to storage and therefore requires additional energy for the heat transfer. POMs in Malaysia currently employ the wet separation system for the separation of kernel from the shell. The drawbacks associated with the wet process were successfully eliminated in the current system, which is based on a dry technique. A few attempts have been made to introduce a dry technique for the separation of the kernel from its cracked mixture on a commercial and pilot scale. One method in particular uses an experimental rotary separator, as reported by Olasumboye and Koya [17]. Though the objective of achieving dry separation was a success, the study also reported a significant amount of kernel losses, ranging from 4% to 5%, due to attachment to the shell during the separation process [17].

The cracked mixtures produced by the nutcracker consist of kernel, broken kernel, half-cracked nuts, small shell and large shell [13]. The separation process of the shell and kernel fragments depends on the physical characteristics of the components of the mixture. If the fragments are large, the separation process is simple. The separation process becomes more complex when there are different sizes and shapes of the shell and kernel [18]. Therefore, a single- or double-stage separating system will not perform efficiently without incurring high kernel loss [14]. There is an urgent need by the industry to improve the conventional separation system as well as to maximize the recovery of dry kernel and shell. Therefore, the aim of this study is to develop an efficient system that can recover the maximum amount of kernel irrespective of their sizes and at the same time produce no waste effluent.

The system developed in this study is designed to separate shell and kernels based on their respective mass and densities. More specifically, the kernel with a greater density compared to other fractions (shell and its fragments) will fall vertically onto the outlet chutes of the columns, while the shell fragments will be blown upwards towards the top opening of each column and then to the cyclone member, which reduces the air velocity to enable the lighter material to settle [14]. Some of the kernel particles with less mass may be blown upwards together with the lightweight shell fragments and discharge as waste material, leading to inefficient separation in the single-stage winnowing column. The purpose of this study, therefore, is to design a separation system consisting of multiple stages of a winnowing column using a suitable airflow rate to extract kernel fragments from the cracked nut mixture without necessarily using a conventional method, such as the claybath or hydrocyclone.

2. Experimental Sections

2.1. Materials

The feeding material for the five-stage winnowing column system was obtained by cracking the palm nuts produced after the milling process. The cracked mixture consists of palm kernel (whole kernel and broken kernel), palm shell and other foreign materials, such as sand and stones.

Rolek nutcracker (MPOB & Hur Far, Selangor, Malaysia) was used to crack the palm nuts due to its cracking efficiency of more than 98%, irrespective of the nut size [13].

2.2. Development of the Separation System

In this study, a winnowing column is used for the dry separation of kernel and shell from the nut's cracked mixture. The system consists of five winnowing columns that can cater about 10 t/h of the cracked nut mixture throughput was developed. In addition to aiming for a totally dry process, it is also designed to improve the separation efficiency of the cracked nut mixture to provide better kernel recovery and shell removal. Figure 1 (left) shows a single-stage winnowing column, whereas Figure 1 (right) shows a dry separation system—the five-stage winnowing column system. Each column is designed with different specifications, *viz.* airflow rate, column height, inlet level and outlet level, in order to accommodate the feeding materials and to achieve the desired shell and kernel ratio at each outlet point.

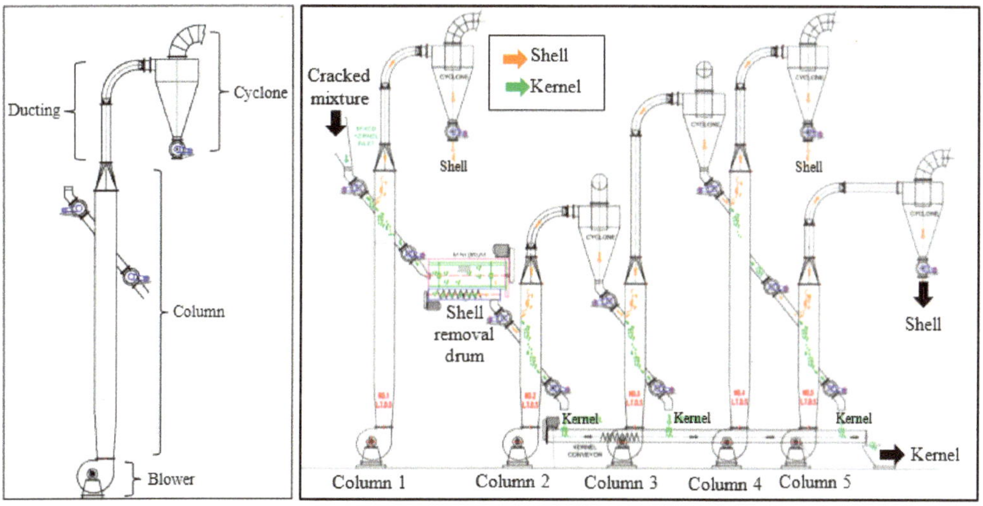

Figure 1. (**left**) A single stage consists of a blower, a column, a ducting and a cyclone unit; and (**right**) a five-stage winnowing column system for the separation of the kernel from the cracked nut mixture.

The winnowing column is designed with two special features to enhance the separation process. The first feature is that the internal column is equipped with a special device responsible for diverting the direction of the airflow away from the cracked mixture feed flow path, thus creating a laminar airflow through the separating column for better separation of the cracked mixture. The second feature is an adjustable damper for regulating the airflow rate in the separating column by altering the throat diameter of the column. This feature can also be enhanced by column design that uses forced draught instead of induced draught where its airflow rate is powered by an electrical blower and adjusted via the damper located at the ground or at an elevated level as shown in Figure 2. Theoretically, a forced draught puts the system under positive pressure (above atmospheric pressure) and therefore consumes less power compared to an induced draught system that operates under negative pressure (below atmospheric pressure). The forced draught system is due to the placement of the fan at the entry end of the path flow where the outside air is drawn and forced into the system. The throat geometry is easily adjustable on each separating column to cater for the differing characteristics of the cracked nut mixture. The air velocities at each column are optimized by varying the fan speed from the control panel.

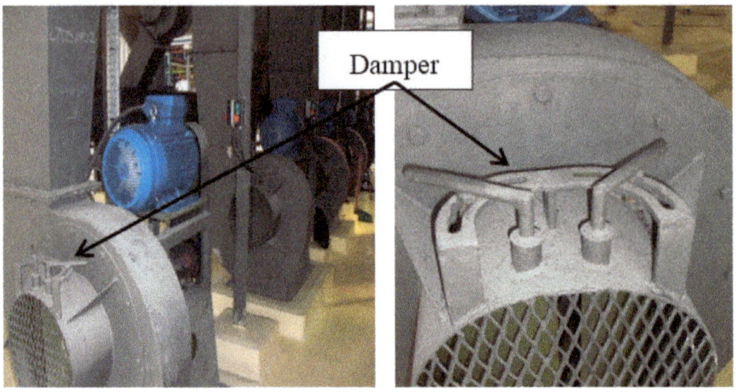

Figure 2. Blower damper installed at the bottom of the column.

Air velocity was measured using an air velocity meter (PCE-AM81, Southampton, UK). The principle of the separation is based on the airflow rate, which was calculated as follows:

$$V_i = v_i \times A_i \tag{1}$$

where V_i is the volumetric airflow rate (m^3/min), v_i is the air velocity (m/min) and A_i is the throat area (m^2), for the respective column i.

The separation of shell and kernel in this system works on the principle of the specific gravity differential of the materials in the presence of a high velocity airflow caused by a blower powered by an electric motor. Each stage, as shown in Figure 1 (left), consists of a blower, a winnowing column, a ducting and a cyclone, whereby the separation occurs at each column but the separated material is not necessarily discharged. Circular ducting connects the upper end of the column to the cyclone.

The first column discharges most of the shell after the cyclone unit where the kernel and heavy shell are sent to the second column. The second column discharges most of the kernel and the remaining shells and light kernel are sent to the third column. The light material that goes to the cyclone still remains as a mixture of kernel and shell that will again be subjected to more rigorous separation in the following stages; its quality is not critical. In Stage 5 of the system, the separating column will lift almost all the shells and deposit them in the cyclone due to the reduced height of the cyclone and the ducting. If some small and broken kernels and small shell still remain, these will be discharged from the bottom chute of the fifth separating column. A unit of the shell removal drum (Figure 3) with a mesh size between 14 and 16 mm is placed between the first and second columns to remove any contaminant, specifically thick shell (large and small shell) and uncracked nut from continuing to the next separation column. The removal of these shell fragments at an early stage reduces the load of the cracked mixture to be separated in the subsequent stages.

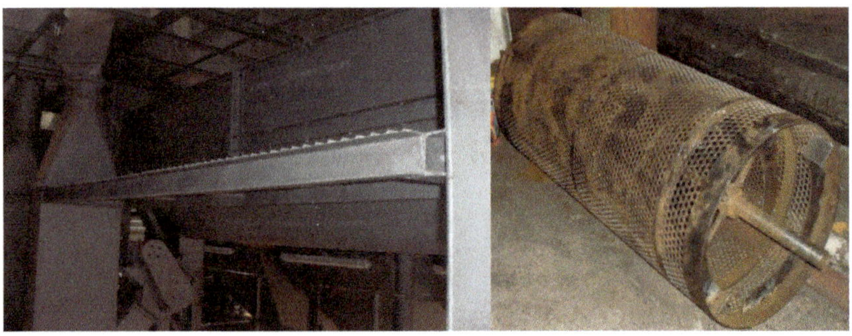

Figure 3. Shell removal drum placed between the first and the second stage.

In all stages of the separation, fan speeds, airflow damper control positions and heights of the cyclone inlet from the top outlet of the separating column are critical because efficient separation of the different components of the cracked mixture rests solely on these settings. These settings were obtained by conducting a number of expensive research trials and as such were safeguarded by means of intellectual property rights and patent.

2.3. Evaluation of System Performance

The five-stage winnowing column system was installed and commercially evaluated at Air Tawar POM, Johor, with a mill capacity of 45 t/h processed FFB. The performance of the separation process was carried out for 8 months, and the system was continuously monitored for another 12 months. Parameters that were closely monitored during the commercial trials were kernel throughput, kernel recovery rate (KER), dirt or shell content in the recovered kernel and kernel losses to ensure meeting the targeted performance as follows: total kernel throughput in the range of 35 to 45 wt %, dirt and shell content in the kernel <5.5 wt %, kernel losses to FFB <0.3 wt % and KER >6.0 wt %. Results were analyzed and compared with the targeted performance of the system and other sister mills in the same region that used a conventional separation system.

2.4. Determination of the Total Dirt Content

We collected 1 kg of the recovered palm kernel (from the final discharge outlet to determine its dirt content. The kernel sample was sorted as whole nut (d_1), half-cracked nuts (d_2), loose shells (d_3) and stones and debris (d_4). Prior to weighing, the whole nuts (d_1) and half-cracked nuts (d_2) were carefully cracked and the shells were weighed separately. Total dirt content was calculated as follows:

$$D_t = [(d_1 + d_2 + d_3 + d_1) \; / \; PK]_t \times 100\% \tag{2}$$

where D is the total dirt content (wt %), d_1 is the weight of shells of whole nuts (kg), d_2 is the weight of shells of half-cracked nuts (kg), d_3 is the weight of loose shells (kg), d_4 is the weight of stones and debris (kg) and PK is the weight of the palm kernel sample (kg), for a t period of evaluation.

2.5. Determination of Kernel Losses

We collected 1 kg of shell fragment containing mostly shells along with kernel carryover from the shell bunker. The kernel loss was calculated as follows:

$$KL_t = K_t \times [(S_t N_t)] \times (N_t F_t) \times 100\% \tag{3}$$

where KL_t is the kernel loss over FFB processed (wt %), K_t is the weight of kernel carryover in the dirt sample (kg), S_t is the weight of shell (kg), N_t is the weight of nut (kg) and F_t is the weight of FFB (kg), for a t period of evaluation. Shell to nut (S/N) and nut to FFB (N/F) ratios were obtained from actual milling activity.

2.6. Determination of the Kernel Recovery Rate

The KER was calculated by the weight ratio of kernel and fresh fruit bunch as follows:

$$KER_t = (K_t/F_t) \times 100\% \tag{4}$$

where KER_t is the kernel recovery rate (wt %), K_t is the weight of kernel (kg) and F_t is the weight of FFB processed (kg), for a t period of evaluation.

2.7. *Determination of Nut and Kernel Size Distribution*

We collected 2.0 kg of nut and kernel sample from the feeding to the Rolek nutcracker and the kernel discharge outlet, respectively, for size measurement. Both nut and kernel sizes were measured using a perforated metal plate consisting of different sizes, as shown in Figure 4 (left). The horizontal cross-sectional diameter of the nuts and kernels was measured using the plate, and its size distribution was determined.

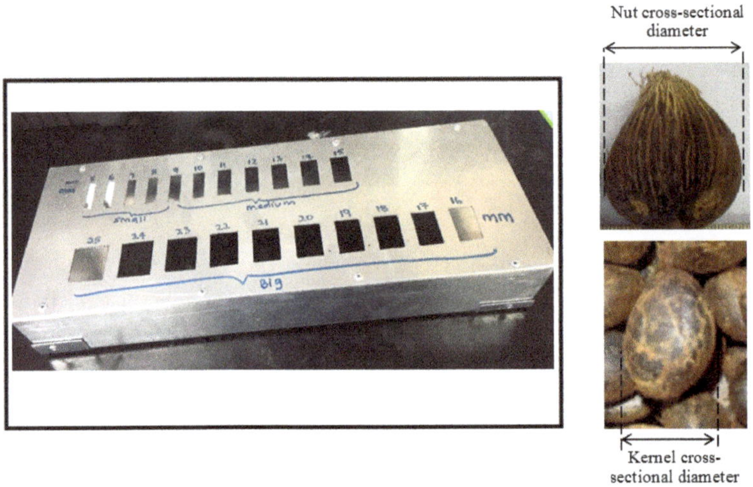

Figure 4. **(left)** Perforated plate to measure different nut and kernel size; and **(right)** horizontal cross-sectional diameter of nut and kernel.

3. Results and Discussion

3.1. *Nut and Kernel Size Distribution*

The Rolek nutcracker cracks palm nuts to produce a cracked mixture consisting of more than 35 wt % whole kernel, less than 12 wt % broken kernel, less than 1 wt % uncracked nuts and less than 2 wt % half-cracked nuts. Conventional nutcrackers used at POMs, such as ripple mill and super cracker, often crack both the nuts and the kernel; the cracked mixture then consists of a high broken kernel of *ca.* 15–25 wt % [19]. Rolek nutcracker differs from other commercial crackers in terms of its principle of operation, which relies on dynamic interactions of adjustable rods varied to break all sizes of nuts without causing high kernel breakage [13,19].

Currently, there is no established method to measure nut and kernel size. Nut and kernel are not perfectly round in shape; thus, the horizontal cross-section of the nut and kernel, as shown in Figure 4 (right), is deemed appropriate to represent the size of the nut and kernel. To determine the size distribution of the collective samples, the sizes are generally grouped into the acceptable range of nut and kernel size, ranging from 5–25 mm. A perforated metal plate consisting of holes with different sizes ranging from 5–25 mm was therefore developed to facilitate size measurement of the collective samples, as shown in Figure 4 (left). This method of measurement is common in POMs in Malaysia but has not been properly documented elsewhere.

Measurement of nut and kernel size distribution prior to cracking was carried out to determine the exact composition of nut type processed in the mill, such as *Dura*, *Pisifera* and *Tenera*. *Pisifera* is a pollen-type nut, while *Dura* has a thicker shell and a lower oil content compared to *Tenera*, which is a hybrid between *Pisifera* and *Dura*. *Tenera* has a high oil content with a thin shell. Figure 5 shows nut size distribution prior to cracking and kernel size distribution after separation. It shows that Air Tawar POM processed about 96.7% *Tenera* nuts and 3.3% *Dura* nuts. Typical *Tenera* nuts may be 20 mm or less in length, and *Dura* nuts may be 20 to 30 mm in length. In terms of horizontal cross-sectional diameter,

Tenera nut size is less than 16 mm, while *Dura* nuts are often more than 16 mm. The size distribution of nuts ranged from 10 to 18 mm in diameter; the largest nut sizes ranged from 13 to 16 mm with kernel sizes between 10 and 12 mm. Determination of the nut and kernel size distribution was important prior to optimization of kernel separation using the five-stage winnowing column system.

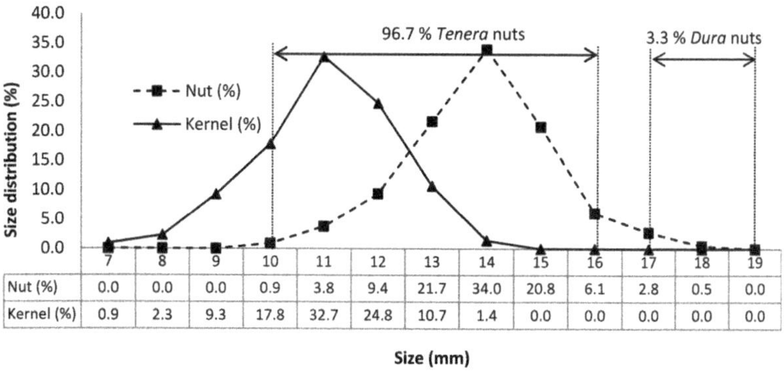

Figure 5. Nut and kernel size distribution of the feeding materials.

3.2. Kernel Throughput and Separation Efficiency

The determination of the kernel and shell throughput was carried out at three kernel-discharge points and three shell-discharge outlets. Table 1 shows the optimized setup for the five-stage winnowing column system for the dry kernel separation process in terms of airflow rate, damper opening and column height to achieve the desired 45 wt % of total kernel throughput. The cracked mixture fed into the system contained *ca.* 55 wt % shells and 45 wt % kernels; thus, the maximum recovery of kernel was at 45 wt % and maximum shell removal at 55 wt %. As mentioned previously, the system was designed so that shell and kernel would be discharged at Stages 1, 4, and 5 and Stages 2, 3, and 5, respectively. Separation of shell and kernel in a column is driven by the specific gravity differential of materials and in the presence of high velocity airflow blown by an electrical fan. Furthermore, different openings of the damper drive the variation in air velocity to suit the characteristic of the feed material.

Table 1. Optimized setup for the five-stage winnowing column system for the dry separation of kernel and its product throughput.

Stage	Blower Power (HP)	Airflow Rate (m^3/min)	Damper Open-up (%)	Column Height (m)	Products Discharged	Products Description	Products Throughput (wt %)	
							Min.	Max.
1	30	257	50	8.64	Shell	Small shell	30	35
2	25	202	50	5.03	Kernel	Big whole kernel Big broken kernel	20	28
3	25	196	40	5.03	Kernel	Small whole kernel Small broken kernel	5	8
4	25	173	30–35	8.64	Shell	Big and thin shell	10	12
5	25	173	30–35	5.03	Shell Kernel	Big and thick shell Small kernel	5 5	8 7
			Total shell throughput (wt %) Total kernel throughput (wt %)				45 30	55 43

We found most of the small shells (30–35 wt %) were successfully discharged at the first stage. Large and thin shells (10–12 wt %) were discharged at the fourth stage, whereas large and thick shells

(5–8 wt %) were discharged at the fifth stage. Overall, the system showed excellent shell removal efficiency by more than 82 wt % (min. 45 wt % shell throughput), as shown in Table 1.

The highest removal of shells in the first stage was due to the highest airflow rate utilized at 257 m^3/min in the first column compared to 202, 196, 173 and 173 m^3/min in the second, third, fourth, and fifth columns, respectively. Technically, a high airflow rate blows the small light shells upwards and leaves the heavy kernels and thick shell fraction deposited at the bottom of the column. However, due to the high air velocity, this is also associated with a loss of light broken kernel that is blown away with the small shells, amounting to 0.1–0.2 wt % (kernel to FFB) (Table 2). For comparison, this loss is higher than the kernel losses at the fourth and the fifth stage, which are both in the range of 0.06–0.08 wt % (kernel to FFB) (Table 2). This stage makes use of a high column height (8.64 m) to allow longer separation time in order to effectively push up most of the small shells (Table 1). A similar case was also observed at the fourth stage where a high column design enables good removal of shells in the range of 10–12 wt % from the total throughput.

Table 2. Contaminants in terms of [1] kernel losses and [2] free shells in the recovered products.

Stage	Products Discharged	Contaminants	Throughput (wt %)	
			Min.	Max.
1	Shell	Kernel losses	0.1	0.2
2	Kernel	Free shells	1.0	2.0
3	Kernel	Free shells	2.0	3.0
4	Shell	Kernel losses	0.06	0.08
5	Shell	Kernel losses	0.06	0.08
5	Kernel	Free shells	2.0	3.0
	Total kernel losses (wt %)		0.22	0.36
	Total free shells (wt %)		5.0	8.0

[1] Kernel losses are reported based on FFB processed. [2] Free shells are reported based on (kernel) sample collected.

For the separation and recovery of the kernel, 20–28 wt % large whole kernels and large broken kernels were discharged at the second stage. The remaining 5–8 wt % and 5–7 wt % small whole kernels and small broken kernels were recovered at the third and fifth stages, respectively. In general, the system showed excellent kernel recovery efficiency by more than 67% (min. 30 wt % kernel throughput), as shown in Table 1.

The system was intentionally designed for the second stage to carry out most of the separation of large whole kernels. The discharge at the second-stage fraction was kept as clean as possible by using a high separating velocity in the separating column, represented by the highest airflow rate (202 m^3/min) compared to the airflow rate at the third and fifth columns where kernel was also discharged (196 m^3/min and 173 m^3/min, respectively). This was achieved through the opening of the damper to 50% at the second stage, which provides a higher air velocity compared to opening the damper at the third and fifth columns to 40% and 30%, respectively. Nevertheless, we found that the recovered kernel from the second stage was contaminated with 1–2 wt % of free shells, which was still within the acceptable limit (Table 2). We note that the second and the third stages, which both accounted for almost 84% of the kernel recovery (max. 36 wt % of total throughput), had a lower column height at 5.03 m compared to 8.64 m for the first and fourth stages, which are meant for shell removal. In contrast to the shell removal principle, which requires a high column to allow a sufficient floating duration of shells, recovery of the kernel requires a short column to avoid kernel from being lifted to the upper end of the column.

The third stage, also designed for kernel recovery, used a lower separating velocity represented by a lower airflow rate (from 202 to 196 m^3/min) by reducing the damper opening from 50% to 40% (Table 1). As the separation column throat area remained the same as for the first stage, the lower separating velocity was able to separate out the light kernel fraction in the form of small whole kernel and large broken kernel, both having almost the same specific gravity. About 5–8 wt % kernels (small whole and large broken kernels) were successfully discharged through the bottom outlet. Consequently, the remaining broken kernel and shell were lifted and deposited in the third-stage discharge cyclone for further separation. In addition to the optimum separating velocity, the height of the circular ducting from the column is much higher (Figure 1 (right)) to prevent kernel from being lifted to the cyclone. The height of the circular ducting was in the order of: third column > fourth column = first column > fifth column > second column (Figure 1 (right)). Results obtained confirm that kernel loss in the third-stage shell discharge outlet was in the range of 0.06–0.08 wt %, which was less than in the shell discharge outlet at the first stage (Table 2).

Small kernels and shells from the third-stage cyclone were then transferred by a screw conveyor to the feed inlet of the fourth-stage separating column where the large and thin shells (10–12 wt %) and smashed kernels (0.06–0.08 wt %) were lifted and delivered to the discharge cyclone for discharging into the boiler as fuel. The airflow rate here was further reduced from 196 to 173 m^3/min by adjusting the setting of the damper position from 40% to 30%–35% opening. Apart from the designated air velocity, the reduced height of the cyclone from the column also cut-short the travelling distance of the lightweight shell along the ducting, therefore lifted almost all the shell and deposited it in the cyclone.

The fifth stage recovered the remaining 5–7 wt % small kernel and removed another 5–8 wt % large and thick shells. The fifth column also served a cleaning purpose, which helps the system to maximize kernel recovery, achieve a dry kernel separation process and generates zero effluent waste. The predecessor of this system that uses a four-stage separation technique still requires a wet separation process using claybath and hydrocyclone, indicating the remarkable concept and performance improvement of our design [14].

Another work reported by Olasumboye and Koya [17] demonstrated a separation technique using a rotary separator that employs a multi-cyclic-separation process for the dry mixture. The work reports the highest kernel recovery over feeding cracked mixture of 92%. The kernel purity was 73%, which indicates contamination of other foreign materials of up to 27% [17]. Our study, however, as presented in Table 2, showed better separation efficiency with maximum kernel losses over FFB processed of 0.36 wt %, equivalent to 97.5% of kernel recovery over feeding cracked mixture. In a later section, we discuss the KER over FFB processed instead of feeding cracked mixture. This is because the KER is a parameter commonly used among millers and therefore provides convenient data to actual end users. In addition to kernel recovery, purity of the recovered kernel in this study was much higher, ranging from 92 to 95 wt % (5–8 wt % shell contaminated in the kernel fraction), as shown in Table 2.

3.3. Kernel Recovery Rate

KER is commonly used to measure the kernel recovery performance of a POM calculated over the processed FFB. The palm oil milling sector makes use of this parameter to evaluate the production performance of POMs in addition to its main CPO production. In this study, the system performance in terms of KER and kernel losses between three commercial mills using different kernel separation system was evaluated on a monthly basis.

Figure 6 illustrates a comparison of the trend of KER in Air Tawar POM with other sister mills in the Johor region over a period of 20 months. Air Tawar POM, mill A and mill B show their KER values in the range of 5.69–6.89 wt %, 4.84–6.26 wt % and 4.85–6.21 wt %, respectively. In general, Air Tawar POM, which adopted a totally dry kernel separation process using the five-stage winnowing column system, showed a higher KER with an average KER of 6.34 $\pm$ 0.30 wt % compared to mill A and mill B, which had average KERs of 5.78 $\pm$ 0.34 wt % and 5.46 $\pm$ 0.28 wt %, respectively. Air Tawar POM maintained its KER values above 6.00 wt % at a 90% consistency rate, whereas the other two mills were only able to maintain their KER above the targeted limit of 6.00 wt % at a 35% and 5% consistency rate, respectively. In an attempt to reduce the effluent generated from the kernel separation system, mill A and mill B improved their system by using a combination of a two-stage winnowing column with a wet separation system using either hydrocyclone or claybath. However, as shown in Figure 6, their KER values were still below 6.00 wt % in most cases.

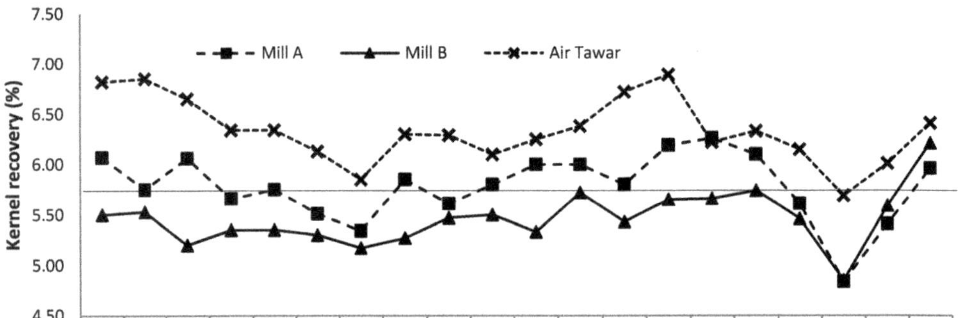

	Jan	Feb	Mar	Apr	May	Jun	Jul	Aug	Sep	Oct	Nov	Dec	Jan	Feb	Mar	Apr	May	Jun	Jul	Aug
Mill A	6.07	5.75	6.06	5.66	5.75	5.51	5.34	5.85	5.61	5.80	6.00	6.00	5.80	6.19	6.26	6.10	5.61	4.84	5.41	5.96
Mill B	5.50	5.53	5.20	5.35	5.35	5.30	5.17	5.27	5.47	5.50	5.33	5.72	5.43	5.65	5.66	5.74	5.46	4.85	5.59	6.21
Air Tawar	6.82	6.85	6.65	6.34	6.34	6.13	5.85	6.30	6.29	6.10	6.25	6.38	6.72	6.89	6.22	6.33	6.15	5.69	6.01	6.41

Month

Figure 6. Monthly KER for the three evaluated mills.

On a yearly basis for two consecutive years, Air Tawar POM recorded the highest average KER of 6.32 wt % and 6.27 wt % compared to mill A (5.77 wt % and 5.70 wt %) and mill B (5.39 wt % and 5.58 wt %) (Figure 7).

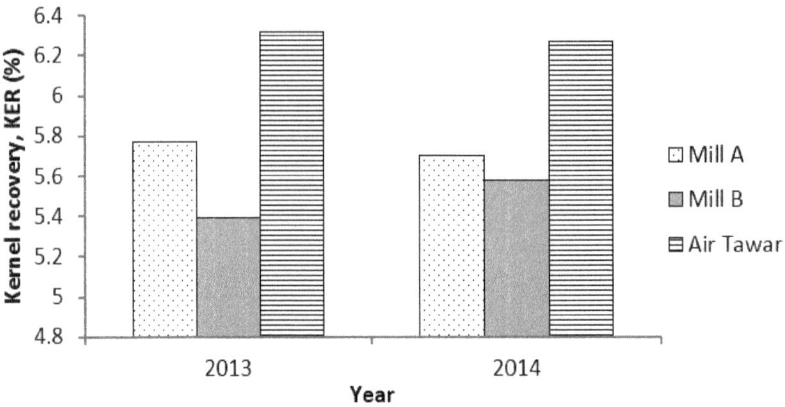

Figure 7. Yearly average kernel recovery rate for the three evaluated mills.

These results imply better KER associated with the five-stage winnowing column system for the kernel separation evaluated in this study. The conventional wet separation system, e.g., the claybath, comprises a cylindrical tank with an inverted cone shape to allow settlement of the heavy phase. A clay solution with a relative density of 1.12 is maintained in the claybath and slowly facilitates the sinking of denser shells and the floating of lighter kernels, both having a relative density in the range of 1.15 to 1.20 and about 1.07, respectively [16]. While the claybath separation is subjected to, and limited to, the density of the clay solution used, the five-stage winnowing column system optimizes the separation mechanism at each column based on the characteristics of the feed into the column, thereby reducing kernel losses and improving kernel recovery.

3.4. Kernel Losses

Similar to KER, measurement of kernel loss is also common in the POM sector so as to compare mill production performance. In general, the average total kernel loss in the shell bunker varied from 0.11 to 0.30 wt % to FFB with almost all the values found to be below 0.30 wt %. These values are far below the set limit of 0.30 wt % to FFB, and these losses do not contribute to the high dirt in the production of kernel.

Kernel losses at these three mills were maintained below 0.30 wt % (Figure 8). The kernel losses varied from 0.11 to 0.30 wt %, and Air Tawar POM was able to maintain the separation efficiency with minimal kernel losses below 0.20 wt % to FFB with an average of 0.17 ± 0.05 wt %. However, mill A and mill B experienced average kernel losses of 0.21 ± 0.05 wt % and 0.20 ± 0.02 wt % to FFB, respectively, both being below 0.20 wt % at a consistency rate of 35% (Figure 8).

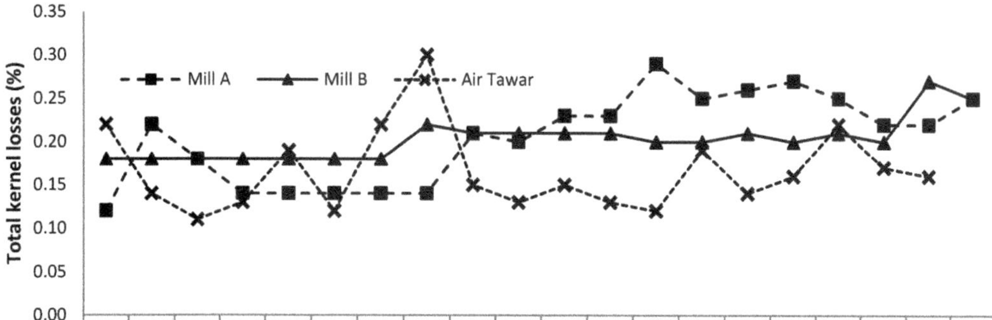

	Jan	Feb	Mar	Apr	May	Jun	Jul	Aug	Sep	Oct	Nov	Dec	Jan	Feb	Mar	Apr	May	Jun	Jul	Aug
Mill A	0.12	0.22	0.18	0.14	0.14	0.14	0.14	0.14	0.21	0.20	0.23	0.23	0.29	0.25	0.26	0.27	0.25	0.22	0.22	0.25
Mill B	0.18	0.18	0.18	0.18	0.18	0.18	0.18	0.22	0.21	0.21	0.21	0.21	0.20	0.20	0.21	0.20	0.21	0.20	0.27	0.25
Air Tawar	0.22	0.14	0.11	0.13	0.19	0.12	0.22	0.30	0.15	0.13	0.15	0.13	0.12	0.19	0.14	0.16	0.22	0.17	0.16	

Month

Figure 8. Monthly kernel losses to FFB for the three evaluated mills.

On a yearly basis for two consecutive years, Air Tawar POM recorded low average kernel losses of 0.17 wt % compared to mill A (0.17 and 0.25 wt %) and mill B (0.19 and 0.22 wt %), respectively (Figure 9).

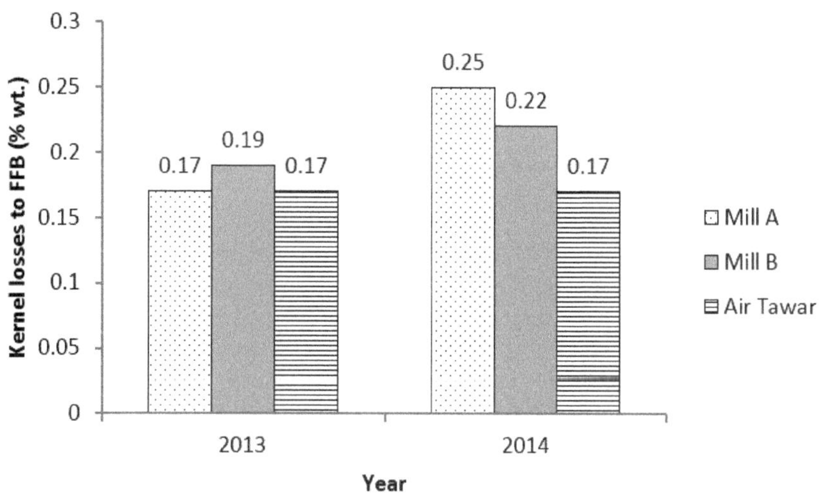

Figure 9. Yearly average kernel losses to FFB for the three evaluated mills.

3.5. Maintenance Activity and Its Effect on Dirt Content

Maintenance activity is scheduled for every 2500 h and 5000 h of operation, equivalent to once for every 5–6-month interval. For every 2400 to 2600 h operation, the system required minor services and parts replacement, including the carbon steel liner plate for the cyclone and transfer trunking of the first and second columns. The replacement of the trunking column was due to the wearing from blown shells that hit the trunking.

There is a major service and parts replacement after 5000 h of operation, including replacement of the liner plate for the cyclone and trunking of all columns as well as servicing of the motor, bearing and impeller of the fan. The system eases the maintenance needs and requires less supervision in order to promote a user-friendly technology concept.

The wear and tear of the system's parts influences its performance in terms of the dirt content in the recovered kernel. Figure 10 shows the trends of dirt content in the recovered kernel for the three mills over the 20 months of evaluation. For the Air Tawar POM, which utilized the five-stage dry kernel separation system, the total dirt and shell content in the kernel sampled at the end of the kernel conveyor varied from 4.56 to 6.03 wt %, and in most cases (80% consistency rate) the dirt content was below the set limit of 5.5 wt %. The high dirt content was observed for three different months as 5.82 wt %, 6.03 wt % and 5.97 wt %. In the first six months, during which the system starts its operation, the dirt content slowly increased from 4.68 wt % in the first month to 5.82 wt % in the sixth month. Maintenance was done gradually from Month 7 until Month 10, resulting in better (reduced) dirt content but still fluctuations. From Month 11 until Month 15, the system retained its dirt content below 5.5 wt % because of the new parts from maintenance. This trend was repeated for the next six-month cycle, thus justifying the need to undertake maintenance every 2500 h and 5000 h of operation.

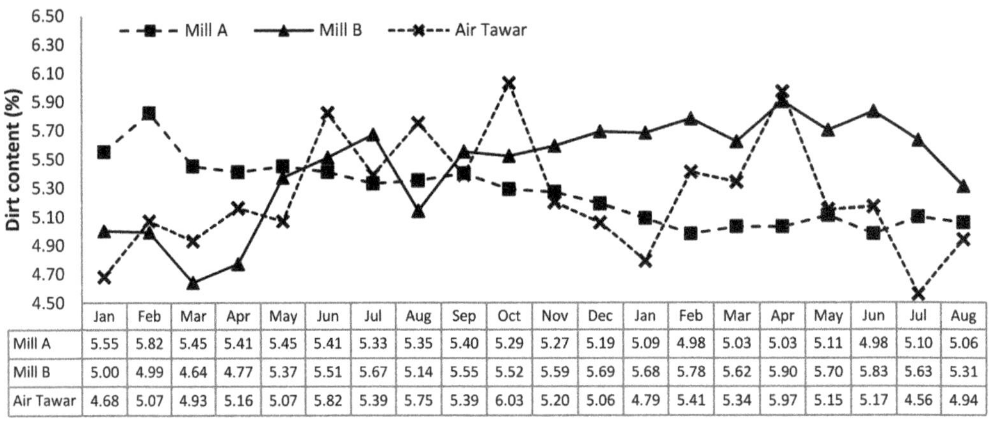

	Jan	Feb	Mar	Apr	May	Jun	Jul	Aug	Sep	Oct	Nov	Dec	Jan	Feb	Mar	Apr	May	Jun	Jul	Aug
Mill A	5.55	5.82	5.45	5.41	5.45	5.41	5.33	5.35	5.40	5.29	5.27	5.19	5.09	4.98	5.03	5.03	5.11	4.98	5.10	5.06
Mill B	5.00	4.99	4.64	4.77	5.37	5.51	5.67	5.14	5.55	5.52	5.59	5.69	5.68	5.78	5.62	5.90	5.70	5.83	5.63	5.31
Air Tawar	4.68	5.07	4.93	5.16	5.07	5.82	5.39	5.75	5.39	6.03	5.20	5.06	4.79	5.41	5.34	5.97	5.15	5.17	4.56	4.94

Month

Figure 10. Monthly dirt content in the shell bunker for the three evaluated mills.

Air Tawar POM experienced slightly lower average dirt content compared to other mills even with a totally dry separation system to separate kernel and shell. As shown in Figure 10, Air Tawar POM experienced an average dirt content of 5.24 $\pm$ 0.40 wt %, slightly lower than the dirt content experienced by Mill A and Mill B at 5.27 $\pm$ 0.22 wt % and 5.44 $\pm$ 0.36 wt %, respectively. Generally, thick shell could be separated efficiently by the wet separation system with either claybath or hydrocyclone, but the use of a shell removal drum in the five-stage winnowing column system eliminates the presence of thick shell and contaminant carryover as the dirt in the production kernel.

Over two successive years, the average dirt content in Air Tawar POM was 5.33 wt % and 5.12 wt %, whereas the dirt content in Mill A was 5.38 and 5.05 wt % and Mill B was 5.39 wt % and 5.65 wt % (Figure 11).

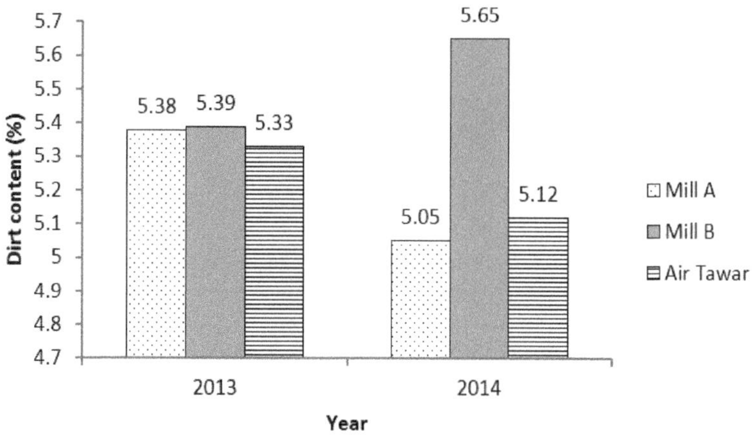

Figure 11. Yearly average dirt content for the three evaluated mills.

3.6. Power Consumption and Economic Analysis

The five-stage winnowing column system requires about 245 HP or 170 kW of power consumption, as tabulated in Table 3. The use of five columns to enhance the separation process and to maximize kernel recovery resulted in high power consumption, with nearly half the power (130 HP) for the operation of the columns, namely to blow air.

Table 3. Power consumption for the five-stage winnowing column system.

Machine	Quantity	Power	
		(HP)	(kW)
Cracked mixture conveyor (top and bottom)	2	11	8
Cracked mixture elevator	1	7.5	5.5
Winnowing column	5	130	100
Air lock	17	51	37.4
Blower	2	40	30
Kernel conveyor	1	5.5	4
Total	-	245	170

The cost of maintenance was calculated for an eight-month period over which the total FFB processed was 122,890 t in 2889 running hours. Two units of impeller for the first and third stages with a total cost of MYR9000 were replaced, and the cost of maintenance was less than MYR 0.10/t FFB processed. An additional cost incurred in the second year of operation is expected since more parts will be replaced after more operational hours. However, this cost will be very low compared to the conventional separation system. The only drawback of the five-stage winnowing column system is that it requires a higher capital cost of nearly MYR one million compare to the wet conventional system, which costs less than MYR500,000.

The kernel plant in Air Tawar POM has been in operation for nearly two years and has recorded an average of an additional min. of 0.3 wt % KER. With the actual FFB processed of 184,335 t/year, this will generate an additional production of palm kernel of 533 t/year. Calculated at the average price of palm kernel in 2015 at MYR1564 per tonne of palm kernel, an additional revenue of MYR844,560 per year could be generated [20]. Therefore, the expected payback period for the capital investment of MYR950,000 is less than two years.

4. Conclusions

A dry process to separate palm kernel and palm shell was developed using a series of five winnowing columns, each powered by a single blower unit. The blower was installed at the bottom of the column to create a forced-draught system with positive pressure along the column. Additionally, the blower system was incorporated with a damper flap to enable airflow rate adjustment to achieve good separation of kernel and shell. The separation in the column works on the principle of the different specific gravity of materials, where the air blown upwards causes the light material, usually palm shell, to rise. A partition was created inside the column to divert the airflow rate and consequently prevent a turbulent flow that would restrain the separation of materials. The five-stage winnowing column system was installed and evaluated at a local POM. Parameters evaluated included size distribution of feeding materials, airflow rate, column sizing, kernel recovery rate, dirt content and kernel losses. The system was able to recover up to 97.5% kernel from the mixture of palm kernel and shell, translated from maximum kernel losses of 0.36 wt %. The system produced kernel products of high purity, *ca.* 92–95 wt %, or 5–8 wt % shell contamination, in the recovered kernel. The KER throughout 20 months of continuous monitoring was in the range of 5.69–6.89 wt %, with most of the values achieving the minimum target limit of 6.00 wt %. Moreover, the system used no water in its separation process, therefore generated zero waste effluent. The novelty of the system could benefit millers in terms of profit, kernel quality, operational performance and maintenance as well as promote user-friendly operations by reducing the water and chemicals used by conventional wet separation.

Acknowledgments: We thank Felda Palm Industries Sdn Bhd (FPISB) and Air Tawar POM for their valuable support and cooperation in this project. We are grateful to all staffs of the milling group of MPOB who were involved in conducting this study.

Author Contributions: Rohaya Mohamed Halim designed and conceived the experiments, designed the system, analyzed the data and wrote the paper; Ridzuan Ramli designed the system; Che Rahmat Che Mat analyzed the data; Choo Yuen May reviewed the paper; Nasrin Abu Bakar designed, analyzed and documented the experimental results; and Nu'man Abdul Hadi analyzed the data and wrote the paper. All authors have read and approved the final manuscript.

Conflicts of Interest: The authors declare no conflict of interest.

Abbreviations

The following abbreviations are used in this manuscript:

POM	Palm oil mill
FFB	Fresh fruit bunch
CPO	Crude palm oil
KER	Kernel recovery rate
MYR	Malaysian ringgit

References

1. Lin, S.W. Palm oil. In *Vegetable Oils in Food Technology: Composition, Properties, and Uses*, 2nd ed.; Gunstone, F.D., Ed.; Wiley-Blackwell: Oxford, UK, 2011; pp. 25–58.
2. Norhaizan, M.E.; Hosseini, S.; Gangadaran, S.; Lee, S.T.; Ramezani, F.; Moghadasian, M.H. Palm oil: Features and applications. *Lipid Technol.* **2013**, *25*, 2011–2014. [CrossRef]
3. Traitler, H.; Dieffenbacher, A. Palm oil and palm kernel oil in food products. *J. Am. Oil Chem. Soc.* **1985**, *62*, 417–421. [CrossRef]
4. Sreedhara, N.; Arumughan, C.; Narayanan, C.S. Dehulling of palm kernel of oil palm (*Elaeis guineensis*) to obtain superior-grade palm kernel flour and oil. *J. Am. Oil Chem. Soc.* **1992**, *69*, 1015–1018. [CrossRef]
5. Zainal, Z.; Yusoff, M.S.A. Enzymatic interesterification of palm stearin and palm kernel olein. *J. Am. Oil Chem. Soc.* **1999**, *76*, 1003–1008.

6. Vaidya, R.; Chaudhary, G.; Raut, N.; Shinde, G.; Deshmukh, N. Synthesis of Biocomposites from Natural Oils—A Review. Available online: http://ipcbee.com/vol41/013-ICEBB2012-H00035.pdf (accessed on 15 April 2016).
7. Haron, M.J.; Jahangirian, H.; Silong, S.; Yusof, N.A.; Kassim, A.; Rafiee-Moghaddam, R.; Mahdavi, B.; Peyda, M.; Abdollahi, Y.; Amin, J. Benzyl and methyl fatty hydroxamic acids based on palm kernel oil as chelating agent for liquid-liquid iron(III) extraction. *Int. J. Mol. Sci.* **2012**, *13*, 2148–2159. [CrossRef] [PubMed]
8. Palm Oil Research Institute of Malaysia. Part 1. General description of the palm oil milling process. In *Palm Oil Factory Process Handbook*; Palm Oil Research Institute of Malaysia: Kuala Lumpur, Malaysia, 1985.
9. Vijaya, S.; Choo, Y.M.; Halimah, M.; Zulkifli, H.; Tan, Y.A.; Puah, C.W. Life cycle assessment of the production of crude palm oil (Part 3). *J. Oil Palm Res.* **2010**, *22*, 895–903.
10. Nizamuddin, S.; Jayakumar, N.S.; Sahu, J.N.; Ganesan, P.; Bhutto, A.W.; Mubarak, N.M. Hydrothermal carbonization of oil palm shell. *Korean J. Chem. Eng.* **2015**, *32*, 1789–1797. [CrossRef]
11. Rahman, M.M.; Adil, M.; Yusof, A.M.; Kamaruzzaman, Y.B.; Ansary, R.H. Removal of heavy metal ions with acid activated carbons derived from oil palm and coconut shells. *Materials* **2014**, *7*, 3634–3650. [CrossRef]
12. Ismaiel, A.A.; Aroua, M.K.; Yusoff, R. Cadmium (II)-selective electrode based on palm shell activated carbon modified with task-specific ionic liquid: Kinetics and analytical applications. *Int. J. Environ. Sci. Technol.* **2014**, *11*, 1115–1126. [CrossRef]
13. Rohaya, M.H.; Nasrin, A.B.; Choo, Y.M.; Ma, A.N.; Ravi, N. A commercial scale implementation of Rolek™ palm nut cracker: Techno-economic viability study for production of shell-free kernel. *J. Oil Palm Res.* **2006**, *18*, 153–167.
14. Rohaya, M.H.; Nasrin, A.B.; Mohd Basri, W.; Choo, Y.M.; Ridzuan, R.; Ma, A.N.; Ravi, N. Maximizing the recovery of dry kernel & shell via a 4-stage winnowing column. In Proceedings of the Malaysian Palm Oil Board (MPOB) International Palm Oil Congress, Kuala Lumpur, Malaysia, 9–12 November 2009.
15. Andoh, P.Y.; Agyare, W.A.; Dadzie, J. Selection of an ideal mesh size for the cracking unit of a palm kernel processing plant. *J. Sci. Technol.* **2010**, *30*, 109–118. [CrossRef]
16. Okoronkwo, C.A.; Ngozi-Olehi, L.C.; Nwufo, O.C. The adaptation of clay-bath and hydro-cyclones in palm nut cracked-mixture separation to small—Scale oil palm processing industry. *Int. J. Mod. Eng. Res.* **2013**, *3*, 2023–2026.
17. Olasumboye, A.; Koya, O. A rotary separator for the dry mixture of palm kernel and shell. *Innov. Syst. Des. Eng.* **2014**, *5*, 32–42.
18. Rohaya, M.H.; Osman, A. The quality of Malaysian palm kernel: Effect of shell and broken kernel on the quality of final products. In Proceeddings of the National Seminar on Palm Oil Milling, Refining Technology, Quality and Environment (Malaysian Palm Oil Board), Sabah, Malaysia, 27–28 November 2002.
19. Shafii, A.F. Nut cracking using ripple mill. *PORIM Bull.* **1989**, *19*, 8–18.
20. Price of Palm Products. Available online: http://bepi.mpob.gov.my/index.php/statistics/price/monthly.html (accessed on 1 April 2016).

9

Terahertz and Cultural Heritage Science: Examination of Art and Archaeology

Antonino Cosentino

Cultural Heritage Science Open Source (CHSOS), Viagrande 95029, Italy; antoninocose@gmail.com

Academic Editor: Manoj Gupta

Abstract: Cultural Heritage scientists need methodologies to examine Art and Archaeology in order to understand artistic materials and techniques and devise better conservation procedures. This review discusses the most successful and promising applications of Terahertz (THz) technology in Cultural Heritage Science. THz is used in homeland security and for plenty of other industrial sectors and it presents a number of valuable features specifically for the investigation of Art and Archaeology: No radiation risk, low power, non-contact and reflection mode. Recent technical advancements are also making its application fast, mobile and relatively affordable creating a potential for its diffused implementation in museums. While THz is most promising for the investigation of multilayered art, such as paintings, it has been tested on a very large range of artifacts, from manuscripts to mummies and lacquered historical furniture.

Keywords: Terahertz; art conservation; conservation science; cultural heritage science

1. Introduction

Cultural Heritage Science (CHS) is a multidisciplinary academic science which encompasses sectors as Physics, Chemistry, Biology, and Engineering, and bridges them with Art conservation and Archaeology [1,2]. Cultural Heritage scientists examine works of art and archaeology by means of technical and scientific methodologies to understand when and how these artifacts were made, and, as well important, how are they to be preserved, what conservation treatment represents the best option and why.

A large number of imaging and spectroscopy techniques are commonly used by scientists involved in CHS. Among the most used imaging methods there are technical photography [3–7], ultraviolet photography [8], reflectance transformation imaging [9], infrared reflectography [10], multispectral imaging [11–13] and X-radiography [14]. Imaging methods are largely used since they do not need any sampling or treatment of the objects. On the other hand, analytical methods can be invasive and destructive. In many cases, however, sampling cultural artifacts is not permitted and, therefore, the application of non-invasive analytical techniques, such as reflectance spectroscopy [15], Raman spectroscopy [16], XRF spectroscopy [17] or neutron techniques [18,19] as well as mobile atomic force microscopy [20], is mandatory.

Terahertz (THz) radiation lies between the infrared and microwave regions (frequency range about 0.1–10 THz), Figure 1. With the advent of stable, sub-picosecond-pulse-duration lasers, researchers at Bell Laboratories in the mid-1980's explored the generation and detection of short transients of radiation that possessed a broad spectrum ranging from tens of gigahertz up to several THz [21]. These transients—which were generated by the laser excitation of DC-biased, fast-responding photoconductive gaps or non-centrosymmetric electro-optic crystals and detected using laser-triggered photoconductive sampling gates or electro-optic sampling—became known as

THz pulses [22]. Terahertz technology was used sparingly by few researchers for spectroscopy [23,24] and astronomy [25] until recent advances in instrumentation allow now to generate and detect terahertz with commercial equipment [26–28]. Research papers exploring THz applications are growing considerably and THz radiation is currently used in quality-control [29,30] and homeland-security [31,32] to screen for hidden weapons and explosives, since it can penetrate clothing. Its applications are also explored in medical-diagnostic [33,34] (mainly for cancer diagnosis [35]) and pharmaceuticals [36,37] for tablet inspection, process improvement and polymorph screening as well as to investigate coatings integrity [38].

This paper introduces THz technology and discusses its applications for Art and Archaeology. Its first test in this field dates back to 1998 when THz was investigated as a tool for dating wooden objects (dendrochronology) [39–41]. The same instrumental advancements that led the growing interest into this technology also generated research into a number of cultural heritage applications [42]. THz as a tool for Art diagonstics is still in its infancy because the technology is relatively less explored than other parts of the electro-magnetic spectrum, and many THz devices are relatively bulky and have not yet been optimized for use on-site in museums and in field sites. Yet the studies presented in this review point to a promising imaging and analytical method. THz waves can characterize many materials at different layers from the surface providing a visual understanding along the depth axis. THz data could yield more precise information about what is beneath the layer of an artifact – to better understand its materials and its current conservation condition.

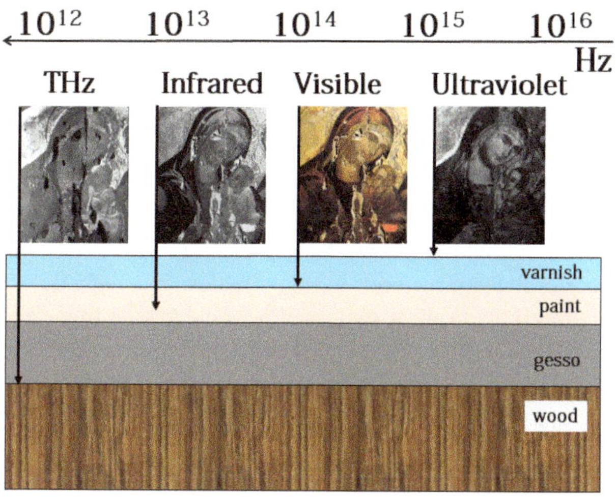

Figure 1. Terahertz radiation (10^{12} Hz) is often applied for the study of paintings. Terahertz (THz) and X-ray radiography can penetrate a panel painting deeper than other radiations traditionally used for paintings examination, such as infrared, but only THz allows for safe (non-ionizing) in deep 3D imaging of its layers.

2. THz Unique Features

THz is complementary to the other techniques commonly used for art examination and it provides different and unique information.

2.1. Optical Opaque Materials Become Transparent

THz radiation penetrates non-polar, non-metallic optically opaque materials (such as plastics, paper, painting materials and textiles) and it can be used for subsurface imaging since it penetrates much further into the sample than infrared techniques, commonly used in art examination.

2.2. *Sensitive to Water*

Unlike infrared radiation which induces molecular bending and stretching, THz excites collective motions, particularly in water molecules. THz radiation is so sensitive to water that it is used for biomedical studies of tissues whose water content is linked to important information. Humidity and water content are important information also in art conservation [43].

2.3. *Safe*

THz pulses have low energy and power and THz radiation is non-ionizing. In accordance with international safety guidelines moderate exposure to THz radiation is safe [44] and it doesn't heat the sample or cause damage [45]. The other nondestructive methods commonly used for sub-surface imaging in art and archaeology (X-ray, γ-ray, protons and neutrons) are ionizing and consequently their accessibility is limited.

3. Instrumentations

3.1. *THz Range*

Different wavelengths are appropriate for analyzing different types of cultural assets: art (notably easel paintings), for which THz radiation at the higher frequency range is preferred (0.5–13 THz), and historic architecture and archaeology, for which lower frequencies (0.01–0.5 THz) can be successfully applied for the analysis of thicker objects such as wall paintings and architectural structures.

3.2. *Scanning*

By focusing the terahertz beam with a short focal length lens, spatial resolution can be scaled down to tens of micrometers much better than sub-surface radars. THz waves lie between the infrared and microwave spectral regions, and therefore they combine the superior lateral resolution of infrared radiation (as compared with microwaves) with the higher penetration depth of electro-magnetic GHz-frequency fields (as compared with infrared). The high spatial resolution allows to scan the THz probes across the surface of the sample both in transmission and reflection mode. The probes are coupled with umbilical optical fiber cables, making it possible to scan the sample rather than move it and to quickly configure the scanner for transmission or reflection mode. The THz emitter usually delivers a picosecond duration THz pulse, allowing high-speed, scanned images to be easily produced acquiring a waveform at each position.

3.3. *Time-Domain or Frequency-Domain*

There are two main type of THz instruments and the use of each one depends on the specific goals of the examination [46], Figure 2. Time-domain based spectrometers (TDS) use short pulses (few picoseconds) to create an instantaneous wide frequency spectrum covering usually up to 3 THz. The time domain information is acquired with a Fast-Fourier transform analyzer and the absorption or reflection spectrum is then studied. This is a very powerful technique but is limited in its dynamic range due to the very low power provided by the ultra-short pulses. By gating the time domain signal, reflections from sub-surface discontinuities within a layered sample can be revealed, both in reflection and transmission mode.

Frequency-Domain is another type of analysis consisting in frequency measurements. In this case, a continuous or relatively short (ms) THz wave is sent and a frequency domain analyzer (THz receiver) is used to measure the reflected signal. The data is taken one frequency point at a time, each with an integration period of few ms. This method provides a much larger dynamic range but with a longer measurement time. Since the current systems are built using waveguides, they have limited bandwidth (around 100–150 GHz) and it is necessary to exchange waveguide bands in order to cover different frequency regions.

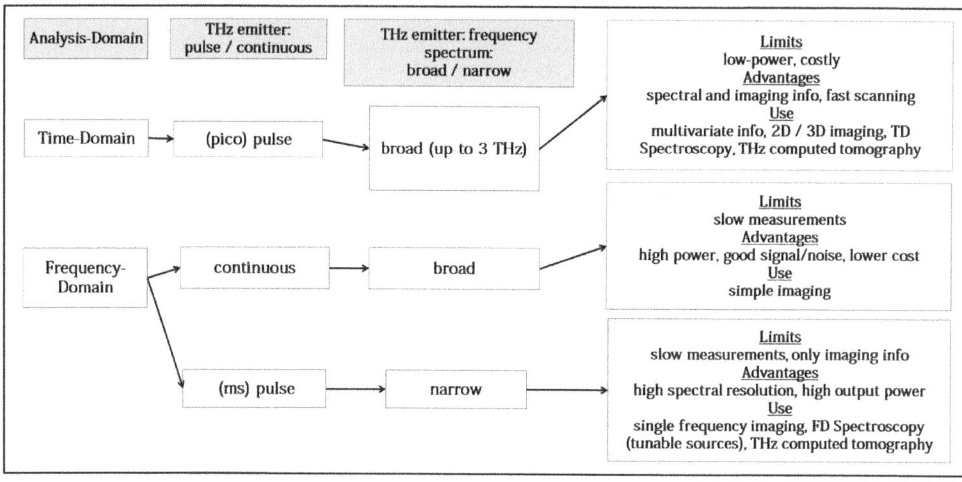

Figure 2. THz imaging and spectroscopy methods used in Cultural Heritage Science.

3.4. Pulse or Continuous Wave

Pulsed terahertz imaging is the most diffused method since it provides multivariate information on the sample. THz picoseconds pulses can be used in time-domain systems to generate images based on the pulse's peak amplitude or on its integration over a time window, or even its time delay [47–50]. On the other hand, continuous-wave systems are more affordable, have higher peak power and better image signal to noise. They are preferred for samples with simple subsurface structure.

3.5. Broad or Narrow Bandwidth

Most THz systems use broad bandwidth ultrafast THz pulses in Time-Domain for spectroscopy or imaging. A Fourier transformation of the time domain signal provides the reflected or transmitted spectrum of the signal, its amplitude or its phase. On the other hand, some THz scanners use narrow bandwidth THz emitters. They offer high spectral resolution and high output power but in order to collect spectra it is necessary to return the output frequency and therefore they are used mostly for single frequency imaging.

3.6. THz Spectroscopy

Spectral features in the terahertz region are due to molecular and inter-molecular interactions, weak bonds and phonon absorptions. Terahertz spectroscopy is implemented as time-domain (TDS) or frequency-domain (FDS). TDS systems use a broadband pulse signal and the transient terahertz electric field is sampled using an optical delay line while amplitude and phase spectra are delivered by Fourier transformation. In TDS it is possible to retrieve with a better signal-to-noise ratio the materials' complex index of refraction directly from the Fresnel equations rather than from the more complex Kramer-Kroenig relations. FDS is performed mostly by FTIR using broad-band and continuous sources and a two-beam interferometric spectrometer [51]. Regarding art materials there is an online THz spectra database by The National Institute of Information and Telecommunications Technology (NICT) [52,53].

THz spectroscopy of art materials is challenging for a number of reasons. Pigments' particle size fall within the terahertz wavelength range and Rayleigh and Mie scattering affect spectra significantly [54]. Also, most historical pigments have fillers and crystal polymorphs that impact the resulting THz spectra [55].

3.7. THz 2D and 3D Imaging

THz scanning can provide 2D imaging which is used for fast analysis of large surfaces but it is 3D imaging of sub-surface structures that makes THz so interesting for art examination. After a Time Domain Imaging scanning is acquired, it is possible to plot the amplitude at each point as a false color image [56,57]. Similarly to ultrasound scans, this THz image is called B-scan, when it shows just one scanning line, where the x-axis represents the position on the line and the y-axis indicates the time increment. The false color scale is related to the signal's amplitude. B-scan representation is used to show the sub-surface layers as a cross-section.

On the other hand, a THz image is called C-scan, if it shows an area, rather than only a scan line. The false color scale shows the intensity of the THz reflection at any desired subsurface depth (time delay). Signal processing such as apodisation [58], wavelet denoising [59] and deconvolution [60] is usually applied to enhance both B-scan and C-scan images.

3.8. THz Tomography

Terahertz computed tomography (THz-CT) features poor spatial resolution but in some cases it is preferred to X-ray since it doesn't affect thermoluminescence dating and it is sensitive to organic materials [61]. THz-CT uses mostly narrow bandwidth sources, multiple projection angles and back projection [62,63]. It is also possible to use broadband sources in time domain mode and to represent in the XY-plane for each pixel the time series (as depth Z) [64]. THz-CT imaging can also be performed with pulsed terahertz radiation in the broadband Time-Domain and even using one single projection it is possible to reconstruct the 3D internal structure of the sample thanks to the time-of-flight delay of the terahertz pulse [65].

4. THz for Art and Archaeology

4.1. Historical Documents

Often, historical documents on papyrus as well as on parchment and on paper cannot be read for a number of reasons. Sometimes the fragile sheets just cannot be separated because they are stuck together as a result of deterioration and damages. In other cases, the sheets have been reused as supports or covers for newer documents. There is considerable interest in reading this hidden information and a number of techniques have been tested to pursue this goal while preserving the documents. So far, X-ray computed tomography [66] has been the most successful method. Since THz-TD imaging was already evaluated for the inspection of postal envelopes [67], it was also tested with encouraging results for stacked papyrus layers written with carbon black ink [68]. Recently, even more successful results were obtained with a new and more sophisticated THz method called tomosynthesis [69] which was applied to image pencil writing on a stack of 50 paper sheets. THz tomography has been applied to resolve text on both sides of a single papyrus sheet [70,71]. THz was tested also for those cases where writing is obscured by stains and other inks in old purchament manuscripts [72] and it seems successful to characterize and evaluate conservation of iron gall inks [73] and parchment [74].

4.2. Panel Paintings

Diagnostics is preliminary to any conservation treatment in order to identify painting materials and technique, to detect previous restorations and additions as well as the nature and extent of alterations. Ultraviolet radiation is used for the examination of the varnish, infrared to reveal underdrawing and X-rays to study underpaint and the wooden support. THz has been successful to investigate all of these layers, even reaching the wood support, providing cross-section (B-scan) and top-view (C-scan) images of the artworks [75].

Gilded panel paintings are a particular kind of art developed in the early medieval age. They were made [76] applying a thin primer on the wooden panel support and then the gold background on a layer of red bole. First attempts to apply THz for the examination of this technique dates back to 2009 when mock-up panels consisting of gold leaf and pigments on wood panel partially obscured by a gesso layer (chalk or gypsum preparation) were tested successfully [77].

THz is used mostly to locate the gilding leaves under the paint layer. Figure 3 shows a detail from a 14th century icon which was analyzed with both infrared imaging and THz [75]. In the THz-TDI image the square gilding leaves are clearly visible at the contours of the figures' faces while infrared reflectography (IRR) imaging cannot reveal them [78]. Before cleaning treatment, THz can be used for the inspection of the condition of the gilding leaves and of the decorative motives made with gold powder (shell gold), because THz can reveal gold even under a thick and dark varnish. Furthermore, the THz image of a panel painting at its different interfaces (wood, primer, paint) can reveal structural problems or paint's flacking. THz imaging of the wood grain reveals the cause of the cracks in the paint documented with infrared-RTI [78], Figure 3. Thanks to Terahertz's non-invasive cross-section imaging it is possible to evaluate the extent of cracks in the paint that are visible to the naked eye but whose depth and extent into the other sublayers cannot be determined otherwise [79]. It is also possible to visualize the traditional complex preparation, consisting in a canvas between two gesso layers [79].

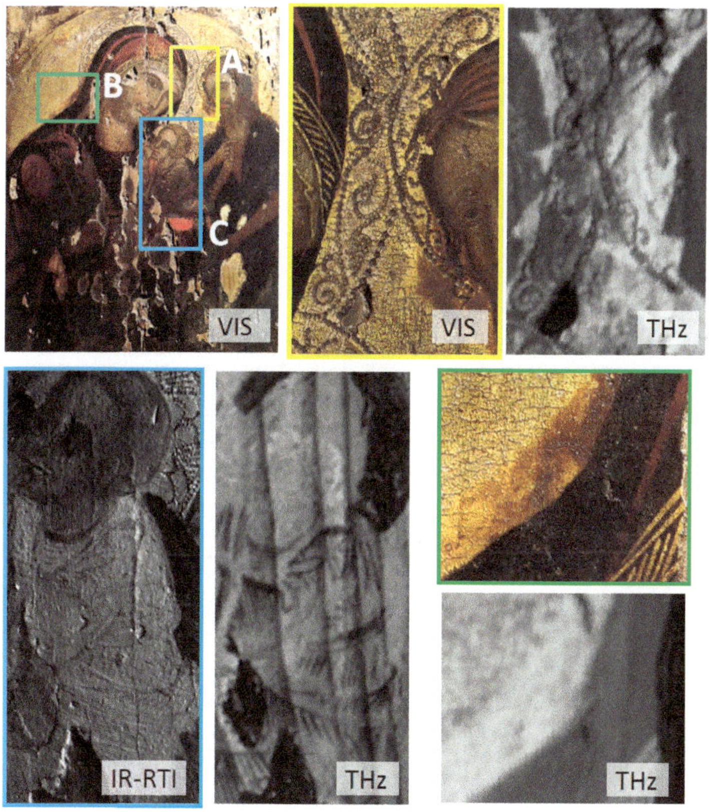

Figure 3. Virgin with Child and a Saint, 14th century icon (32 cm × 39 cm × 0.8 cm), Public Library in Taormina, Sicily. The icon was affected by extensive damages. Before its conservation treatment, it underwent a diagnostic campaign using THz and other imaging techniques (VIS (photography), IR-RTI (Infrared-Reflectance Transformation Imaging)). Each insert illustrates a specific application of THz for panel paintings examination: (A) gilding leaves' edges localization; (B) pre-cleaning visualization of gilding and gold decoration under dirt and varnish; (C) structural condition evaluation (interface wood-preparation). Courtesy of Danish Technical University [78].

4.3. Canvas Paintings

During the early Renaissance, painters started to prefer canvases to panels as supports for their paintings because they were much affordable and could be folded for more convenient transportation. Differently than wood panels, canvases are more transparent to THz and, indeed, the first test of THz for the inspection of paintings on canvas in 2006 [80] was performed with a transmission THz-TDS system. THz radiation has been used to detect underdrawings, changes ("pentimenti"), or previous compositions that are now hidden. While infrared imaging and X-ray radiography are the preferred methods for these kinds of cases, THz has the advantage of further penetration (compared to infrared radiation) and depth resolution not offered by standard X-radiography. The first test of THz for underdrawing observation dates back to 2008 when a THz-TDS in reflection mode could detect graphite pencil drawing under different pigments [81]. Underdrawing with umber was also tested in 2009 with a reflection THz-TDI system [82]. While graphite can be detected with infrared imaging, umber is transparent to the infrared and THz inspection is superior.

4.4. Wall Paintings

In 2008 it was demonstrated that THz imaging can reach through a depth of almost 1 cm into plaster [83]. This announcement started up a prolific research line dedicated to wall paintings that were whitewashed or even covered with a thicker layer of plaster [84]. Mobile and compact up-to-date THz-TDI scanners can be used for in-situ analysis of wall paintings. Their picosecond pulses are reflected back to the detector from discontinuities in the refractive index caused by cracks, covered paint layers or objects inserted in the plaster such as nails. In particular, THz scanning has been used to detect wall paintings whitewashed with lead white, which is very opaque to infrared imaging [84]. As for canvas paintings, THz can also map pigments on wall paintings [85] but the most important application on this specific kind of art is to evaluate the extent of cracks into the plaster [84].

4.5. Textiles

The interaction of THz with textiles has been studied extensively for homeland security applications [86,87]. This research has been instrumental to apply THz to study art and archaeology textile-wrapped objects such as mummies which are otherwise examined routinely with X-ray radiography or computed tomography systems in order to image their internal anatomical features [88,89]. THz adds the advantage of using non-ionizing radiation, portability and complementarity since in THz imaging contrast depends on the refractive index of the materials rather than the elemental density as in X-radiography. THz radiation does not have so much depth penetration as radiography but it provides high contrast images close to the surface useful to reveal the structure and condition of the bandages and objects placed within them [90]. On the other hand, THz systems are also used on small animal mummies and on mummified human body parts to study soft-tissues close to the surface [65].

4.6. Other Applications

THz studies on art and archaeology have been focused on historical documents, textiles and multilayered art such as paintings and lacquer furniture [91,92], Figure 4. There are other promising applications where THz has been tested even if less frequently. There is an increasing interest regarding THz application to detect and measure corrosion under paint for industrial applications [93–96]. This research could develop new THz methods for conservation of painted metal historical objects, such as furniture and sculptures.

THz spectroscopy has been tested on art made with plastics [97] since it is sensitive to their plasticizers and fillers, while THz 3D imaging can distinguish coatings and layers of degradation in historical plastics. There is interest into using THz imaging to obtain stratigraphic information regarding corrosion and encrustations layers of archaeological metals [98] and for historical stone objects conservation [43]. THz tomography with specialized equipment has been used to examine clay Egyptian vessels [64].

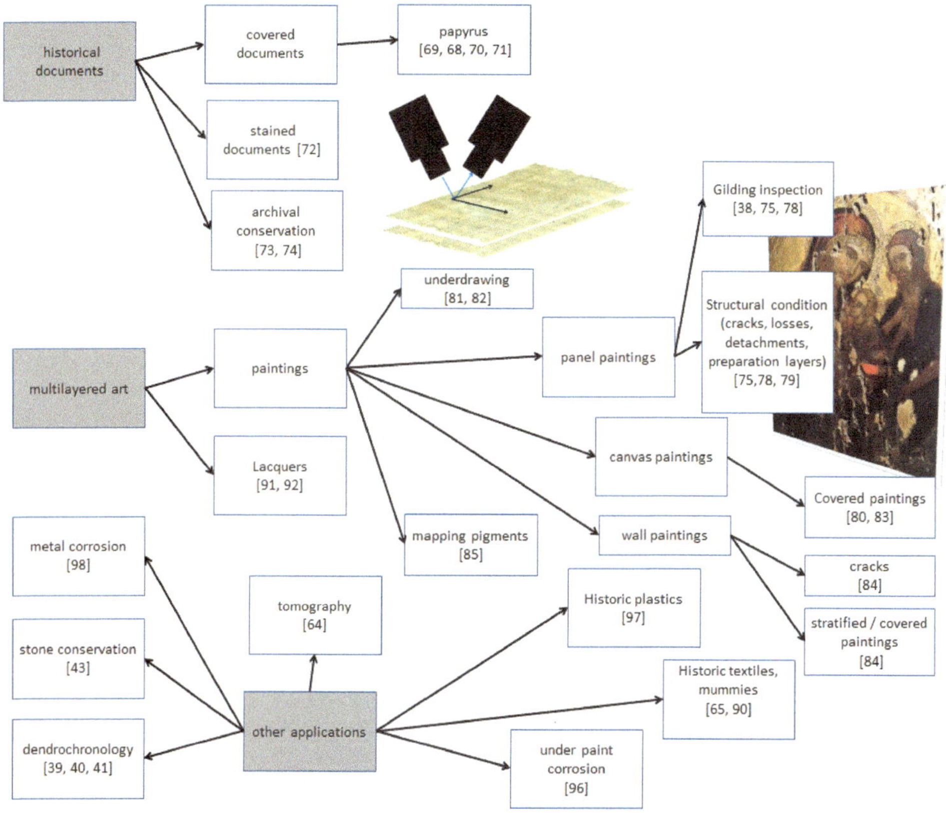

Figure 4. Applications of Terahertz in cultural heritage science.

5. Conclusions and Perspectives

THz technologies have experienced intense development, mainly because of their applications in homeland security, where depth profiling can be used to distinguish weapons and other dangerous items under clothing, and spectroscopy can be employed to sense explosive materials. Still a somewhat unexplored spectral range, Terahertz radiation offers a potentially important new tool for understanding art, searching for lost works, characterizing materials, authenticating artifacts, understanding decay processes, as well as improving conservation of paintings, murals and other cultural objects. Specifically to art investigations, use of the THz portion of the electromagnetic spectrum presents a number of valuable technical features: No radiation risk, THz radiation is non-ionizing and does not pose risk to health. Even if X-ray radiography and spectroscopy are used quite widely in art investigation, it is evident that the potential health and safety ramifications for the users of these methods would provide a strong incentive for the wide spread adoption of THz methods that would not require skilled personnel and specific facilities such as radiation-shielded rooms. Another advantage is that THz is a very low-power technique: The energy in a single THz pulse is typically less than the background radiation present, yet the signal-to-noise ratio can be very high due to the use of gated detection. Other advantages are that THz is non-contact, works in reflection mode, it is relatively fast and affordable. The historically decreasing costs of pulsed lasers and other optical components, and thus of the entire THz systems, creates a potential for broad use in museums. Furthermore, the relatively bulky THz experimental stations common to research facilities have evolved into commercial, suitcase-sized instrumentation that can be easily brought into museums—for *in situ* examinations—and also easily moved from one conservation department to another. Tomography methods, such as CT-scan, have been used successfully for some exceptional initiatives, and even for investigations of mummies, but they cannot be used routinely for practical reasons (for instance, it is not likely that curators will habitually move art objects to external facilities like hospitals, due to limited budgets and security issues). THz tomography allows conservators to investigate thin layers, such those in a painting, as well as thick bulk samples, such as wood artifacts. This versatility is mandatory for the success of the method for practical budget reasons. If it finds applications in more than one conservation department, it has more chances to be implemented for the broadest benefit.

Recent technical developments are providing THz commercial equipment relatively small and lightweight. They are also becoming sturdy for transportation and less affected by environmental condition. Their portability is an extremely important feature for their actual use in the cultural heritage field.

Acknowledgments: No funds were received for this review paper.

Conflicts of Interest: The author declares no conflict of interest.

References

1. Gates, G.A. Discovering the material secrets of art: Tools of cultural heritage science. *Am. Ceram. Soc. Bull.* **2014**, *93*, 20–27.
2. Creagh, D.C.; Bradley, D.A. *Radiation in Art and Archaeometry*; Elsevier Science: Philadelphia, PA, USA, 2000.
3. Cosentino, A. Identification of pigments by multispectral imaging a flowchart method. *Herit. Sci.* **2014**, *2*. [CrossRef]
4. Cosentino, A. A practical guide to Panoramic Multispectral Imaging. *E-Conserv. Mag.* **2013**, *25*, 64–73.
5. Cosentino, A.; Gil, M.; Ribeiro, M.; Di Mauro, R. Technical Photography for mural paintings: The newly discovered frescoes in Aci Sant'Antonio (Sicily, Italy). *Conserv. Patrim.* **2014**, *20*, 23–33. [CrossRef]
6. Cosentino, A.; Stout, S. Photoshop and multispectral imaging for art documentation. *E-Preserv. Sci.* **2014**, *11*, 91–98.
7. Cosentino, A. Effects of different binders on technical photography and infrared reflectography of 54 historical pigments. *Int. J. Conserv. Sci.* **2015**, *6*, 287–298.
8. Cosentino, A. Practical notes on ultraviolet technical photography for art examination. *Conserv. Patrim.* **2015**, *21*, 53–62. [CrossRef]
9. Cosentino, A. Macro photography for reflectance transformation imaging: A practical guide to the highlights method. *E-Conserv. J.* **2013**, *1*, 70–85. [CrossRef]
10. Cosentino, A. Panoramic infrared reflectography. technical recommendations. *Int. J. Conserv. Sci.* **2014**, *5*, 51–60.
11. Cosentino, A. Panoramic, macro and micro multispectral imaging: An affordable system for mapping pigments on artworks. *J. Conserv. Mus. Stud.* **2015**, *13*, 1–17. [CrossRef]
12. Cosentino, A. Multispectral imaging of pigments with a digital camera and 12 interferential filters. *e-Preserv. Sci.* **2015**, *12*, 1–7.
13. Cosentino, A. Multispectral imaging system using 12 interference filters for mapping pigments. *Conserv. Patrim.* **2015**, *21*, 25–38. [CrossRef]
14. Gilardoni, A.; Orsini, R.A.; Taccani, S. *X-rays in Art*; Gilardoni Spa: Mandello Lario, Italy, 1977.
15. Cosentino, A. FORS spectral database of historical pigments in different binders. *E-Conserv. J.* **2014**, *2*, 57–68. [CrossRef]
16. Burrafato, G.; Calabrese, M.; Cosentino, A.; Gueli, A.M.; Troja, S.O.; Zuccarello, A. ColoRaman project: Raman and fluorescence spectroscopy of oil, tempera and fresco paint pigments. *J. Raman Spectrosc.* **2004**, *35*, 879–886. [CrossRef]
17. Cosentino, A.; Stout, S.; Scandurra, C. Innovative imaging techniques for examination and documentation of mural paintings and historical graffiti in the catacombs of San Giovanni, Syracuse. *Int. J. Conserv. Sci.* **2015**, *6*, 23–34.

18. Bom, V.R.; Cosentino, A.; Seracini, M.; Rosa, R. Neutron back scattering for the search of the Battle of Anghiari. *Appl. Radiat. Isot.* **2010**, *68*, 66–70. [CrossRef] [PubMed]
19. Kuznetsov, A.V.; Gorshkov, I.Y.; Evsenin, A.V.; Osetrov, O.I.; Vakhtin, D.N.; Cosentino, A.; Seracini, M. Nanosecond Neutron Analysis for the search of the lost Leonardo's masterpiece, the Battle of Anghiari. *Nucl. Instrum. Methods Phys. Res. B* **2009**, *267*, 3694–3697. [CrossRef]
20. Kampasakali, E.; Bronwyn, O.; Cosentino, A.; Miliani, C.; Learner, T. A preliminary evaluation of the surfaces of acrylic emulsion paint films and the effects of wet-cleaning treatment by atomic force microscopy (AFM). *Stud. Conserv.* **2011**, *56*, 216–230. [CrossRef]
21. Auston, D.H.; Cheung, K.P.; Valdmanis, J.A.; Kleinman, D.A. Cherenkov radiation from femtosecond optical pulses in electro-optic media. *Phys. Rev. Lett.* **1984**, *53*, 1555–1558. [CrossRef]
22. Auston, D.H.; Cheung, K.P.; Smith, P.R. Picosecond photoconducting Hertzian dipoles. *Appl. Phys. Lett.* **1984**, *45*, 284–286. [CrossRef]
23. Chamberlain, J.M.; Kimmitt, M.F.; Crompton, A.; Havenith, M.; Smith, G.; Mittleman, D.M. Where optics meets electronics: Recent progress in decreasing the terahertz gap. *Philosoph. Trans. Ser. A Math. Phys. Eng. Sci.* **2004**, *362*, 199–211. [CrossRef] [PubMed]
24. Karr, C.; Kovach, J.J. Far-infrared spectroscopy of minerals and inorganics. *Appl. Spectrosc.* **1969**, *23*, 219–223. [CrossRef]
25. Dragoman, D. Terahertz fields and applications. *Prog. Quantum Electron.* **2004**, *28*, 1–66. [CrossRef]
26. Hosako, I.; Sekine, N.; Patrashin, M.; Saito, S.; Fukunaga, K.; Kasai, Y.; Baron, P.; Seta, T.; Mendrok, J.; Ochiai, S.; *et al.* At the dawn of a new era in terahertz technology. *Proc. IEEE* **2007**, *95*, 1611–1623. [CrossRef]
27. Tonouchi, M. Galore new applications of terahertz science and tech-nology. *Terahertz Sci. Technol.* **2009**, *2*, 90–101.
28. Schmutternmaer, C. Exploring dynamics in the far-infrared with terahertz spectroscopy. *Chem. Rev.* **2004**, *104*, 1759–1779. [CrossRef] [PubMed]
29. Frank, R.; Koch, M.; Khare, S.; Moneke, M.; Richter, H.; Ewert, U. Terahertz quality control of polymeric products. *Int. J. Infrared Millim. Waves* **2006**, *27*, 547–556. [CrossRef]
30. Dong, J.; Kim, B.; Locquet, A.; McKeon, P.; Declercq, N.; Citrin, D.S. Nondestructive evaluation of forced delamination in glass fiber-reinforced composites by terahertz and ultrasonic waves. *J. Compos. Part B Eng.* **2015**, *79*, 667–675. [CrossRef]
31. Shen, Y.C. Terahertz pulsed spectroscopy and imaging for pharmaceutical applications: A review. *Int. J. Pharm.* **2011**, *417*, 48–60. [CrossRef] [PubMed]
32. Liu, H.B.; Zhang, X.C. Terahertz Spectroscopy for Explosive, Pharmaceutical, and Biological Sensing Applications. In *Terahertz Frequency Detection and Identification of Materials and Objects*; Miles, R.E., Zhang, X.-C., Eisele, H., Krotkus, A., Eds.; Springer: New York, NY, USA, 2007; pp. 251–323.
33. Fitzgerald, A.J.; Berry, E.; Zinovev, N.N.; Walker, G.C.; Smith, M.A.; Chamberlain, G.M. An introduction to medical imaging with coherent terahertz frequency radiation. *Phys. Med. Biol.* **2002**, *47*, R67–R84. [CrossRef] [PubMed]
34. Jacoby, M. Medical imaging turns to oft-neglected part of light spectrum. *Chem. Eng. News* **2015**, *93*, 10–14.
35. Yu, C.; Fan, S.; Sun, Y.; Pickwell-MacPherson, E. The potential of terahertz imaging for cancer diagnosis: A review of investigations to date. *Quant. Imaging Med. Surg.* **2012**, *2*, 33–45. [PubMed]
36. Nowak, K.; Pliński, E.F.; Karolewicz, B.; Jarząb, P.; Plińska, S.; Fuglewicz, B.; Walczakowski, M.; Augustyn, Ł.; Sterczewski, Ł.; Grzelczak, M.; *et al.* Selected aspects of terahertz spectroscopy in pharmaceutical sciences. *Acta Pol. Pharm. Drug Res.* **2015**, *75*, 851–866.
37. Taday, P.F. Applications of terahertz spectroscopy to pharmaceutical sciences. *Phil. Trans. R. Soc. Lond. A* **2004**, *362*, 351–364. [CrossRef] [PubMed]
38. Wagh, M.P.; Sonawane, Y.H.; Joshi, O.U. Terahertz technology: A boon to tablet analysis. *Indian J. Pharm. Sci.* **2009**, *71*, 235–241. [PubMed]
39. Koch, M.; Hunsche, S.; Schumacher, P.; Nuss, M.C.; Feldmann, J.; Fromm, J. THz-imaging: A new method for density mapping of wood. *Wood Sci. Technol.* **1998**, *32*, 421–427. [CrossRef]
40. Groves, R.M.; Pradarutti, B.; Kouloumpi, E.; Osten, W.; Notni, G. 2D and 3D non-destructive evaluation of a wooden panel painting using shearography and terahertz imaging. *NDT E Int.* **2009**, *42*, 543–549. [CrossRef]

41. Jackson, J.B.; Mourou, M.; Labaune, J.; Whitaker, J.F.; Duling, I.N. Terahertz pulse imaging for tree-ring analysis: A preliminary study for dendrochronology applications. *Meas. Sci. Technol.* **2009**, *20*, 075502. [CrossRef]
42. Jackson, J.B.; Bowen, J.; Walker, G.; Labaune, J.; Mourou, G.; Menu, M.; Fukunaga, K. A survey of terahertz applications in cultural heritage conservation science. *IEEE Trans. Terahertz Sci. Technol.* **2011**, *1*, 220–231. [CrossRef]
43. Krügener, K.; Schwerdtfeger, M.; Busch, S.F.; Castro-Camus, E.; Koch, M.; Viöl, W. Terahertz meets sculptural and architectural art: Evaluation and conservation of stone objects with T-ray technology. *Sci. Rep.* **2015**, *5*, 14842. [CrossRef] [PubMed]
44. Berry, E.; Walker, G.C.; Fitzgerald, A.J.; Zinov'ev, N.N.; Chamberlain, M.; Smye, S.W.; Miles, R.E.; Smith, M.A. Do *in vivo* terahertz imaging systems comply with safety guidelines? *J. Laser Appl.* **2003**, *15*, 192–198. [CrossRef]
45. Kristensen, T.; Withayachumnankul, W.; Jepsen, P.U.; Abbott, D. Modeling terahertz heating effects on water. *Opt. Express* **2010**, *18*, 4727–4739. [CrossRef] [PubMed]
46. Karpowicz, N.; Zhong, H.; Xu, J.; Lin, K.I.; Hwang, J.S.; Zhang, X.C. Comparison between pulsed terahertz time-domain imaging and continuous wave terahertz imaging. *Semicond. Sci. Technol.* **2005**, *20*, S293–S299. [CrossRef]
47. Herrmann, M.; Tani, M.; Sakai, K. Display modes in time-resolved terahertz imaging. *Jpn. J. Appl. Phys.* **2000**, *39*, 6254–6258. [CrossRef]
48. Shen, X.; Dietlein, C.R.; Grossman, E.; Popovic, Z.; Meyer, F.G. Detection and Segmentation of Concealed Objects in Terahertz Images. *IEEE Trans. Image Process.* **2008**, *17*, 2465–2475. [CrossRef] [PubMed]
49. Jackson, J.B.; Mourou, M.R.; Whitaker, J.F.; Duling, I.N.; Williamson, S.L.; Menu, M.; Mourou, G. Terahertz time-domain reflectometry applied to the investigation of hidden mural paintings. In Proceedings of the 2008 Conference on Quantum Electronics and Laser Science, Lasers and Electro-Optics, CLEO/QELS 2008, San Jose, CA, USA, 4–9 May 2008.
50. Jackson, J.B.; Mourou, M.R.; Whitaker, J.F.; Duling, I.N.; Labaune, J.; Mourou, G.A. Reflection pulse imaging of hidden fresco paintings. In Proceedings of the EOSAM 2008 TOM2 THz Science Technology, Paris, France, 29 September–2 October 2008; pp. 3–4.
51. Fukunaga, K.; Hosako, I.; Duling, I.N.; Picollo, M. Terahertz imaging systems: A non-invasive technique for the analysis of paintings. In Proceedings of the SPIE 7391 Optics for Art, Archiecture, and Archaeology II, Munich, Germany, 15 June 2009; p. 73910.
52. Fukunaga, K. Terahertz spectral database: Construction of open terahertz spectral database. *J. Nat. Inst. Inf. Commun. Technol.* **2008**, *55*, 61–66.
53. Fukunaga, K.; Picollo, M. Terahertz spectroscopy applied to the analysis of artists' materials. *Appl. Phys. A* **2010**, *100*, 591–597. [CrossRef]
54. Pearce, J.; Mittleman, D.M. Scale model experimentation: Using terahertz pulses to study light scattering. *Phys. Med. Biol.* **2002**, *47*, 3823–3830. [CrossRef] [PubMed]
55. Mizumo, M.; Fukunaga, K.; Saito, S.; Hosako, I. Analysis of calcium carbonate for differentiating between pigments using terahertz spectroscopy. *J. Eur. Opt. Soc. Rapid Publ.* **2009**, *4*, 09044. [CrossRef]
56. Fukunaga, K.; Hosako, I. Innovative non-invasive analysis techniques for cultural heritage using terahertz technology. *Comptes Rendus Phys.* **2010**, *11*, 519–526. [CrossRef]
57. Wenliang, L.; Argyros, A. Terahertz spectroscopy and imaging with flexible tube-lattice fiber probe. *J. Lightwave Technol.* **2014**, *32*, 4621–4627. [CrossRef]
58. Hadjiloucas, S.; Galvão, R.K.H.; Zafiropoulos, A.; Walker, G.C.; Dudley, R.; Bowen, J.W. Optimization of apodization functions in THz transient spectrometry. *Opt. Lett.* **2007**, *32*, 3008–3010.
59. Berry, E.; Boyle, R.D.; Fitzgerald, A.J.; Handley, J.W. *Computer Vision Beyond the Visible Spectrum*; Springer: London, UK, 2004; Chapter 9; pp. 271–311.
60. Walker, G.C.; Bowen, J.W.; Jackson, J.B.; Labaune, J.; Hadjiloucas, S.; Mourou, G.; Menu, M. Terahertz Deconvolution. *Opt. Express* **2012**, *20*, 27230–27241. [CrossRef] [PubMed]
61. Tite, M.S. Thermoluminescent dating of ancient ceramics: A reassessment. *Archaeometry* **1966**, *9*, 155–169. [CrossRef]

62. Recur, B.; Younus, A.; Salort, S.; Mounaix, P.; Chassagne, B.; Desbarats, P.; Caumes, J.-P.; Abraham, E. Investigation on reconstruction methods applied to 3D terahertz computed tomography. *Opt. Express* **2011**, *19*, 5105–5117. [CrossRef] [PubMed]
63. Younus, A.; Mounaix, P.; Salort, S.; Caumes, J.P. Fresnel losses in terahertz computed tomography. In Proceedings of the EOSAM 2010 TOM—THz Science Technology, Paris, France, 26–29 October 2010; pp. 26–27.
64. Labaune, J.; Jackson, J.B.; Fukunaga, K.; White, J.; d'Alessandro, L.; White, A.; Menu, M.; Mourou, G. Investigation of terra cotta artefacts with terahertz. *Appl. Phys. A* **2011**, *105*, 5–9. [CrossRef]
65. Ohrstrom, L.; Bitzer, A.; Walther, M.; Rhli, F.J. Technical note: Terahertz imaging of ancient mummies and bone. *Am. J. Phys. Anthropol.* **2010**, *142*, 497–500. [CrossRef] [PubMed]
66. Lin, Y.; Seales, W.B. Opaque document imaging: Building images of inaccessible texts. In Proceedings of the Tenth IEEE International Conference on Computer Vision (ICCV'05), Washington, DC, USA, 17–21 October 2005; pp. 662–669.
67. Sasaki, Y.; Hoshina, H.; Yamashita, M.; Okazaki, G.; Otani, C.; Kawase, K. Inspection system of hidden drugs in sealed envelopes using terahertz waves. *Conf. Infrared Millim. Waves* **2007**, *1–2*, 266–267.
68. Labaune, J.; Jackson, J.B.; Pages-Camagna, S.; Menu, M.; Mourou, G.A. Terahertz investigation of Egyptian artifacts. In 35th International Conference on Infrared Millimeter and Terahertz Waves (IRMMW-THz), Piscataway, NJ, USA, 5–10 September 2010; pp. 1–3.
69. Sunaguchi, N.; Sasaki, Y.; Maikusa, N.; Kawai, M.; Yuasa, T.; Otani, C. Depth-resolving THz imaging with tomosynthesis. *Opt. Express* **2011**, *17*, 9558–9570. [CrossRef]
70. Labaune, J.; Jackson, J.B.; Pagès-Camagna, S.; Mourou, G.A.; Duling, I.N.; Menu, M. Papyrus imaging with terahertz time domain spectroscopy. *Appl. Phys. A* **2010**, *100*, 607–612. [CrossRef]
71. Walker, G.C.; Labaune, J.; Bowen, J.W.; Jackson, J.; Hadjiloucas, S.; Mourou, G.; Menu, M. Deconvolution: Imaging the unturned page. In Proceedings of the 2011 36th International Conference on Infrared, Millimeter and Terahertz Waves (IRMMW-THz), Houston, TX, USA, 2–7 October 2011; pp. 1–2.
72. Fukunaga, K.; Ogawa, Y.; Hayashi, S.; Hayashi, S. Application of terahertz spectroscopy for character recognition in a medieval manuscript. *IEICE Electron. Express* **2008**, *5*, 223–228. [CrossRef]
73. Bardon, T.; May, R.K.; Taday, P.F.; Strlič, M. Systematic study of terahertz time-domain spectra of historically informed black inks. *Analyst* **2013**, *138*, 4859–4869. [CrossRef] [PubMed]
74. Bardon, T.; May, R.K.; Taday, P.F.; Strlic, M. Material characterization of historical parchment using terahertz time-domain spectroscopy. In Proceedings of the 2014 39th International Conference on Infrared, Millimeter, and Terahertz waves (IRMMW-THz), Tucson, AZ, USA, 14–19 September 2014; pp. 1–2.
75. Koch Dandolo, C.L.; Cosentino, A.; Uhd Jepsen, P. Inspection of panel paintings beneath gilded finishes using terahertz time-domain imaging. *Stud. Conserv.* **2015**, *60*, S159–S166. [CrossRef]
76. Cennini, C. *The Craftsman's Handbook: The Italian "Il libro Dell' Arte"*; Thompson, D.V., Translator; Dover Publications: New York, NY, USA, 1960.
77. Gallerano, G.P.; Doria, A.; Germini, M.; Giovenale, E.; Messina, G.; Spassovsky, I.P. Phase-sensitive reflective imaging device in the mm-wave and terahertz regions. *J. Infrared Millim. Terahertz Waves* **2009**, *30*, 1351–1361. [CrossRef]
78. Cosentino, A.; Koch Dandolo, C.L.; Cristaudo, A.; Uhd Jepsen, P. Diagnostics pre and post Conservation on a 14th Century Gilded Icon from Taormina, Sicily. Available online: http://www.e-conservation.org/issue-3/49-Diagnostics-on-a-14th-Century-Gilded-Icon-from-Taormina (accessed on 18 February 2016).
79. Picollo, M.; Fukunaga, K.; Labaune, J. Obtaining noninvasive stratigraphic details of panel paintings using terahertz time domain spectroscopy imaging system. *J. Cult. Herit.* **2015**, *16*, 73–80. [CrossRef]
80. Köhler, W.; Panzner, M.; Klotzbach, U.; Beyer, E.; Winnerl, S.S.; Helm, M.; Rutz, F.; Jördens, C.; Koch, M.; Leitner, H. Non-destructive investigation of paintings with THz-radiation. In Proceedings of the 9th European Conference on NDT, ECNDT, Berlin, Germany, 11 September 2006. Poster 181.
81. Abraham, E.; Younus, A.; El Fatimy, A.; Delagnes, J.C.; Nguéma, E.; Mounaix, P. Broadband terahertz imaging of documents written with lead pencils. *Opt. Commun.* **2009**, *282*, 3104–3107. [CrossRef]
82. Adam, A.J.L.; Planken, P.C.M.; Meloni, S.; Dik, J. Terahertz imaging of hidden paint layers on canvas. *Opt. Express* **2009**, *17*, 3407–3416. [PubMed]

83. Jackson, J.B.; Mourou, M.; Whitaker, J.F.; Durling, I.N., III; Williamson, S.L.; Menu, M.; Mourou, G.A. Terahertz imaging for non-destructive evaluation of mural paintings. *Opt. Commun.* **2008**, *281*, 527–532. [CrossRef]
84. Walker, G.C.; Jackson, J.B.; Giovannacci, D.; Bowen, J.W.; Delandes, B.; Labaune, J.; Mourou, G.; Menu, M.; Detalle, V. Terahertz analysis of stratified wall plaster at buildings of cultural importance across Europe. In Proceedings of the SPIE 8790, Optics for Arts, Architecture, and Archaeology IV, Munich, Germany, 12 May 2013; Volume 8790, p. 87900H.
85. Fukunaga, K.; Hosako, I.; Kohdzuma, Y.; Koezuka, T.; Kim, M.-J.; Ikari, T.; Du, X. Terahertz analysis of an East Asian historical mural painting. *J. Eur. Opt. Soc. Rapid Publ.* **2010**, *5*. [CrossRef]
86. Fletcher, J.R.; Swift, G.P.; Dai, D.C.; Levitt, J.A.; Chamberlain, J.M. Propagation of terahertz radiation through random structures: An alternative theoretical approach and experimental validation. *J. Appl. Phys.* **2007**, *101*, 01310. [CrossRef]
87. Kurabayashi, T.; Kikuchi, N.; Tanno, T.; Watanabe, M. Significance of terahertz spectrometry for textile article of wool. In Proceedings of the 34th International Conference on Infrared, Millimeter and Terahertz Waves, Busan, Korea, 21–25 September 2009; pp. 1–2.
88. Cesarani, F.; Martina, M.C.; Ferraris, A.; Grilletto, R.; Boano, R.; Fiore Marochetti, E.; Donadoni, A.M.; Gandini, G. Whole-Body Three-Dimensional Multidetector CT of 13 Egyptian Human Mummies. *Am. J. Roentgenol.* **2002**, *180*, 597–606. [CrossRef] [PubMed]
89. Hoffman, H.; Torres, W.E.; Ernst, R.D. Paleoradiology: Advanced CT in the evaluation of nine Egyptian mummies. *Radiographics* **2002**, *22*, 377–385. [CrossRef] [PubMed]
90. Cortes, E.; Cosentino, A.; Duling, I.N.; Fukunaga, K.; Mininberg, D.T.; Stuenkel, I.; Leona, M. Investigating the use of terahertz pulsed time domain reflection imaging for the study of fabric layers of an Egyptian mummy. *J. Eur. Opt. Soc. Rapid Publ.* **2011**, *6*, 11040.
91. Koch Dandolo, C.L.; Jepsen, P.U.; Christensen, M.C. Characterization of european lacquers by terahertz (THz) reflectometric imaging. In Proceedings of the IEEE Proceedings of the 1st Digital Heritage International Congress (DigitalHeritage), Marseille, France, 28 October–1 November 2013; Volume 1, pp. 89–94.
92. Dandolo, C.L.K.; Cattersel, V.; Jepsen, P.U. Terahertz time-domain imaging of a 17th century lacquered cabinet: A contribution to European lacquerwares characterization. In Proceedings of the IRMMW-THz, Hong Kong, China, 23–28 August 2015.
93. Chen, C.C.; Lee, D.J.; Pollock, T.; Whitaker, J.F. Pulsed-terahertz reflectometry for health monitoring of ceramic thermal barrier coatings. *Opt. Express* **2010**, *18*, 3477–3486. [CrossRef] [PubMed]
94. Kurabayashi, T.; Sakai, S.; Fujino, K. Sub-terahertz imaging of painted steel. In Proceedings of the 35th International Conference on Infrared, Millimeter and Terahertz, Rome, Italy, 5–10 September 2010; pp. 1–2.
95. Anastasi, R.F.; Madaras, E.I. Terahertz NDE for under paint corrosion detection and evaluation. *AIP Conf. Proc.* **2006**, *820*, 515–522.
96. Zhao, H.; Wu, D.-B.; Zhan, H.-L.; Sun, Q.; Zhao, K. Detection of iron corrosion by terahertz time-domain spectroscopy. In Proceedings of the SPIE 9795, Selected Papers of the Photoelectronic Technology Committee Conferences, Hefei, Suzhou, and Harbin, China, 14 June 2015; p. 97953.
97. Pastorelli, G.; Trafela, T.; Taday, P.F.; Portieri, A.; Lowe, D.; Fukunaga, K.; Strlič, M. Characterisation of historic plastics using terahertz time-domain spectroscopy and pulsed imaging. *Anal. Bioanal. Chem.* **2012**, *403*, 1405–1414. [CrossRef] [PubMed]
98. Cacciari, I.; Agresti, J.; Siano, S. Combined THz and LIPS analysis of corroded archaeological bronzes. *Microchem. J.* **2016**, *126*, 76–82.

10

Environmental Performance of Hypothetical Canadian Pre-Combustion Carbon Dioxide Capture Processes Using Life-Cycle Techniques

Lakkana Piewkhaow [1,†], Anastassia Manuilova [2,†], Christine W. Chan [1,*], Malcolm Wilson [2,†] and Paitoon Tontiwachwuthikul [1,†]

1 Faculty of Engineering and Applied Science, University of Regina, Regina, SK S4S 0A2, Canada; ying_gf@hotmail.com (L.P.); Paitoon.Tontiwachwuthikul@uregina.ca (P.T.)

2 ArticCan Energy Services, Regina, SK S4P 3N6, Canada; amanuilova@articcanenergy.com (A.M.); malcolm.alan.wilson@gmail.com (M.W.)

* Correspondence: christine.chan@uregina.ca

† These authors contributed equally to this work.

Academic Editor: Gustavo A. Fimbres Weihs

Abstract: The methodology of life-cycle assessment was applied in order to evaluate the environmental performance of a hypothetical Saskatchewan lignite-fueled Integrated Gasification Combined Cycle (IGCC) electricity generation, with and without pre-combustion carbon dioxide (CO_2) capture from a full life-cycle perspective. The emphasis here is placed on environmental performance associated with air contaminants of the comparison between IGCC systems (with and without CO_2 capture) and a competing lignite pulverized coal-fired electricity generating station in order to reveal which technology offers the most positive environmental effects. Moreover, ambient air pollutant modeling was also conducted by using American Meteorological Society/Environmental Protection Agency Regulatory Model (AERMOD) air dispersion modeling to determine the ground-level concentration of pollutants emitted from four different electricity generating stations. This study assumes that all stations are located close to Estevan. The results showed a significant reduction in greenhouse gas (GHG) emissions and acidification potential by applying both post-combustion and pre-combustion CO_2 capture processes. The GHG emissions were found to have reduced by 27%–86%, and IGCC systems were found to compare favorably to pulverized coal systems. However, in other environmental impact categories, there are multiple environmental trade-offs depending on the capture technology used. In the case of post-combustion capture, it was observed that the environmental impact category of eutrophication potential, summer smog, and ozone depletion increased due to the application of the CO_2 capture process and the surface mining coal operation. IGCC systems, on the other hand, showed the same tendency as the conventional coal-fired electricity generation systems, but to a lesser degree. This is because the IGCC system is a cleaner technology that produces lower pollutant emission levels than the electricity generating station; thus, the benefits of capture are reduced on a comparative basis. The results from air dispersion analysis showed that the maximum ground-level concentrations of pollutants from all electricity generating stations are in compliance with all air quality standards, except for Co, Pb and Ni. The IGCC with capture revealed the lowest nitrogen dioxide (NO_2) ground-level concentration compared to all other scenarios. Moreover, IGCC systems both with and without pre-combustion CO_2 capture revealed no ground-level concentration of trace elements. This is because the IGCC system operates with an acid gas cleaning process that removes most of the trace contaminants from the syngas.

Keywords: IGCC; life-cycle assessment (LCA); AERMOD; Saskatchewan lignite; coal; pre-combustion CO_2 capture processes

1. Introduction

Currently, CO_2 emissions are the primary concern in the electrical generation industry, as carbon dioxide emissions from electrical generation accounted for 12% of the total Canadian emissions in 2013, Figure 1 [1]. Coal-fired electrical generation is the predominant source of electricity, accounting for 47% of generation in Saskatchewan, while natural gas accounts for 28%, hydropower 19%, and renewable and others 6% [2]. Efforts are being made by engineers and scientists to overcome anthropogenically induced climate change by researching and developing electrical generation systems, including clean coal technology, which produces lower CO_2 emissions than conventional plants. One promising approach to reduce the emission of CO_2 into the atmosphere is carbon capture and storage (CCS).

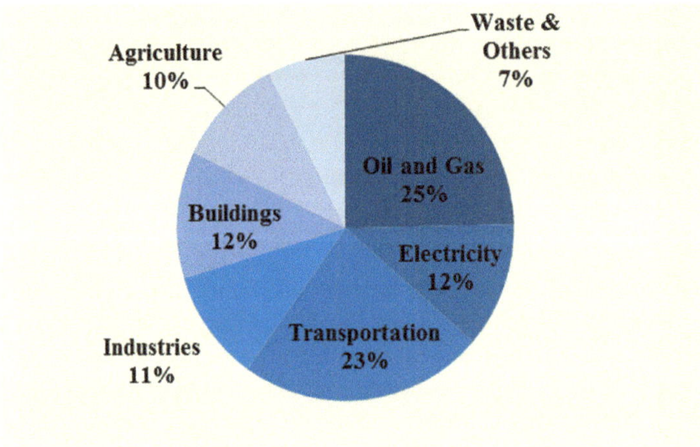

Figure 1. Canada carbon dioxide emissions by source in 2013 [1].

Coal gasification is a proven technology that is being adopted in commercial Integrated Gasification Combined Cycle (IGCC) electricity generation. The IGCC is an advanced generation concept that combines modern coal gasification technology with gas turbines (Brayton cycle) and steam turbines (Rankine cycle). IGCC technology has proven to be one of the cleanest coal-based electrical generation technologies, emitting the least of several key pollutants of NO_x, SO_x, CO_x and PM_{10} compared to any other coal-based technology [3–5]. The IGCC can be combined with the pre-combustion CO_2 capture technology for further gas purification, which enables efficient separation of the CO_2 by means of pressure-swing absorption or physical solvents such as chilled methanol (Rectisol®, Munich, Germany) or a mixture of dimethyl ethers of polyethylene glycol (Selexol®, Des Plaines, IL, USA) [6,7]. With CO_2 concentration of 15%–60% by volume in a pre-combustion process and only 15% by volume in a post-combustion process, the CO_2 partial pressure can be significantly higher in the pre-combustion CO_2 capture process. These factors help to improve the driving force for the separation process. As a consequence, the required equipment for the pre-combustion CO_2 capture process can be smaller than that for the post-combustion CO_2 capture process [8].

Although there are many environmental and performance-enhancing advantages associated with CO_2 capture with IGCC technology, it must be recognized that CO_2 capture is an energy-intensive process which significantly decreases the overall efficiency of the electrical generation station; hence, a larger amount of coal is needed to compensate for the efficiency loss. This will lead to further emissions associated with the additional coal. Moreover, the upstream emissions from fuel/material acquisition and the downstream emissions from waste disposal cannot be readily captured. The conversion of uncontrolled emissions release from the generating station itself to the atmosphere to a solid or liquid stream offers the potential for well-designed landfills to store waste safely. The CO_2 emissions produced at the plant can be captured for underground storage. In a traditional plant, the emissions from upstream and downstream processes are small when compared to the emissions from the combustion process [9]. However, the upstream and downstream emissions become more relevant when the CO_2 capture process is applied with the reduction in the "mid-stream" emissions. Therefore, it is very important to conduct a life-cycle assessment of an electricity generation station with CO_2 capture and evaluate its environmental performance. Given the current lack of commercial experience in IGCC, it is necessary to predict its environmental performance from a full life-cycle perspective to comprehensively understand its environmental impacts. It is also important to understand the full ramifications of the process when comparisons are made to other alternative technologies for greenhouse gas emissions reduction. Moreover, air dispersion modeling was used to evaluate the ground-level concentration of sixteen air contaminant components including sulfur dioxide (SO_2), nitrous oxide (N_2O), particulate matter ($PM_{2.5}$), arsenic (As), cadmium (Cd), chromium (Cr), mercury (Hg), manganese (Mn), lead (Pb), antimony (Sb), cobalt (Co), copper (Cu), molybdenum (Mo), selenium (Se), vanadium (V), and nickel (Ni).

At present, there are only fourteen IGCC plants operating worldwide with a capacity size ranging from 40 MWe to 582 MWe, most of them are in the United States and Europe [10]. None of these plants currently capture CO_2. However, the Kemper County IGCC is nearing commercial operations and will be operating in 2016. This plant will capture 65% of the CO_2 and send it to oil companies for enhanced oil recovery. According to [11], there are ten IGCC plants (exclude Nuon IGCC, Buggenum, which was closed down in April 2013) in operation or advanced construction in Europe. Four of them use coal (accounts for over 1500 MWe or over 47% of total electricity generated from the IGCC in Europe). The other six plants are fueled by using gasified refinery by products. In the United States there are currently four IGCC coal-based plants in operation including Tampa Electric, Polk County (250 MWe), Wabash, West Terre Haute (265 MWe), Edwardsport IGCC Station, Indiana (618 MWe) and Kemper County IGCC, Mississippi (582 MWe). In addition, there are about eighteen plants that are being planned, six plants that are in the initial stage will be located in Europe, twelve plants will be located in North America; ten plants will be located in the United States and two in Canada [12–14].

Based on the database of operating plants, the IGCC electrical generating station is still in the early phase which makes for a significant limitation on actual technical data and experiences. The life-cycle assessment evaluation of electrical generation had never been applied and studied in detail, in Saskatchewan, until the year 2006 when Suebsiri conducted the CO_2 emission accounting study of the Weyburn project in Saskatchewan. So far, there have only been two studies on LCA-CCS electrical generating stations based in Saskatchewan [15] and [16]. Suebsiri [15] conducted the study of an environmental model of carbon capture and storage for SaskPower's Boundary Dam. In 2011, the full life-cycle assessment of SaskPower's Boundary Dam power generation station was studied in detail. Both of the studies focused on the post-combustion technology coupled with the existing electrical generating station in Saskatchewan. Up to this point, there has been no research on LCA-IGCC coal-based generation coupled with pre-combustion technology in Saskatchewan. This research is the starting point for the evaluation of LCA of IGCC systems when fueled by Saskatchewan's unique lignite coal. This research also focuses on air dispersion modeling associated with an IGCC electrical generating station. The database resulting from this research will benefit future LCA-IGCC Saskatchewan lignite-based projects.

This study applies life-cycle assessment (LCA) methodology to evaluate the environmental performance associated with air contaminants from the IGCC, with and without pre-combustion CO_2 capture from a full life-cycle perspective. Moreover, the air dispersion modeling method is also conducted to predict the impact of atmospheric emissions associated with electrical generating operations in the SaskPower Boundary Dam (Estevan) area. The results from both LCA methodology and the air dispersion modeling of IGCC electricity generation, with and without pre-combustion CO_2 capture, are compared to conventional lignite coal-fired electricity generation, with and without post-combustion CO_2 capture.

The paper is organized as follows. Section 2 describes the methodology for life-cycle assessment and air dispersion modeling. Section 3 provides a detailed description and inventories of IGCC and conventional lignite coal-fired electricity generation. Section 4 presents the results and discussion of the life-cycle impact assessment and air dispersion modeling of both technologies. Section 5 presents a conclusion and recommendation for the future work.

2. Methodology

2.1. Environmental Life-Cycle Assessment Methodology

2.1.1. Goal and Scope Definition

The goal of this LCA study is to evaluate the environmental performance of Saskatchewan lignite-fueled IGCC electrical generation, with and without CO_2 capture. This research was intended to provide results specific to Western Canadian lignite used in an IGCC facility. The emphasis was thereby put on the comparison between IGCC systems (with and without CO_2 capture) and the competing lignite-pulverized coal electrical generation stations (with and without CO_2 capture) in order to reveal which technology offers the most positive environmental effects. The composition of Saskatchewan lignite coal used in all case scenarios is listed in Table 1, and the heating value of lignite coal was 15,119 kJ/kg.

Table 1. Composition of Saskatchewan lignite coal [16].

Parameter	Unit	Value
Moisture	%	35.00
Carbon	%	41.70
Hydrogen	%	2.61
Nitrogen	%	0.79
Sulfur	%	0.54
Ash	%	9.47
Oxygen	%	9.90
Mercury	ppb	79.00
Chlorine	ppm	10.20

The assessment of environmental impacts performance from IGCC and PC systems was conducted using the "cradle-to-gate" approach, where the activities stop at transport entry point to the field. Four system options were analyzed and compared:

1. conventional lignite coal-fired electricity generation without a post-combustion CO_2 capture process;
2. conventional lignite coal-fired electricity generation with an amine post-combustion CO_2 capture process;
3. integrated Gasification Combined Cycle (IGCC) without a pre-combustion CO_2 capture process;
4. integrated Gasification Combined Cycle (IGCC) with Selexol® as the chosen pre-combustion CO_2 capture process.

Functional Unit

The functional unit provides a standard reference metric for all input and output variables (*i.e.*, energy, materials, emissions, and wastes). The functional unit for this study was selected as 1 MWh of electricity generated, considering that this is a product of the electrical generation by any means.

System Boundaries

In this LCA study, the full life cycle of electricity generating stations was considered.

The "cradle-to-gate" approach is one in which the environmental effects associated with all life-cycle activities, from resource extraction (e.g., coal, limestone and iron) and the production of materials (e.g., steel, concrete and monoethanolamine (MEA)) to the generation of electrical energy at the generating station, as well as the CO_2 capture process and all transportation stages, are taken into account; however, no accounting is made for the utilization of any oil produced under enhanced oil recovery (EOR) nor for underground CO_2 sequestration, for example. The material and energy inputs needed for equipment manufacture are not included in this study. Exclusion of these activities is common practice where it is expected that these processes have far smaller environmental implications than the operations of the facilities [17]. MEA solvent consumption and production for the post-combustion process is included; this is the amine with abundant information in the literature. For pre-combustion, the Selexol® solvent production was not included because of a lack of data. It was thought to have no relevant effects on the environmental impacts since the Selexol® consumption is two orders of magnitude less than amine consumption [18]. For downstream processes, waste transport and disposal in a nearby landfill were considered. The system boundary for lignite-IGCC electrical generating stations with pre-combustion capture is presented in Figure 2.

Temporal and geographical boundaries: The temporal boundary was defined as 34 years, and it was assumed for the purpose of this study that plant construction (IGCC and PC) would be completed by 2015. This duration included the time needed for coal mine development, plant construction, commissioning, and operation. The construction and commissioning phase was assumed to be four years before the operation starts. This number was based on the report from the Polk Electricity Generating Station [19]. Although the system boundary of this research does not include downstream processes after CO_2 is captured, it is assumed that the CO_2 will eventually be used for enhanced oil recovery. Therefore, the lifespan of the electrical generating station was assumed to be 34 years to accommodate CO_2-EOR projects, with an expected requirement for new CO_2 purchase of 15 years, where the first four years is the plant construction phase and the rest (30 years) is the operational phase.

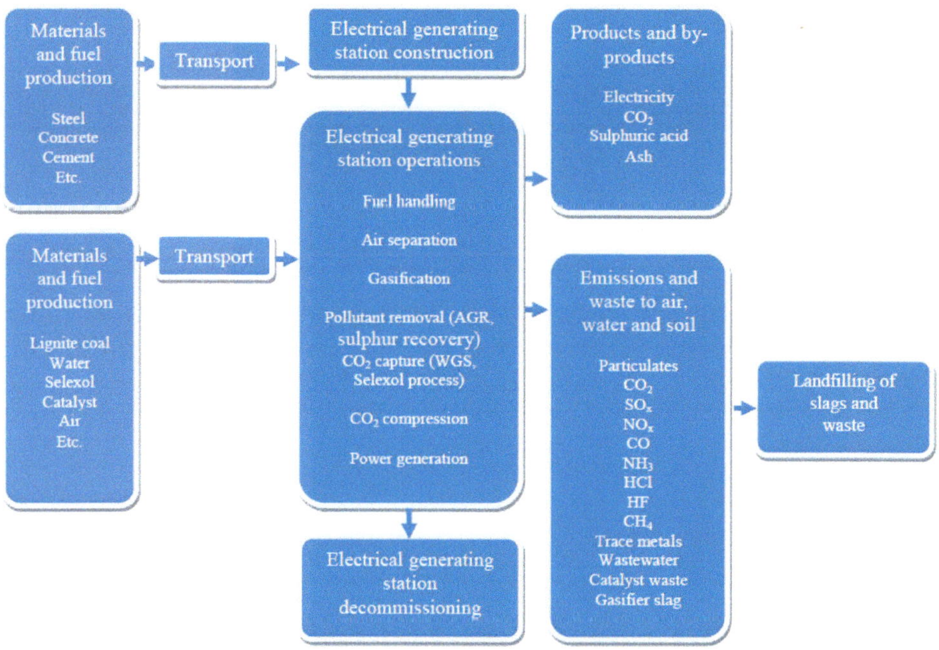

Figure 2. System boundaries for lignite-IGCC electrical generating station with pre-combustion capture.

2.1.2. Life-Cycle Inventory Data Quality, Sources and Assumptions

The majority of the data used in this study is specific to Western Canada. In the absence of Western Canadian data, information was drawn from other Canadian sources, followed by North American and European sources and, finally, from pre-existing life-cycle databases, most notably Ecoinvent v2.1.

The majority of operational data for IGCC plants came from the technical documentation of the IGCC with carbon dioxide capture and storage (CCS), [20]. Other sources of data came from US Department of Energy (DOE) reports.

In the case of post-combustion CO_2 capture, refer to [21] and [16], since the conventional lignite coal-fired electricity generation (PC) in both cases, with and without CO_2 capture, were adopted from the previous work of [16].

2.1.3. Life-Cycle Assessment Modeling

All unit processes at the electrical generating plant were modeled so that emissions associated with each unit process could be tracked. The IGCC model was created using an Excel™ (2010) spreadsheet and AspenPlus® (Version 2004.1) simulation software. The modeling results were transferred to the LCA software GaBi™ (Version 5.0), which is a software product of PE International, Germany.

2.1.4. Life-Cycle Impact Assessment

The life-cycle impact assessment was carried out in order to identify environmental impacts that arise from this LCA study. In this study, the TRACI life-cycle impact assessment (LCIA) method was adopted for the analysis. Since this study is specific to Saskatchewan/Canada, it was thought that the selected impact assessment methodology should represent Canadian or US environmental conditions for accurate impacts evaluation. LUCAS, the Canadian LCIA method, is still in the development stage and so was not usable for this purpose at this time. Therefore, TRACI impact assessment methodology was applied in this study. The environmental impacts were categorized into five categories. Brief descriptions of each environmental impact category are summarized in Table 2.

Table 2. TRACI impact categories considered in the study.

Impact Category	Abbreviation	Short Description	Relevant LCI Data	Characterization Factor
Global warming potential	GWP	Radiative forcing of atmosphere, causing a temperature rise	CO_2, N_2O, CH_4, SF_6, $CHCL_3$, CF_4, CFCs, HCFCs, CH_3Br	kg CO_2-equivalent
Acidification potential	AP	Emission of acid-forming substances	SO_2, NO_x, NH_3, HCL, HF, H_2S	kg SO_2-equivalent
Eutrophication potential	EP	Excessive supply of nutrients	NO_x, NH_3, PO_4^{3-}	kg N-equivalent
Ozone depletion potential	ODP	The relative amount of degradation to the ozone layer	R 11, R 114, R12, and R13	kg CFC 11-Equivalent
Photochemical oxidation potential (Summer smog)	POCP	The complex reactions during the formation of reactive chemical compounds by action of sunlight on primary pollutants	NOx, PAH, NMVOC, CH_4	kg ethylene-equivalent

2.2. Air Dispersion Modeling Methodology.

Air dispersion modeling assessment was conducted for the purpose of predicting the impact of air contaminant emissions to the atmosphere associated with electrical generation operations in the SaskPower Boundary Dam area. The assessment was conducted to evaluate and compare the atmospheric emissions of an IGCC-based electricity generating plant, with and without pre-combustion CO_2 capture, with a competing pulverized lignite coal-fired electricity generating stations (with and without CO_2 capture). The atmospheric emissions associated with a pulverized coal-fired electricity generating station without CO_2 capture process was used for the base case scenario, and the rest, the atmospheric emissions of pulverized lignite coal-fired electricity generating station with CO_2 capture and the IGCC, with and without CO_2 capture, are compared to the base case scenario. The location of all plants was assumed to be the same site as SaskPower's Boundary Dam, approximately 5 kilometers southwest of the city of Estevan, Saskatchewan.

In this study, the U.S. EPA AERMOD dispersion modeling (Version 13350) was selected to predict the maximum ground-level air concentrations of SO_2, nitrogen dioxide (NO_2), $PM_{2.5}$, Hg, Cd as well as other heavy metals that could have an impact on human health resulting from the operation of electricity from both pulverized and IGCC plants. The dispersion modeling was performed based on Saskatchewan Air Quality Modeling Guidelines [22]. The primary input to AERMOD was the "run stream setup file" and contains source location, parameter data, receptor location, pre-processor meteorological data file (American Meteorological Society/Environmental Protection Agency Regulatory Meteorological (AERMET)), and output options. The background concentration values and ambient air quality standard values for air contaminants were adopted from available regional ambient air quality monitoring data provided by the Saskatchewan Ministry of Environment [22–25]. The cumulative assessment of each air contaminant was addressed by the inclusion of ambient background concentrations of an air contaminant to the predicted value from AERMOD to obtain total concentrations before making a comparison to Saskatchewan ambient air quality standards. In the case that Saskatchewan air quality standards were not available, the comparison was made to Ontario.

2.2.1. Facility Regional Setting and Study Area

In this paper, the study area of 20 kilometers by 20 kilometers centered on the SaskPower Boundary Dam electricity generating station was selected to conduct the air quality modeling assessment. As the plant is located approximately 5 kilometers southwest of the city of Estevan, it was thought that a 20 km $\times$ 20 km area was a reasonable size for covering Estevan and the surrounding area, as most pollution concentration occurs within this range based on the reports from [25] and [26]. Figure 2 presents the location of air dispersion quality study area. The City of Estevan is the major population concentration in the region.

2.2.2. Meteorology, Receptors and Terrain

In order to assess air quality impacts, meteorology data are necessary for air dispersion modeling, as the contaminants being studied migrate through the environment. AERMOD uses hour-by-hour meteorological data that is processed through AERMET, where the meteorology data is organized into a suitable format for the AERMOD dispersion model. To run the air dispersion model, five consecutive years of representative meteorology data should be used to adequately predict the pollutant concentrations at ground level [25]. AERMET requires at least one surface meteorological station and one station where the data of wind speed, wind direction and temperature are collected at standard heights [22]. The surface parameters include albedo (the percentage of radiation returning from a surface compared to that which strikes it [27]), Bowen ratio (to determine the surface moisture), and surface roughness (the measure of how efficiently momentum can be transferred to the ground at a given wind speed [28]). Preparing AERMET input files is time-consuming; therefore, to accommodate the air quality modeling, the Saskatchewan Ministry of Environment has developed regional meteorological data sets, and these are available on the ministry's website. Saskatchewan's meteorological data sets have been developed based on geographical area and are categorized into five zones; Northern, North Central, Central, Southwestern and Southeastern. Each regional metrological data set provided by the Ministry contains five years of hourly meteorological data for the years 2003 to 2007, as this AERMET input file is the most recent data available and is representative of the area under the study. In the AERMET file, the standard height used in the sampling at the meteorological station is set as 9 meters.

In this study, the flat terrain option was chosen to model the effects of terrain above stack base in the rural setting as the geography of Estevan and the surrounding area in the southwest of Saskatchewan is relatively flat. The origin receptor grid defined the locations where the model would predict ground-level concentration or where the stack was located. The distances along each directional radial at which the receptor was located were defined as 2000 meters, 4000 meters, 6000 meters and 10,000 meters for all case scenarios. The model also generated direction radials for the receptor network; in this case, there were 36 directions, beginning with the 10 degree flow vector and incrementing every 10 degrees clockwise. The receptor network is shown in Figure 3.

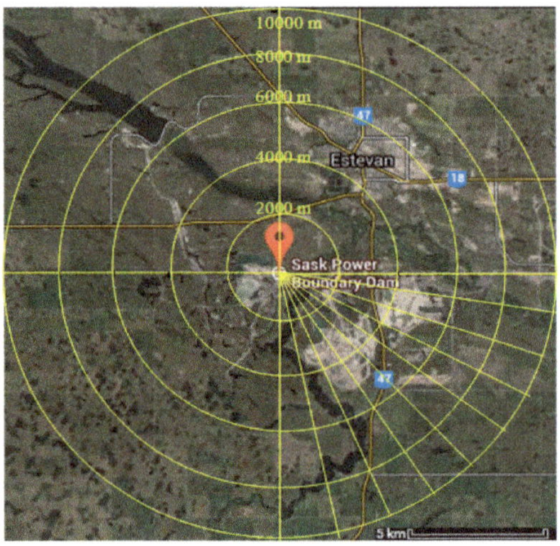

Figure 3. Location of air dispersion study area and receptor network specification.

2.2.3. Emission Source Characterization

In this study, a prediction of maximum ground-level air concentrations of emissions from a Saskatchewan lignite IGCC-based electricity generating plant, with and without the pre-combustion CO_2 capture process, and from the competing pulverized lignite coal-fired electricity generating stations, with and without CO_2 capture process, was made. The specific air contaminants released into the atmosphere associated with the operation of all electricity generating stations include NO_2, SO_2, $PM_{2.5}$, As, Cd, Cr, Hg, Mn, Sb, Co, Cu, Pb, Mo, Se, V, and Ni.

The emission data related to the operation of a pulverized coal-fired electricity generating station were adopted from the previous work of [16]. The emissions data of IGCC systems were conducted with the engineering simulation software AspenPlus®, Excel™ spreadsheets and life-cycle assessment simulation software GaBi5™. The emission rates are based on each MWh net electricity produced. Physical dimensions of the emission source (stack height and stack diameter) as well as emission parameters characterizing of the exit conditions (exit velocity and exit temperature) were taken from [1] and [29] for Pulverized coal-fired and IGCC systems, respectively. It has to be noted that, due to the limitation and confidentiality of the data, the stack height and stack diameter of pulverized coal-fired with post-combustion CO_2 capture process are assumed based on [30]. Table 3 shows the emission rate and stack parameters used for air dispersion modeling.

Table 3. Emission rates and stack parameters for AERMOD.

Parameters	PC	PC with Capture	IGCC	IGCC with Capture
Net Electricity Output (MW)	820	552	342	261
Stack Height (m)	92	55	99.06	99.06
Stack Diameter (m)	5.18	3	7.3	7.3
Stack Exit Velocity (m/s)	24.4	24.4	14.02	14.02
Exit Gas Temperature (K)	436.15	340	373.7	373.7
Emission rate (g/s) per MWh				
NO_2	3.1×10^{-2}	3.7×10^{-2}	2.2×10^{-2}	1.9×10^{-2}
SO_2	1.8×10^{-2}	1.42×10^{-3}	2.6×10^{-2}	2.9×10^{-2}
$PM_{2.5}$	2.4×10^{-2}	1.9×10^{-2}	2.3×10^{-2}	1.4×10^{-2}
As	1.2×10^{-2}	2.09×10^{-4}	8.33×10^{-7}	1.43×10^{-8}
Cd	1.1×10^{-3}	2.42×10^{-4}	5.94×10^{-8}	7.0×10^{-9}
Cr	1.6×10^{-2}	7.06×10^{-5}	1.44×10^{-8}	6.11×10^{-12}
Hg	1.3×10^{-3}	1.29×10^{-6}	1.39×10^{-7}	1.39×10^{-11}
Mn	1.2×10^{-3}	1.42×10^{-6}	1.83×10^{-7}	3.11×10^{-9}
Sb	1.1×10^{-3}	2.28×10^{-7}	7.22×10^{-8}	1.44×10^{-9}
Co	1.9×10^{-3}	1.92×10^{-6}	4.03×10^{-8}	4.03×10^{-9}
Cu	6.4×10^{-4}	3.89×10^{-6}	1.53×10^{-7}	3.06×10^{-9}
Pb	8.3×10^{-3}	1.25×10^{-5}	1.08×10^{-6}	1.62×10^{-8}
Mo	1.1×10^{-2}	2.11×10^{-5}	8.89×10^{-9}	1.78×10^{-10}
Se	1.1×10^{-1}	2.4×10^{-2}	2.0×10^{-6}	9.89×10^{-7}
V	2.4×10^{-2}	4.06×10^{-5}	3.14×10^{-6}	6.28×10^{-8}
Ni	1.6×10^{-1}	5.64×10^{-4}	2.94×10^{-7}	7.36×10^{-9}

3. Data Inventory Modeling and Systems Descriptions

3.1. Integrated Gasification Combined Cycle (IGCC) with Pre—Combustion CO_2 Capture Process

The IGCC system equipped with a CO_2 capture process includes the following: an Air Separation Unit (ASU), a Gasification Unit, a Gas Cleaning/Acid Gas Removal unit, a Sulfur Recovery Unit, a Water Gas Shift Unit (WGS), a CO_2 Removal Unit, and a Combined Cycle Electrical Generation Unit (Gas turbine, Steam turbine, and Heat Recovery Steam Generator). The simplified flow diagram of the IGCC with a pre-combustion CO_2 capture process is shown in Figure 4.

The processes mentioned above (ASU, Gasifier, Gas cleaning and WGS) were simulated with AspenPlus®, while the CO_2 removal and combined cycle electrical generation unit were simulated using a spreadsheet. The specific design and detailed description of the process unit in this study are given below.

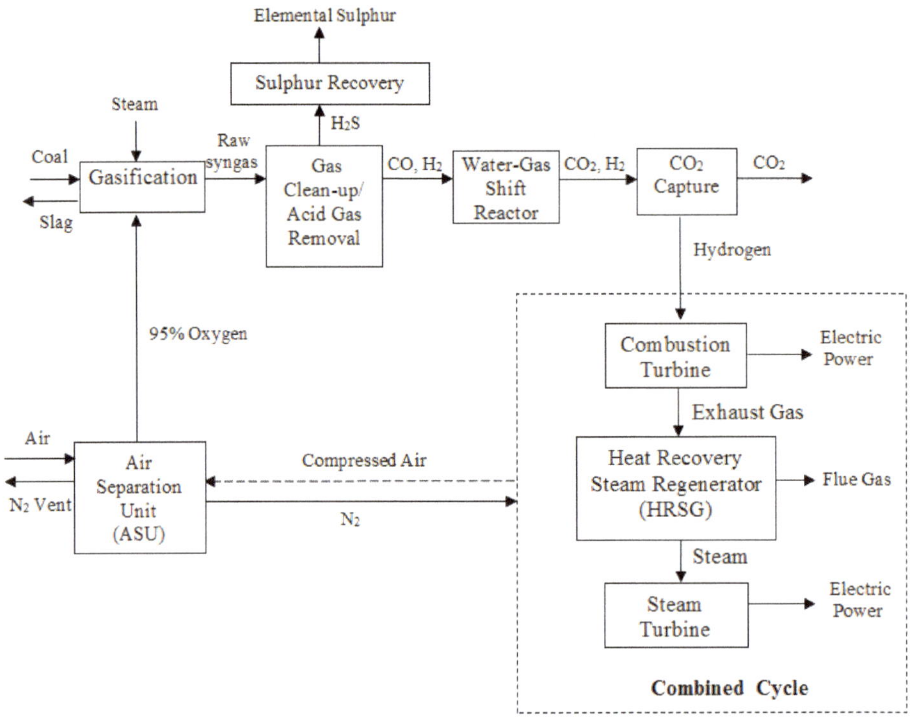

Figure 4. Simplified flow diagram of IGCC with CO_2 capture process.

3.1.1. Air Separation Unit (ASU)

In this study, cryogenic separation was applied since it is the most common method and widely used for medium and large scale oxygen production with high purity requirements. Ambient air is first sent to a multistage air compressor unit. The air is compressed to approximately 6.3 bar (630 kPa). The compressed air is then cooled in the main heat exchangers. The cooled air is then fed into the cryogenic separation unit. As a result, the air stream is liquefied and split into three streams: a nitrogen stream, a 95% oxygen stream, and a water stream. In this study, 70% of nitrogen was sent to gas turbines to moderate combustion conditions and minimize or reduce NO_x formation. In the meantime, the 95% oxygen was compressed and sent to the gasifier to further the fuel reaction process.

3.1.2. Gasification

The feed coal slurry was pumped into the gasifier (reactor) together with a high purity oxidant (95% O_2) to partially oxidize the fuel, by controlling the amount of oxygen and steam. The reaction of coal slurry and oxygen took place at a high temperature of between 1250 °C and 1427 °C.

In the simulation, an entrained flow gasifier (GE) was modeled at a pressure of 580 psia (4 MPa) and operated on a 35 kg/s coal mass flow. The entrained-flow gasifier was a plug type reactor. This reactor is suitable for fine feedstock fuel, which, in this case, had a particle diameter less than 0.1 mm. The fuel flows co-current and reacts with oxygen and with or without steam (depending on whether the coal was added as dry feed or water-based slurry). The highly pure oxygen (95% O_2) at an ambient temperature of 25 °C and slurry coal were introduced at the top of the gasifier. The mass ratio between injected water and coal was set at 0.54 and the oxygen-fuel ratio was set at 0.85, based on the report from [31] for the optimum steam-fuel ratio and oxygen-fuel ratio, respectively. The fuel (slurry-coal) flows co-currently and reacts with the oxygen. The fuel selected in this study was Saskatchewan lignite coal.

3.1.3. Acid Gas Removal and Sulfur Recovery Process

The raw syngas from the gasification process mainly consisted of carbon monoxide (CO) and hydrogen (H_2), and a small quantity of contaminant compounds such as hydrogen-sulfide (H_2S) and carbonyl sulfide (COS), oxides of carbon and sulfur such as carbon dioxide (CO_2) and sulfur dioxide (SO_2), nitrogen compounds such as ammonia (NH_3), hydrogen cyanine (HCN), Hg, and particulates from the mineral fraction of the coal such as unconverted carbon and ash [32]. Some of the corrosive components such as sulfide, nitride, and dust were removed in this cleaning section.

The H_2S rich gas from the Acid Gas Removal (AGR) process was then fed into the two-step Claus Plant Process (sulfur recovery) to produce sulfur [33]. In this process, 96% of H_2S in the acid gas feed stream was converted into elemental sulfur [33]. Sulfur is a valuable product, which can be used for manufacturing sulfuric acid and fertilizer, *etc.* It has to be noted that SaskPower converts the captured SO_x to sulfuric acid (H_2SO_4). The alternative of IGCC would be to take the H_2S gas stream and re-inject it with the captured CO_2 stream, as there is no impact of H_2S on EOR (Weyburn receives a CO_2 stream at 1% H_2S) or to inject it separately into a disposal well (Alberta has over 40 acid gas disposal wells).

3.1.4. Water Gas Shift and CO_2 Removal Unit

After the H_2S scrubbing process, the syngas stream containing mainly CO and H_2 was fed to the shift reactors. In order to meet the steam-to-carbon ratio requirement, steam was added to the clean syngas before passing through the shift reactor.

In the first shift reactor, the steam reacted with syngas at high temperature. The reaction took place with the ratio of 1.0 mole of water per mole of carbon monoxide. The water gas shift reaction was an exothermic reaction that produced sensible heat, and the syngas from the high temperature reactor had to be cooled before being fed into the low temperature reactor. The shift reaction was repeated in the second reactor with a lower temperature than the first reactor. The syngas that left the WGS reactor had a much higher H_2 composition than the raw syngas with the CO conversion of 90% [32,33].

The CO_2 and H_2 rich syngas from water-gas-shift process was then fed to a Selexol® process, licensed by UOP LLC, to remove CO_2. The Selexol® solvent was a mixture of Dimethylethers and Polyethylene glycol with the formulation of $CH_3(CH_2O)_nCH_3$, where n is between 3 and 9. The Selexol® solvent absorbed 95% of the CO_2 from the feed syngas when at the relatively high pressure of between 300 and 2000 psia (2.068 and 13.8 MPa) [34]. The rich solvent containing acid gas (CO_2) was then sent to a regenerator to strip off the solvent and recover the CO_2. The resulting gases, containing 95% by mass CO_2, was then sent to a dehydrator and compressor and compressed to 80 bar (8 MPa) in an intercooled staged compressor and liquefied. The CO_2 was then ready to be transported and used for enhanced oil recovery. One of the advantages of Selexol® process over an amine absorption process is that Selexol® requires less energy than amine-based processes. This is because there is no chemical reaction involved between the solvent and the acid gas. However, the solvent capacity is reduced when the feed syngas pressure is less than 300 psia (2.068 MPa). With high pressure syngas, the physical absorption using Selexol® has proven to be favorable for CO_2 removal [20].

3.1.5. Electrical Generation Unit

After the CO_2 removal process, the hydrogen rich syngas was then introduced to the combined cycle electricity generation unit to produce electricity. The combined cycle electricity generation consists of a gas turbine system and steam turbine system. The inventory modeling of combined cycle was conducted based on the Excel™ spreadsheet of mass and energy balance model of a gas turbine combined cycle from [35,36] and are described in detail in the following section. The main inputs and assumptions of the combined cycle are also presented in Table 4.

Table 4. Input assumptions for combined cycle system [36].

Parameter	Value
Ambient Pressure (psia)	14.7 (101.35 kPa)
Ambient Temperature (K)	288
Compressor Pressure Ratio	15.7
Combustor Pressure Drop (psia)	4 (27.5 kPa)
Turbine Backpressure (psia)	2 (13.8 kPa)
Turbine Inlet Temperature (K)	1600
Turbine Inlet Reference Mass flow (kg/s)	455
Cooling Air Extraction Fraction (%)	12
Nitrogen Injection (%)	70
Nitrogen/Syngas molar ratio	0.6–0.8
Adiabatic Compressor Efficiency	0.77
Adiabatic Turbine Efficiency	0.87
Shaft/generator Efficiency	0.98
Steam Cycle Heat Rate (10^6 J/kWh, LHV)	9.653
HRSG Outlet Temperature (K)	386

3.1.6. Combined Cycle Electricity Generation

After the CO_2 scrubbing process, the hydrogen rich syngas was fed into a combined cycle generator to produce electricity. The combined cycle generation consists of two main components; gas turbine system and steam turbine system.

A gas turbine system consists of three main components; the compressor, the combustor, and the turbine. Air at ambient conditions enters the compressor. The compressed air then enters the combustion chamber, where the air is mixed with the high pressure fuel (clean syngas). In the combustion unit, the air and fuel are burned at a constant pressure [35]. The high pressure gases from combustor are sent to the turbine (expander) where the pressure and temperature of the gas is reduced. The difference between the work output of the expander and the work input to the compressor is the net electricity produced from the generator. The high temperature exhaust gas is then passed to the steam turbine cycle process.

The steam turbine cycle consists of a heat recovery steam generator (HRSG), a steam turbine and other auxiliary parts. The gas turbine exhaust flows through the heat recovery steam regenerator to provide superheated steam, saturated steam and boiler feed water [36]. The steam is then sent to the high pressure steam turbine and low pressure steam turbine. Then, the generator converts shaft work into electricity. The cooled flue gas from the HRSG is exhausted from a flue stack.

3.2. Integrated Gasification Combined Cycle (IGCC) Without Pre-Combustion CO_2 Capture

A typical IGCC system without CO_2 capture is similar to an IGCC system with CO_2 capture except that there is no water gas shift (WGS) or Selexol® unit installed in the system. Thus, the turbine is powered by syngas not a relatively pure stream of H_2. In the gasification process, slurry-coal is first reacted with steam and high purity oxygen to produce syngas. The high purity oxygen (95% O_2) is produced in an ASU. After the gasification process, the syngas, which contains a high content of CO and H_2, is cooled and cleaned in the gas cleaning/acid gas removal unit and Sulfur recovery unit. Then, the clean syngas is combusted with excess N_2 (to keep NO_x level down) in the gas turbine combined cycle system. In the combined cycle, the clean syngas (fuel) is first reacted with the compressed air in the combustor. The product gas from the combustor then enters the expander inside the turbine, and the shaft work is produced. The heat from the gas turbine exhaust flows through the HRSG to make steam. The steam is then expanded in the steam turbine. Finally, the electricity is generated by gas turbine and steam turbine. The simplify flow diagram of IGCC without pre-combustion is shown in Figure 5. The process-specific parameters of both IGCC, with and without pre-combustion CO_2 capture, are shown in Table 5, and the results composition of syngas for both IGCC, with and without CO_2 capture along with energy consumptions for each process, are shown in Table 6.

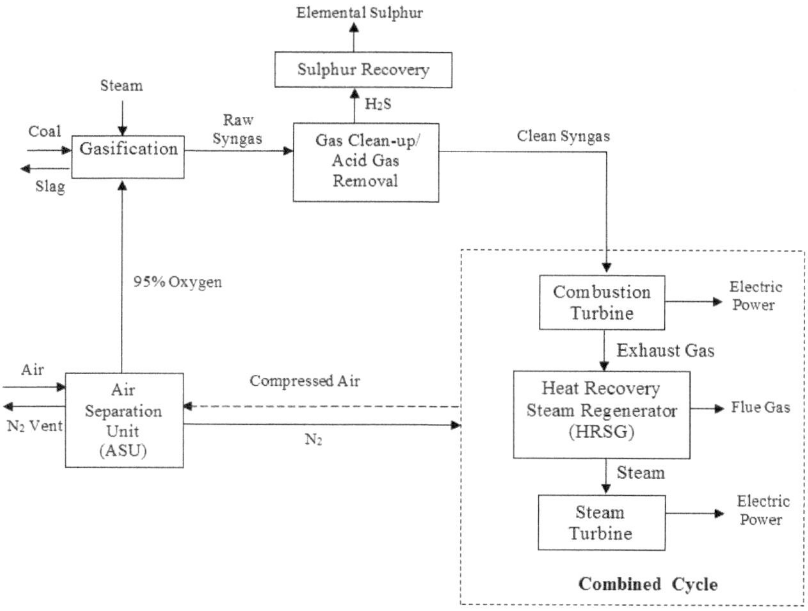

Figure 5. Flow diagram of IGCC without pre-combustion co$_2$ capture process.

Table 5. Main model parameters for IGCC electrical generating station including CO_2 capture process (Selexol®).

Process	Parameter	Value
ASU	Type and number of ASU	One cryogenic separation train
	O_2 purity	95%
	Excess Nitrogen Injection	70%
Gasification	Type and number of Gasifier	One oxygen-blow GE gasifier
	Gasifier temperature, °C	1250
	Gasifier pressure, Mpa	6
	Steam input to gasifier, mol H_2O/mol C	0.54
	Oxidant pressure (at outlet of ASU), MPa	4
	Particulate removal efficiency from syngas,%	50
AGR	COS to H_2S conversion efficiency, %	96
	H_2S removal sufficiency, %	98
	COS removal efficiency, %	40
	Sulfur recovery system	Claus plant and Bravon-Stretford tail gas unit
	Sulfur recovery efficiency, %	95
CO_2 capture	Steam added to shift reactor, mol H_2O/mol CO converted	1
	CO_2 removal efficiency, %	95
Combined cycle electricity output	Combined cycle electricity output, MW net	342.41 (without CO_2 capture),
		261.22 (with CO_2 capture)

Table 6. Model results of IGCC with and without capture.

Syngas Compositions (% by Volume)	IGCC without Pre-Combustion CO_2 Capture	IGCC with Pre-Combustion CO_2 Capture (95% CO_2 Removal)
$AR+N_2$	5.53	6.55
O_2	0.36	0.75
CH_4	0.4	0.45
CO_2	53.63	7.66
H_2S	0.02	0.02
H_2	12.62	68.05
CO	27.62	13.07
H_2O	1.73	3.03
Total	100	100
H_2/CO	0.45	5.22
Fuel heating value (HHV) (J/g)	47,915.4	53,665.3
Gas turbine power output	202.45	184.60
Steam turbine power output (MWh)	252.34	241.1
Combined cycle power output (MWgross)	454.79	425.7
Misc. Power Block Use (MW)	11.35	9.60
Air Separation Unit Use(MW)	77.09	76.77
Gasifier Use (MW)	14.24	14.39
Sulfur Capture Use (MW)	5.90	6.47
Claus Plant Use (MW)	0.49	0.51
Beavon-Stretford Use (MW)	1.49	1.55
WGS and Selexol Use (MW)	-	49.5
Net electricity output (MW_{net})	342	261
Plant Efficiency (%)	64.63	49.3

3.3. The Conventional Lignite Coal-Fired Electricity Generation With and Without Post Combustion CO_2 Capture Process

As mentioned earlier in this paper, the conventional lignite coal-fired electricity generation was adopted from the previous work of [16]. The process description and input data of post-combustion is therefore not explained in extensive detail. For this study, Saskpower's Boundary Dam Power Station (BDPS), a conventional lignite coal-fired power station in Estevan, Saskatchewan was chosen as a case study. Operations of the BDPS were modeled. The plant generates a gross electricity capacity of 882 MW for the purposes of Manuilova's work (part of the station has subsequently been closed).

The amount of lignite coal needed to produce the gross electric capacity of 882 MW was calculated based on the gross cycle heat rate and higher heating value of coal. The amount of oxygen needed for stoichiometric combustion of fuel is the sum of oxygen needed to convert carbon to carbon dioxide, hydrogen to water, and sulfur to sulfur dioxide, minus any oxygen in fuel. The combustion products were calculated using emissions factors for pulverized coal tangentially-fired dry bottom boilers [21]. More detailed information can be found in [16].

In the capture scenario, the flue gas desulfurization (FGDs) and CO_2 capture units were installed for gas cleaning process. The wet flue gas desulfurization unit was installed to remove 99% of SO_2 before the flue gas entered the CO_2 capture unit, in which the post-combustion amine-based absorption of CO_2 from flue gases and CO_2 compression was modeled. 30 wt% monoethanolamine (MEA) was used as the sorbent for removing 90% of CO_2 form the flue gas. The modeling assumptions are presented in Table 7.

Table 7. Model assumptions of the conventional lignite coal-fire electricity generation with and without post-combustion CO_2 capture process [21].

Parameter	Value		
MEA concentration in sorbent, %	30		
Lean sorbent CO_2 loading, mole CO_2/mole MEA	0.2		
Temperature of the flue gas entering the CO_2 absorber, °C	50		
Desired CO_2 product pressure, psig	2000 (13.8 MPa)		
MEA losses, kg MEA/tonne CO_2	1.3607		
Reclaimer waste, kg/tonne CO_2 captured	3.2		
Activated carbon consumption, kg C/tonne CO_2 captured	0.075		
Caustic consumption, kg NaOH/tonne CO_2 captured	0.13		
Ammonia formation, kg NH_3/tonne CO_2	0.136		
Water consumption, tonne/MWh	1.1		
Sorbent regeneration heat requirement, kJ/kg	3600		
Enthalpy of steam, kJ/kg steam	2000		
Reboiler efficiency, %	85		
Steam requirement, kg/MWh	2045		
Parameter	**ESP ***	**FGD ****	**CO_2 Capture and Compression**
Particulates removal efficiency, %	99	70	50
CO_2 removal efficiency, %	-	-	90
SO_2 removal efficiency, %	-	99	99.5
SO_3 removal efficiency, %	25	50	99.5
HCl removal efficiency, %	-	90	95
NO_2 removal efficiency, %	-	-	25
Hg^{2+} removal efficiency, %	7.5	25	80
Auxiliary power requirements, % MW_{gross}	6.9 ***	3.8	34.6

* ESP is installed on both capture and non-capture scenario; **FGD is install on only capture scenario; *** Auxiliary power requirements for the boiler and ESP operations.

4. Results and Discussions

4.1. LCA Results

The electricity outputs for four different scenarios are presented in Figure 6. The net electricity output of pulverized coal without post-combustion CO_2 capture process is 820 MWh and 552 MWh for pulverized coal with post-combustion CO_2 capture process. For the IGCC system, the net electricity outputs are 342 MWh and 261 MWh for the IGCC and the IGCC with pre-combustion CO_2 capture process, respectively. In the case of the IGCC system, there is a 24% reduction in electricity output in the "IGCC with capture" scenario compared to the "IGCC" scenario. For the conventional pulverized coal system, there is a 33% electricity reduction in the "PC with capture" scenario compared to the "PC" scenario. These reductions in electricity output were due to the additional electricity and steam requirements of the CO_2 capture process system. In order to compensate for this loss in electricity output, additional coal needs to be burned. As shown in Figure 7, 36% extra coal is needed for the "IGCC with capture" scenario and 48% extra coal is needed for the "PC with capture" scenario, compared to the "IGCC" and "PC" scenarios, respectively.

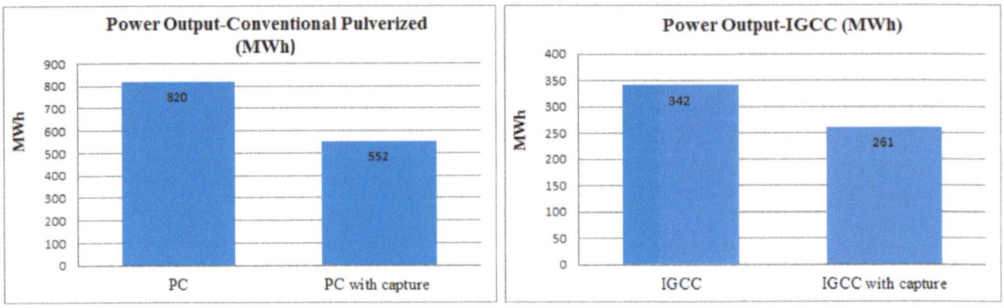

Figure 6. Net power output for four scenarios.

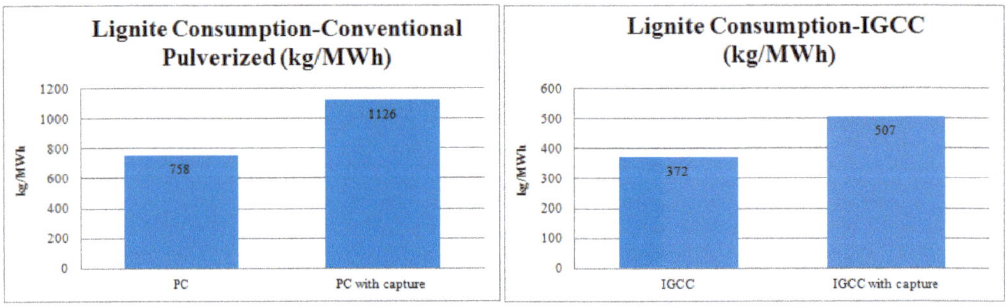

Figure 7. Lignite consumption for four scenarios.

The results for different impact categories per 1 MWh of electricity output as well as the environmental contributions of each process toward the impact categories are shown in Figure 8. The analysis of the environmental impacts for each impact category is presented as follows. The analysis adopts the "conventional pulverized coal-fired electricity generation system without CO_2 capture" or the "PC" scenario as the baseline, and the differences between the "PC" scenario *versus* the "PC with capture," "IGCC" and "IGCC with capture" scenarios are shown as percentage changes. As mentioned earlier, the system boundary of this research does not include the utilization of any oil produced under EOR as well as the CO_2 sequestration underground process. See [16] for the life-cycle assessment of the power generation associated with a CO_2-EOR process.

4.1.1. Global Warming Potential

The results showed a reduction in global warming potential in terms of CO_2 equivalents, achieved through the application of the post-combustion and pre-combustion CO_2 capture and IGCC technologies. Among them, the IGCC with capture performed the best, followed by PC with capture and IGCC without capture. The category of CO_2 emissions from the PC with capture scenario was found to be 233 kg CO_2 equivalent per MWh of electricity produced. For the IGCC with capture scenario, the CO_2 emissions category was found to be 77 kg CO_2 equivalent per MWh of electricity produced. The PC with post-combustion showed a reduction in global warming potential of 80% compared to PC. Compared to PC, the IGCC system, with and without the pre-combustion technology, showed a reduction in global warming potential by 93% and 75%, respectively. Due to the higher CO_2 concentration in the syngas after the WGS in the IGCC process, the removal of CO_2 from the combustion gas stream was more efficient than the pulverized coal-fired electricity generation technology. Additionally, the integrated gasification combined cycle (IGCC) gasifier was operated under higher pressure, which made it easier to remove CO_2 from the gas stream than to capture it from the flue gas at atmospheric pressure.

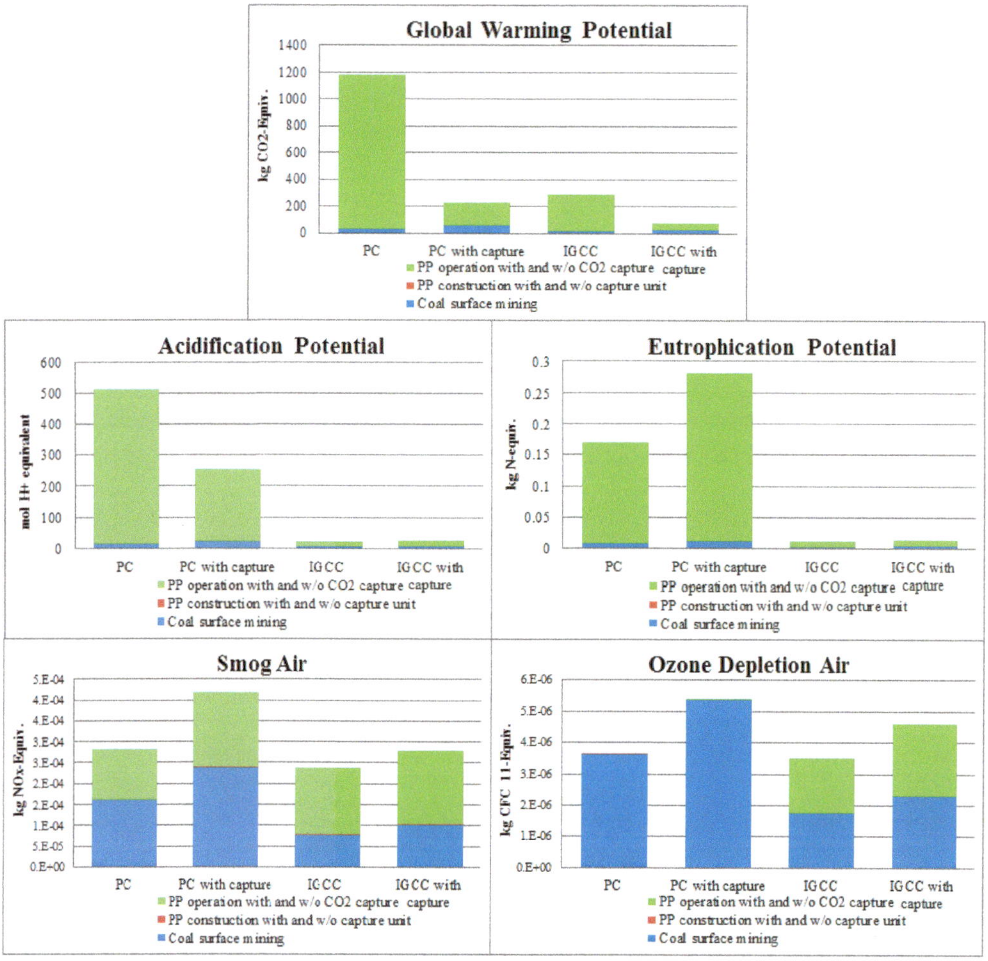

Figure 8. Life-cycle impact assessment.

4.1.2. Acidification

The pollutants that contribute to acidification include: sulfur oxides (SO_x), nitrous oxides (NO_x), hydrochloric acid (HCl), and hydrofluoric acid (HF). These acids can be deposited dry (e.g., dust) or wet (e.g., acid rain). In the acidification category, the impacts are relatively low for the "PC with capture," "IGCC," and "IGCC with capture" scenarios compared to the baseline case of PC.

In the case of the "PC with capture" scenario, the reduction of acidification potential was found to be 50%. This is mainly due to the removal of SO_x in the FGD. The IGCC system performed distinctly better than the conventional electricity generation station. The result showed a reduction in the acidification impact category of 95% and 96% for the IGCC technology, with and without pre-combustion capture, respectively, compared to the baseline case of the coal-fired electricity generation station without CO_2 capture. This was due to the gas cleaning process in the IGCC, which removed most of the pollutants. Only small amounts of sulfur and nitrogen oxides remained in the syngas, and these were converted to SO_2 and NO_x in the combustion turbine and released into the atmosphere. The process of electricity generation was the main contributor to the acidification impact category in all the scenarios.

4.1.3. Eutrophication

The eutrophication category included the impact of nutrients and degradable organic pollution, which led to oxygen consumption in the ecosystem. The release of nutrient and the nitrogen compounds of nitrogen oxide (NO), N_2O, nitrogen dioxide (NO_2) and NH_3 cause eutrophication. In this category, the impacts are relatively high for PC with CO_2 capture compared to the baseline scenario. The observed increase in eutrophication potential impact category is 66% in post-combustion scenarios. This is due to the loss of plant capacity, and NO_x emissions increased with the addition of a capture unit. In addition, the emissions of NH_3, MEA, and ethylene from the production of MEA also contributed to an increase in the eutrophication potential.

Compared to a conventional electricity generating station, the eutrophication potential of the IGCC electricity system, with and without CO_2 capture, showed a significant reduction of 91% and 93%, respectively, compared to the conventional coal-fired electricity generation system without CO_2 capture. The operation of the IGCC system was the primary contributor to the eutrophication impact category. The electricity generation process produced inorganic solid waste, which was collected and landfilled, and with time leached into the soil and ground water. This accounted for 90% of the impact in this category.

4.1.4. Ozone Depletion

The electricity generation systems contribute to ozone depletion because of their production of halogenated organic, volatile organic compounds (VOC), trichlorofluoromethane (R11), and dichlorotetrafluoroethane (R114), which are released into the air. The results show that the ozone depletion air impact category was increased by 48% in the "PC with capture" scenario and by 27% in the "IGCC with capture scenario", compared to the "PC" scenario. There was also an observed reduction by 4% in the "IGCC without capture" scenario. Coal mining and electricity operation are mainly responsible for the ozone depletion. Since the results are presented in the form of percentages and not in absolute numbers, small absolute increases can be exaggerated when viewed as percentage increases.

4.1.5. Smog Air

Coal surface mining and the electrical operation processes produce NO_x, which contributes directly to smog air or summer smog. The results from the "PC with capture" scenario showed an increase of 48.6% on the impact category of smog air compared to the baseline scenario. This increase was mainly due to NOx from coal mining and NO_2 from the operation of the electrical generating station. In the case of the IGCC systems, the results also showed the reduction in the smog air impact category. The results showed a reduction of 15.8% and 1.1% on the "IGCC" and "IGCC with capture" scenarios, respectively, compared to the baseline scenario. The reduction was due to less NO_x content in the flue gas from syngas combustion. The main contributions to the category of smog air of IGCC are the operation of the electrical generating station and coal surface mining.

4.2. Air Dispersion Modeling Results

The results of air dispersion assessment from electricity generating stations are presented in Tables 8–11. The assessment considers air emissions of NO_2, SO_2, $PM_{2.5}$, As, Cd, Cr, Hg, Mn, Sb, Co, Cu, Pb, Mo, Se, V, and Ni. These air emission substances have an impact on human health, which will be further analyzed in the next section of this paper. The maximum air pollution concentrations of each substance are compared to Saskatchewan air quality standards unless otherwise noted.

The total maximum concentration results are reported with the inclusion of ambient background concentrations to air contaminant predicted values as per the methodology outline in Saskatchewan Air Dispersion Modeling guideline. The maximum predicted ground-level concentrations per MWh of net electricity produced on both not including and including the ambient ground-level concentration are summarized in Tables 8 and 9 respectively. The location of predicted ground-level concentration of each element for each averaging time is shown in Table 10. The maximum predicted ground-level concentration with the inclusion of ambient ground-level concentrations on total electricity produced are summarized in Table 11.

Table 8. AERMOD results: maximum predicted ground-level concentration per MWe produced.

Substance	Averaging Period	Ambient Background Concentration ($\mu g/m^3$)	Limited Ambient Air Quality Standard ($\mu g/m^3$)	AERMOD Results of Maximum Predicted Ground Level Concentration ($\mu g/m^3$) per MW Net Electricity			
				PC	PC with Capture	IGCC	IGCC with Capture
NO_2	One-hour	2.2×10^{-2}	400	3.1×10^{-2}	1.3×10^{-1}	2.3×10^{-2}	2.1×10^{-2}
	Daily	2.0×10^{-2}	200	2.9×10^{-3}	1.4×10^{-2}	2.5×10^{-3}	2.2×10^{-3}
	Annual	1.0×10^{-2}	100	6.1×10^{-4}	9.6×10^{-4}	1.7×10^{-4}	1.6×10^{-4}
SO_2	One-hour	3.0×10^{-3}	450	1.8×10^{-2}	4.9×10^{-3}	2.8×10^{-2}	3.0×10^{-2}
	Daily	2.0×10^{-3}	150	1.7×10^{-3}	5.5×10^{-4}	2.9×10^{-3}	3.2×10^{-3}
	Annual	0	30	3.5×10^{-4}	4.0×10^{-5}	2.1×10^{-4}	2.2×10^{-4}
$PM_{2.5}$	Daily	8.3	30	2.3×10^{-4}	7.5×10^{-3}	2.6×10^{-3}	1.6×10^{-3}
	Annual	3.7	15	2.0×10^{-5}	5.0×10^{-4}	1.8×10^{-4}	1.1×10^{-4}
As	Daily	-	0.3	1.6×10^{-4}	8.0×10^{-5}	0	0
Cd	Daily	-	2	1.1×10^{-4}	9.0×10^{-5}	0	0
Cr	Daily	-	1.5	1.5×10^{-3}	3.0×10^{-5}	0	0
Hg	Daily	-	2	1.3×10^{-3}	0	0	0
Mn	Daily	-	0.15	1.1×10^{-4}	0	0	0
Sb	Daily	-	25	1.1×10^{-4}	0	0	0
Co	Daily	-	0.1	1.8×10^{-4}	0	0	0
Cu	Daily	-	50	6.0×10^{-5}	0	0	0
Pb	Daily	-	0.5	7.8×10^{-4}	0	0	0
Mo	Daily	-	120	9.9×10^{-5}	1.0×10^{-4}	0	0
Se	Daily	-	10	1.1×10^{-2}	9.1×10^{-3}	0	0
V	Daily	-	2	2.3×10^{-3}	0	0	0
Ni	Daily	-	0.2	1.5×10^{-3}	2.2×10^{-4}	0	0

0 means the concentration is very small (less than 6 digits).

Table 9. AERMOD results: maximum predicted ground-level concentration including background concentrations per MWe produced.

Substance	Averaging Period	Ambient Background Concentration ($\mu g/m^3$)	Limited Ambient Air Quality Standard ($\mu g/m^3$)	AERMOD Results of Maximum Predicted Ground Level Concentration Including Ambient Background Concentration ($\mu g/m^3$) Per MW Net Electricity			
				PC	PC with Capture	IGCC	IGCC with Capture
NO_2	One-hour	2.2×10^{-2}	400	5.3×10^{-2}	1.5×10^{-1}	4.5×10^{-2}	4.3×10^{-2}
	Daily	2.0×10^{-2}	200	2.3×10^{-2}	3.4×10^{-2}	2.2×10^{-2}	2.2×10^{-2}
	Annual	1.0×10^{-2}	100	1.1×10^{-2}	1.1×10^{-2}	1.0×10^{-2}	1.0×10^{-2}
SO_2	One-hour	3.0×10^{-3}	450	2.0×10^{-2}	7.9×10^{-3}	3.1×10^{-2}	3.3×10^{-2}
	Daily	2.0×10^{-3}	150	3.7×10^{-3}	2.6×10^{-3}	4.9×10^{-3}	5.2×10^{-3}
	Annual	0	30	3.5×10^{-4}	4.0×10^{-5}	2.1×10^{-4}	2.2×10^{-4}
$PM_{2.5}$	Daily	8.3	30	8.30023	8.30752	8.30259	8.30158
	Annual	3.7	15	3.70002	3.70050	3.70018	3.70011
As	Daily	-	0.3	1.15×10^{-4}	8.0×10^{-4}	0	0
Cd	Daily	-	2	1.1×10^{-4}	9.0×10^{-5}	0	0
Cr	Daily	-	1.5	1.54×10^{-3}	3.0×10^{-5}	0	0
Hg	Daily	-	2	1.3×10^{-3}	0	0	0
Mn	Daily	-	0.15	1.1×10^{-4}	0	0	0
Sb	Daily	-	25	1.1×10^{-4}	0	0	0
Co	Daily	-	0.1	1.8×10^{-4}	0	0	0
Cu	Daily	-	50	6.0×10^{-5}	0	0	0
Pb	Daily	-	0.5	7.8×10^{-4}	0	0	0
Mo	Daily	-	120	9.9×10^{-4}	1.0×10^{-5}	0	0
Se	Daily	-	10	1.1×10^{-2}	9.1×10^{-3}	0	0
V	Daily	-	2	2.3×10^{-3}	0	0	0
Ni	Daily	-	2	1.5×10^{-3}	2.2×10^{-4}	0	0

Table 10. AERMOD results: location of maximum predicted ground-level concentration including background concentrations Per MWe produced.

Substance	Averaging Period	AERMOD Results of Maximum Predicted Ground Level Concentration Including Ambient Background Concentration (μg/m^3) Per MW Net Electricity				Maximum Prediction Location
		PC	PC with Capture	IGCC	IGCC with Capture	
NO_2	One-hour	5.3×10^{-2}	1.5×10^{-1}	4.5×10^{-2}	4.3×10^{-2}	2000 meters, 140 degrees, SW of the facility
	Daily	2.3×10^{-2}	3.5×10^{-2}	2.2×10^{-2}	2.2×10^{-2}	2000 meters, West of the facility
	Annual	1.1×10^{-2}	1.1×10^{-2}	1.0×10^{-2}	1.0×10^{-2}	2000 meters, 140 degrees, SE of the facility
SO_2	One-hour	2.0×10^{-2}	7.9×10^{-3}	3.1×10^{-2}	3.3×10^{-2}	2000 meters, 130 degrees, SW of the facility
	Daily	3.7×10^{-3}	2.6×10^{-2}	4.9×10^{-3}	5.2×10^{-3}	2000 meters, West of the facility
	Annual	3.5×10^{-4}	4.0×10^{-5}	2.1×10^{-4}	2.2×10^{-4}	2000 meters, 140 degrees, SE of the facility
$PM_{2.5}$	Daily	8.30023	8.30752	8.30259	8.30158	2000 meters, West of the facility
	Annual	3.70002	3.70050	3.70018	3.70011	2000 meters, 140 degrees, SE of the facility
As	Daily	1.2×10^{-4}	8.0×10^{-5}	0	0	2000 meters, West of the facility
Cd	Daily	1.1×10^{-4}	9.0×10^{-5}	0	0	2000 meters, West of the facility
Cr	Daily	1.5×10^{-3}	3.0×10^{-5}	0	0	2000 meters, West of the facility
Hg	Daily	1.3×10^{-3}	0	0	0	2000 meters, West of the facility
Mn	Daily	1.1×10^{-4}	0	0	0	2000 meters, West of the facility
Sb	Daily	1.1×10^{-4}	0	0	0	2000 meters, West of the facility
Co	Daily	1.8×10^{-4}	0	0	0	2000 meters, West of the facility
Cu	Daily	6.0×10^{-5}	0	0	0	2000 meters, West of the facility
Pb	Daily	7.8×10^{-4}	0	0	0	2000 meters, West of the facility
Mo	Daily	9.9×10^{-4}	1.0×10^{-5}	0	0	2000 meters, West of the facility
Se	Daily	1.1×10^{-2}	9.1×10^{-3}	0	0	2000 meters, West of the facility
V	Daily	2.3×10^{-3}	0	0	0	2000 meters, West of the facility
Ni	Daily	1.5×10^{-3}	2.2×10^{-4}	0	0	2000 meters, West of the facility

Table 11. AERMOD results: maximum predicted ground-level concentration including background concentrations.

Substance	Averaging Period	Ambient Background Concentration (μg/m³)	Limited Ambient Air Quality Standard (μg/m³)	AERMOD Results of Maximum Predicted Ground Level Concentration Including Ambient Background Concentration (μg/m³)							
				PC	% of Criteria	PC with Capture	% of Criteria	IGCC	% of Criteria	IGCC With Capture	% of Criteria
NO_2	One-hour	2.2×10^{-2}	400	25.0400	6.3%	70.0320	17.5%	8.0110	2.0%	5.5108	1.4%
	Daily	2.0×10^{-2}	200	2.3898	1.2%	7.9522	4.0%	0.8610	0.4%	0.5994	0.3%
	Annual	1.0×10^{-2}	100	0.5102	0.5%	0.5399	0.5%	6.8×10^{-2}	0.1%	5.2×10^{-2}	0.1%
SO_2	One-hour	3.0×10^{-3}	450	14.5330	3.2%	2.7022	0.6%	9.4830	2.1%	7.8878	1.8%
	Daily	2.0×10^{-3}	150	1.3796	0.9%	0.3056	0.2%	1.0000	0.7%	0.8345	0.6%
	Annual	0	30	0.2870	1.0%	2.2×10^{-2}	0.1%	7.1×10^{-2}	0.2%	5.7×10^{-2}	0.2%
$PM_{2.5}$	Daily	8.3	30	8.4886	28.3%	12.4510	41.5%	9.1850	30.6%	8.7123	29.0%
	Annual	3.7	15	3.7164	24.8%	3.9760	26.5%	3.7610	25.1%	3.7287	24.9%
As	Daily	-	0.3	9.4×10^{-2}	31.4%	4.4×10^{-2}	14.7%	0	0.0%	0	0.0%
Cd	Daily	-	2	9.0×10^{-2}	4.51%	4.9×10^{-2}	2.48%	0	0.0%	0	0.0%
Cr	Daily	-	1.5	1.2628	84.2%	1.7×10^{-2}	1.1%	0	0.0%	0	0.0%
Hg	Daily	-	2	1.0438	52.2%	0	0.0%	0	0.0%	0	0.0%
Mn	Daily	-	0.15	9.0×10^{-2}	60.1%	0	0.0%	0	0.0%	0	0.0%
Sb	Daily	-	25	9.0×10^{-2}	0.36%	0	0.0%	0	0.0%	0	0.0%
Co	Daily	-	0.1	0.1476	147.6%	0	0.0%	0	0.0%	0	0.0%
Cu	Daily	-	50	4.9×10^{-2}	0.098%	0	0.0%	0	0.0%	0	0.0%
Pb	Daily	-	0.5	0.6396	127.9%	0	0.0%	0	0.0%	0	0.0%
Mo	Daily	-	120	0.8118	0.6%	5.0×10^{-3}	0.004%	0	0.0%	0	0.0%
Se	Daily	-	10	8.7658	87.66%	5.0450	50.45%	0	0.0%	0	0.0%
V	Daily	-	2	1.8778	93.9%	0	0.0%	0	0.0%	0	0.0%
Ni	Daily	-	0.2	1.2400	620%	0.1214	60.7%	0	0.0%	0	0.0%

4.2.1. NO_2 Predictions

The maximum predicted one-hour average ground-level NO_2 concentrations associated with "PC," "PC with capture," "IGCC" and "IGCC with capture" scenarios are 25, 70, 8 and 5.5 μg/m^3, respectively. The maximum predicted 24-hour average ground-level concentration for NO_2 concentration associated with "PC," "PC with capture," "IGCC" and "IGCC with capture" scenarios are 2.4, 7.9, 0.8 and 0.6 μg/m^3, respectively. For the annual average ground level, the results show NO_2 concentration associated with "PC," "PC with capture," "IGCC" and "IGCC with capture" scenarios are 0.51, 0.54, 0.068 and 0.0517 μg/m^3, respectively. The maximum predicted ground-level concentration of a one-hour averaging period shows the highest NO_2 ground-level concentration of all scenarios.

IGCC with capture scenario shows the lowest NO_2 ground-level concentration compared to all other scenarios. Meanwhile, the "PC with capture" scenario shows the highest ground-level concentration for all averaging periods compared to all case scenarios. The high ground-level concentration of the "PC" scenario is due to the highest NO_2 emission rate that was produced from the post-combustion CO_2 capture process. All maximum predicted NO_2 concentrations associated with all scenarios are lower than the Saskatchewan ambient air quality standard. The maximum predicted 1-hour, 24-hour and annual average ground-level NO_2 concentrations are predicted to occur southwest, west and southeast of the facility, roughly 2 km away.

4.2.2. SO_2 Predictions

The maximum predicted one-hour average ground-level SO_2 concentrations associated with "PC," "PC with capture," "IGCC" and "IGCC with capture" scenarios are 14, 2.7, 9.4 and 7.8 μg/m^3, respectively. The maximum predicted 24-hour average ground-level concentrations for SO_2 concentrations associated with "PC," "PC with capture," "IGCC" and "IGCC with capture" scenarios are 1.4, 0.3, 1 and 0.8 μg/m^3, respectively. For the annual average ground level, the results show SO_2 concentration associated with "PC," "PC with capture," "IGCC" and "IGCC with capture" scenarios to be 0.29, 0.022, 0.071 and 0.0574 μg/m^3, respectively. The maximum predicted ground-level concentration of a one-hour averaging period shows the highest SO_2 ground-level concentration of all scenarios.

The "PC with capture" scenario shows the lowest SO_2 ground-level concentration compared to all other scenarios. Meanwhile, the "PC" scenario shows the highest ground-level concentration for all averaging periods compared to all case scenarios. The high ground-level concentration of the "PC" scenario is due to the highest SO_2 emission rate that was produced from post-combustion CO_2 capture process. All maximum predicted SO_2 concentrations associated with all scenarios are lower than the Saskatchewan ambient air quality standard. The maximum predicted 1-hour, 24-hour and annual average ground-level SO_2 concentrations are predicted to occur southwest, west and southeast of the facility, roughly 2 km away.

4.2.3. $PM_{2.5}$ Predictions

The maximum predicted 24-hour average ground-level $PM_{2.5}$ concentration associated with "PC," "PC with capture," "IGCC" and "IGCC with capture" scenarios are 8.5, 12.5, 9.2 and 8.7 μg/m^3, respectively. For the annual average ground level, the results show $PM_{2.5}$ concentrations associated with "PC", "PC with capture", "IGCC" and "IGCC with capture" scenarios are 3.7, 4.0, 3.76 and 3.72 μg/m^3, respectively. The maximum predicted ground-level concentration of a 24-hour averaging period shows the highest $PM_{2.5}$ ground-level concentration of all scenarios.

The "PC" scenario shows the lowest $PM_{2.5}$ ground-level concentration compared to all other scenarios. Meanwhile, the "PC with capture" scenario shows the highest ground-level concentration for all averaging periods compared to all case scenarios. All maximum predicted $PM_{2.5}$ concentrations associated with all scenarios are lower than the Saskatchewan ambient air quality standard.

The maximum predicted $PM_{2.5}$ concentration is predicted to occur west and southeast (24-hour and annual average) of the facility, roughly 2 km away.

4.2.4. Arsenic (As) Predictions

The maximum predicted 24-hour average ground-level Arsenic concentration associated with the "PC" and "PC with capture" scenarios are 0.0943 and 0.0441 $\mu g/m^3$, respectively. The predicted results show no ground-level concentration in either the "IGCC" or "IGCC with capture" scenarios. This is because the IGCC system generates very minimal amounts of trace elements and heavy metals due to the cleaning process that removes most of the trace element emissions from the syngas.

The maximum predicted 24-hour average As concentrations associated with "PC" and "PC with capture" scenarios are lower than the Ontario air quality standard. The maximum ground-level concentrations are predicted to occur west of the facility.

4.2.5. Cadmium (Cd) Predictions

The maximum predicted 24-hour average ground-level Cd concentration associated with the "PC" and "PC with capture" scenarios are 0.0902 and 0.0496 $\mu g/m^3$, respectively. The predicted results show no ground-level concentration in either the "IGCC" or "IGCC with capture" scenarios due to the cleaning process that removes most of the trace element emissions from the syngas.

The maximum predicted 24-hour average Cd concentrations associated with "PC" and "PC with capture" scenarios are lower than the Saskatchewan standard. The maximum ground-level concentrations are predicted to occur on the west side of the facility.

4.2.6. Chromium (Cr) Predictions

The maximum predicted 24-hour average ground-level Cr concentration associated with the "PC" and "PC with capture" scenarios are 1.2628 and 0.0165 $\mu g/m^3$, respectively. The predicted results also show no ground-level concentration in either the "IGCC" or "IGCC with capture" scenarios. The maximum predicted 24-hour average Cr concentrations associated with "PC" and "PC with capture" scenarios are both lower than Ontario air quality standard. The maximum ground-level concentrations are predicted to occur on the west side of the facility.

4.2.7. Elemental Mercury (Hg) Predictions

The maximum predicted 24-hour average ground-level Hg concentration associated with the "PC" scenario is 1.0438 $\mu g/m^3$. The predicted results show no ground-level concentration in the "PC with capture", "IGCC" and "IGCC with capture" scenario due to acid gas cleaning process on IGCC systems and FGD and ESP process on PC with capture scenario. The maximum predicted 24-hour average Hg concentration associated with the "PC" scenario is lower than the Ontario air quality standard. The maximum ground-level concentrations are predicted to occur on the west side of the facility, roughly 2 km away.

4.2.8. Manganese (Mn) Predictions

The maximum predicted 24-hour average ground-level Mn concentration associated with the "PC" scenario is 0.0902 $\mu g/m^3$. The predicted results show no ground-level concentration in the "PC with capture", "IGCC" and "IGCC with capture" scenarios due to the acid gas cleaning process of IGCC systems and FGD and ESP processes on PC with the capture scenario. The maximum predicted 24-hour average Hg concentration associated with the "PC" scenario is lower than the relevant Ontario air quality standard. The maximum ground-level concentrations are predicted to occur on the west side of the facility, roughly 2 km away.

4.2.9. Lead (Pb) Predictions

The maximum predicted 24-hour average ground-level Pb concentration associated with the "PC" scenario is 0.6396 μg/m^3. Compared to the Limited Ambient Air Quality Standard, the maximum predicted Pb concentration associated with the "PC" scenario is greater than the limited ambient air quality standard. The maximum ground-level concentrations are predicted to occur on the west side of the facility, roughly 2 km away. The predicted results show no ground-level concentration in the "PC with capture", "IGCC" or "IGCC with capture" scenarios.

4.2.10. Other Trace Elements

The maximum predicted ground-level concentrations of other trace elements are lower than the Limited Ambient Air Quality Standard level except for Co and Ni. The maximum predicted 24-hour average ground-level of Co and Ni concentration associated with the "PC" scenario are 0.1476 μg/m^3 and 1.240 μg/m^3, respectively. The exceeding ground-level concentration of Co and Ni indicate that further evaluation may be necessary to determine the possibility of risk associated with human health. The predicted results show no ground-level concentration in the "PC with capture," "IGCC" and "IGCC with capture" scenarios due to the acid gas in the cleaning process on IGCC systems and FGD and ESP process on PC with capture scenario. The maximum ground-level concentrations of other trace elements are predicted to occur on the west side of the facility, roughly 2 km away.

4.3. Discussions

Compared to the amount of lignite coal from an electrical generating station without CO_2 capture, there is an increase in coal used in the "PC with capture" and "IGCC with capture" scenarios due to additional units for CO_2 capture process, which consumed additional auxiliary power. Percentages are used rather than absolute quantities in this study; as a result, the impact categories, particularly under eutrophication potential, appear as significant increases. The results of using percentages as the absolute changes are quite small. Global warming is the most significant, and this impact category demonstrated a major environmental benefit.

The LCA results of this study correlates with the results found on [9,18,37]. Post-combustion CO_2 capture at a conventional pulverized lignite coal-fired electricity generation station resulted in increased impacts in the eutrophication potential, ozone depletion air, and smog air due to the CO_2 capture process, *i.e.*, the production and use of a solvent, as well as coal surface mining operation, in which reductions in the acidification air and global warming impact categories were observed.

IGCC systems in both the capture and non-capture scenarios showed the same tendency of increased impacts in some specific categories as the conventional coal-fired electricity generation system, but to a lesser degree. This is because the IGCC system is a cleaner technology which produces fewer pollutant emissions from the electricity generation station. The environmental impacts of IGCC generally increased only slightly when the IGCC system with CO_2 capture was introduced. On the other hand, the observed increase in the specific impact categories of eutrophication potential was generally small and remained rather low compared to the global warming potential and acidification potential. Ultimately, the IGCC technology was recognized as the cleaner technology, compared to the conventional coal-fired generation process.

The air dispersion analysis in this study involves the assessment of compliance of the predicted ambient air pollution concentrations emitted from four electrical generating stations with a limited ambient air quality standard. The analysis was conducted by following the outline of Saskatchewan air quality modeling guideline. In this study, AERMOD air dispersion modeling was used to predict the ground-level concentration of pollutants in the 20 km $\times$ 20 km modeling domain. The maximum ground-level concentrations of pollutants from all electricity generating stations were found to be in compliance with all limited air quality standards except for Co, Pb and Ni. The predicted maximum ground-level concentration is believed to be occurring 2 km southwest of the facility for a 1-hour averaging time, and west and southeast of the facility for the 24-hour and annual times, respectively.

IGCC with the capture scenario shows the lowest NO_2 ground-level concentration compared to all other scenarios. Meanwhile, the "PC with capture" scenario shows the highest ground-level concentration for all averaging periods on NO_2 substance. For SO_2 ground-level concentration, the "PC with capture" scenario shows the lowest maximum ground-level concentration and the "PC" scenario shows the highest maximum ground-level concentration compared to all other case scenarios. However, in terms of $PM_{2.5}$, "PC" has the lowest maximum ground-level concentration, but the "PC with capture" scenario shows the highest maximum ground-level concentration, compared to all other scenarios. This is because the stack height of the "PC with capture" scenario was the lowest compared to other scenarios, as stack height has an impact on air concentration modeling. The assumption of the "PC with capture" scenario's stack height was made using the best estimation based on SaskPower's available public information. In addition, the maximum ground-level concentrations of most trace elements are also lower than the limited ambient air quality standard. The "PC" scenario shows the highest trace element ground-level concentrations compared to the other scenarios. Nevertheless, IGCC systems in either scenario, with and without the pre-combustion CO_2 capture process, shows no ground-level concentration of the trace elements. This is because the IGCC system operates with an acid gas cleaning process that removes most of the trace contaminants from the syngas.

5. Conclusions and Recommendations for Future Work

5.1. Conclusion

A life-cycle assessment was carried out to evaluate the environmental performance as well as the adverse potential health impact of the hypothetical Saskatchewan lignite-fueled IGCC electricity generation station, with and without a CO_2 capture process. However, the comparison between the capture system and non-capture system might not be completely fair, as the non-capture plant will result in more air pollutants if equipment such as desulfurization and dust filters are not installed. This does not apply when it comes to European electricity generation stations because they do have the desulfurization and dust filters installed. Saskatchewan electricity generation stations operate on low-sulfur coal and are not equipped with NO_x and SO_x capture which is the baseline scenario. Since this study considered a typical Saskatchewan electricity generation station, the comparison was made to a plant equipped only with PM capture. The goal is to compare a typical Western Canadian electricity generation station with and without CO_2 capture.

The results of the study revealed that, by applying both post-combustion and pre-combustion CO_2 capture, a significant reduction in greenhouse gas emissions can be achieved. The greenhouse gas emissions were found to be reduced by 75% and 93% in the "IGCC" and "IGCC with capture" scenarios, respectively. The conventional coal-fired electricity generation plant with post-combustion CO_2 capture resulted in an 80% reduction in greenhouse gas emissions. However, similar improvements in other environmental impact categories are not guaranteed, and the impacts depend on the CO_2 capture technology employed. Additionally, one of the challenges associated with implementing a CO_2 capture process is the decrease in electricity generation efficiency, because additional fuel is needed to compensate for the loss in electricity production. Therefore, implementing a CO_2 capture process will cause a substantial decrease in electricity output or an increase in coal consumption, which has an impact on environmental performance.

In case of air dispersion modeling, the maximum predicted ground-level concentrations are found to be in compliance with all air quality standards except for Co, Pb and Ni. IGCC systems on both with and without the pre-combustion CO_2 capture process show no ground-level concentration on all heavy metals. This is because IGCC systems operate with an acid gas removal unit that works to remove most of the heavy metals from the syngas.

5.2. Future Perspective

The IGCC process is understood to be a clean coal-fired electricity production technology that is cleaner and offers higher efficiency compared to most electricity generation technologies. It provides many advantages in terms of environmental impacts and performance. The results obtained from this LCA study of the IGCC system were consistent with this general perception. However, two weaknesses observed in the current investigation can be addressed in future studies. Firstly, the technology is still in the initial stage of development. Since the IGCC plants are not yet economically feasible, there is a lack of actual technical data about, and experience with, these plants. As a consequence, the technical data used in this study were adopted from the available literature. In the future, data from an actual industrial plant, or at least from a pilot plant, will be used so that the analysis can be more accurate (*i.e.*, once the Kemper County IGCC starts its operation). Secondly, the data on the trace elements emissions used in this study were not obtained via direct measurement because of the lack of available data on Saskatchewan lignite coal and the coal-fired gasification process. The data on the trace elements and heavy metal emissions were taken from a coal-fired power plant in the US. Therefore, it is possible that the environmental performance on some impact categories related to heavy metal and trace elements may be slightly higher than has been evaluated in this study. Therefore, it would improve the accuracy of the LCA analysis in the future if real and directly measured data from an industrial plant can be obtained.

As mentioned earlier, IGCC technology is still in the initial stage of development, and the IGCC plants are not yet economically feasible; therefore, additional research and development is required to identify and evaluate the means by which the technology can be improved to make it economic viable and cost-competitive in comparison with the existing coal-based electrical generating stations. Moreover, future work may include an analysis from the full life-cycle perspective (*i.e.*, a cradle-to-grave approach that includes CO_2-EOR and CO_2-storage).

Acknowledgments: We are grateful for the financial support from the Network of Centres of Excellence of Canada-Carbon Management Canada (CMC-NCE), the Faculty of Engineering and Applied Science and the Faculty of Graduate Studies and Research at the University of Regina.

Author Contributions: Lakkana Piewkhaow, Anastassia Manuilova and Malcolm Wilson conceived and designed the system boundary; Lakkana Piewkhaow performed modelling (Aspenplus®, LCA and AREMOD); Lakkana Piewkhaow, Anastassia Manuilova, Malcolm Wilson, Christine W. Chan and Paitoon Tontiwachwuthikul analyzed the data; Christine W. Chan and Paitoon Tontiwachwuthikul contributed materials and analysis tools; Lakkana Piewkhaow wrote the paper. All authors have read and approved the final manuscript.

Conflicts of Interest: The authors declare no conflict of interest.

References

1. National Inventory Report 1990–2013: Greenhouse Gas Sources and Sinks in Canada—Executive Summary. Available online: https://www.ec.gc.ca/ges-ghg/default.asp?lang=En&n=5B59470C-1&offset=4&toc=show (accessed on 29 February 2016).
2. Economic Review 2013 (Number Sixty-Seven) of Government of Saskatchewan. Available online: http://www.stats.gov.sk.ca/stats/ER2013.pdf (accessed on 29 February 2016).
3. Rai, H.; Bharti, A.; Singh, R.; Prasad, N.K. Energy Efficienct Coal Gasification for IGCC Power Plant. *Int. J. Innov. Technol. Explor. Eng.* **2013**, *2*, 203–206.
4. Ratafia-Brown, J.A.; Manfredo, L.M.; Hoffmann, J.W.; Ramezan, M.; Stigel, G.J. An environmental assessment of IGCC power systems. In Proceedings of the Nineteenth Annual Pittsburgh Coal Conference, Pittsburgh, PA, USA, 23–27 September 2002.
5. Commercial Power Production Based On Gasification: Integrated Gasification Combined Cycle without Carbon Capture and Storage. Available online: http://www.netl.doe.gov/research/Coal/energy-systems/gasification/gasifipedia/igcc (accessed on 2 March 2016).
6. Wall, T.F. Combustion processes for carbon capture. *Proc. Combust. Inst.* **2007**, *31*, 31–47. [CrossRef]
7. Burr, B.; Lyddon, L. *A Comparison of Physical Solvents for Acid Gas Removal*; Bryan Research & Engineering, Inc.: Bryan, TX, USA, 2013.

8. Davison, J.; Thambimuthu, K. Technologies for capture of carbon dioxide. In *Greenhouse Gas Control Technologies 7*; Rubin, E.S., Keith, D.W., Gilboy, C.F., Wilson, M., Morris, T., Gale, J., Thambimuthu, K., Eds.; Elsevier Science Ltd: Oxford, UK, 2005; pp. 3–13.
9. Odeh, N.; Cockerill, T. Life-Cycle GHG Assessment of Fossil Fuel Electrical generating stations with Carbon Capture and Storage. *Energy Policy* **2008**, *36*, 367–380. [CrossRef]
10. Emissions of substances other than CO_2 from power plants with CCS. Available online: https://www.globalccsinstitute.com/publications/emissions-substances-other-co2-power-plants-ccs (accessed on 2 March 2016).
11. Hoffmann, B.S.; Szklo, A. Integrated gasification combined cycle and carbon capture: A risky option to mitigate CO_2 emissions of coal-fired power plants. *Appl. Energy* **2011**, *88*, 3917–3929. [CrossRef]
12. CO_2 Capture and Storage. A key carbon abatement option. Available online: https://www.iea.org/publications/freepublications/publication/CCS_2008.pdf (accessed on 3 March 2016).
13. Carbon capture and storage technologies database at MIT. Available online: http://sequestration.mit.edu/index.html (accessed on 3 March 2016).
14. Major Commercial Examples of IGCC Plants. Available online: http://www.netl.doe.gov/research/coal/energy-systems/gasification/gasifipedia/igcc-config (accessed on 3 March 2016).
15. Suebsiri, J. An Environmental Model of Carbon Capture and Storage with Demonstration to Carbon Footprint and Resource Deletion Evaluation. PhD. Thesis, University of Regina, Regina, SK, Canada, 2010.
16. Manuilova, A. Evaluation of Environmental Performance of Carbon Capture and Storage Project in Canada Using Life-Cycle Assessment Methodology. PhD. Thesis, University of Regina, Regina, SK, Canada, 2011.
17. Baumann, H.; Tillman, A. *The Hitch Hiker's Guide to LCA*; Student Literature: Lund, Sweden, 2014.
18. Pehnt, M.; Henkel, J. Life-cycle assessment of carbon dioxide capture and storage from lignite electrical generating stations. *Int. J. Greenh. Gas Control* **2009**, *3*, 49–66. [CrossRef]
19. Polk Power Station. Available online: http://www.tampaelectric.com/company/ourpowersystem/power generation/polk/ (accessed on 29 February 2016).
20. Rubin, E.S.; Berkenpas, M.B.; Frey, H.C.; Chen, C.; McCoy, S.T.; Zaremsky, C.J. Technical Documentation: Integrated Gasification Combined Cycle Systems (IGCC) with Carbon Capture and Storage (CCS). Available online: http://repository.cmu.edu/cgi/viewcontent.cgi?article=1073&context=epp (accessed on 29 February 2016).
21. Koiwanit, J.; Piewkhaow, L.; Zhou, Q.; Manuilova, A.; Chan, C.W.; Wilson, M.; Tontiwachwuthikul, P. A life-cycle assessment study of a Canadian post-combustion carbon dioxide capture process system. *Int. J. Life-Cycle Assess.* **2014**, *19*, 357–369. [CrossRef]
22. *Saskatchewan Air Quality Modeling Guideline*; Saskatchewan Environment Technical Resources Branch: Regina, SK, Canada, 2012.
23. *Air Monitoring Guideline for Saskatchewan*; Saskatchewan Environment Technical Resources Branch: Regina, SK, Canada, 2012.
24. Air Quality Assessment Report: Rainy River Gold Project. Available online: http://www.ceaa-acee.gc.ca/050/documents_staticpost/80007/97751/Vol3_AppQ-1_RRP_Final_EA_AQ_Assess.pdf (accessed on 29 February 2016).
25. Mokhtar, M.M.; Hassim, M.H.; Taib, R.M. Health risk assessment of emissions from a coal-fired power plant using AERMOD modelling. *Process Saf. Environ. Prot.* **2014**, *92*, 476–485. [CrossRef]
26. Gibson, D.M.; Kundu, S.; Satish, M. Dispersion model evaluation of $PM_{2.5}$, NO_X and SO_2 from point and major line sources in Nova Scotia, Canada using AERMOD Gaussian plume air dispersion model. *Atmos. Pollut. Res.* **2013**, *4*, 157–167. [CrossRef]
27. Climate and Weather Term Glossary. Available online: http://www.wrcc.dri.edu/ams/glossary.html (accessed on 29 February 2016).
28. Al-Jiboori, M.H.; Al-Draji, A.G. Aerodynamic Surface Roughness Length of Baghdad city. *J. Al-Nahrain Univ.* **2010**, *13*, 96–102.
29. Appendix R: Kemper County IGCC Project Hazardous Air Pollutant Risk Screening Analyses. Available online: https://www.netl.doe.gov/File%20Library/Library/Environmental%20Policy/eis%20kemper%20 draft/19_Appendix-R.pdf (accessed on 29 February 2016).
30. SNC-Lavalin Inc. From Engineering to Procurement to Construction of the Boundary Dam Carbon Capture System. In Proceedings of SaskPower CCS Consortium 2013 Information & Planning Symposium, Regina, SK, Canada, 21–23 May 2013.

31. Govin, R.; Shah, J. Modeling and Simulation of An Entrained Flow Coal Gasifier. *AIChE J.* **1984**, *30*, 79–91. [CrossRef]
32. Christou, C.; Hadjipaschalis, I.; Poullikkas, A. Assessment of integrated gasification combined cycle technology competitiveness. *Renew. Sustain. Energy Rev.* **2008**, *12*, 2459–2471. [CrossRef]
33. Fiaschi, D.; Lombardi, L. Integrated Gasifier Combined Cycle Plant with Integrated CO_2-H_2S and Removal: Performance Analysis, Life-Cycle Assessment and Exergetic Life-Cycle Assessment. *Int. J. Appl. Thermodyn.* **2002**, *5*, 13–24.
34. Kohl, A.L.; Nielsen, R.B. *Gas Purification*, 5th ed.; Gulf Publishing Company: Houstun, TX, USA, 1997.
35. Zhu, Y. Evaluation of Gas Turbine and Gasifier-based Power Generation System. PhD. Thesis, North Carolina State University, Raleigh, NC, USA, 2004.
36. Zhu, Y.; Frey, H.C. Simplified Performance Model of Gas Turbine Combined Cycle System. *J. Energy Eng.* **2007**, *133*, 82–90. [CrossRef]
37. Volkart, K.; Bauer, C.; Boulet, C. Life cycle assessment of carbon capture and storage in power generation and industry in europe. *Int. J. Greenh. Gas Control* **2013**, *16*, 91–106. [CrossRef]

11

Fate and Transport of Fire-Born Particles in Porous Media

Prabhakar Sharma [1,2,*], Fritjof Fagerlund [2], Ulrika Iverfeldt [2] and Amanda Eskebaek [2]

Academic Editors: Luciano Feo and Raimondo Luciano

[1] School of Ecology and Environment Studies, Nalanda University, Rajgir, Nalanda, Bihar 803116, India
[2] Department of Earth Sciences, Uppsala University, Uppsala 75236, Sweden; fritjof.fagerlund@geo.uu.se (F.F.); ulrika_iverfeldt@hotmail.com (U.I.); amanda.eskebaek@gmail.com (A.E.)
* Correspondence: psharma@nalandauniv.com

Abstract: A variety of hazardous substances may be generated from the burning materials during fire extinguishing operations, depending on the location, type, and place of the fire. As a result, the fire-extinguishing water may act as a carrier for these nano- and micro-sized fire-born particles, including various types of associated contaminants, and may cause contamination of soil and groundwater resources. While airborne particles from fires have been studied, it is currently not well known what types of nano- and micro-sized contaminants are typically carried by the fire-extinguishing water and how these contaminants can be transported in the natural environment. The main purpose of this study was to increase the understanding about the occurrence and physical and chemical properties of nanoparticles commonly found in discharge water from fire extinguishing operations. The current study was based on collection of original samples from a fire location. A detailed characterization of the particles present in the extinguishing water was performed including both quantification of contaminants associated with the particles (such as metals and polycyclic aromatic hydrocarbons (PAHs) as well as measurement of properties related to the mobility of these particles through porous media. Such mobility properties include size distributions of the particles and the porous media, surface charges and solution chemistry). Results indicate that metals and PAHs are present in both finer and relatively larger fire-born particles. The particles larger than 11 μm were not mobile in porous media. The mobility of the finer particles (<11 μm) was generally high but was dependent on the solution chemistry. Low mobility of large particles in porous media indicates that a large amount of the contamination can likely be trapped in the top soil layer even though the fire extinguishing water infiltrates.

Keywords: fire particles; carbon; transport; retention; straining; porous media

---

1. Introduction

Fires in buildings such as households, and industrial or municipal buildings are common worldwide e.g., more than 10,000 fires in buildings were reported in Sweden alone in 2012 [1], and, in about 40% of these events, the fire was extinguished by the rescue services agencies. In most fire events, water is used as an extinguishing agent. Depending on the type of fire, the burning material and extinguishing method *etc.*, a variety of toxic compounds and particles can be formed. The particles (fire-born particles, FBPs) and associated contaminants can contaminate the extinguishing water. If this water is not collected on-site and disposed of safely, the contaminants carried with it infiltrate soil and groundwater systems or spread in the storm water system.

Anticipating the contamination content in the extinguishing water is a complex process because many factors affect the concentration of contamination and particles. Three different types of sources can contribute to pollution in a fire event: (i) material burning, (ii) items at the fire scene which were not affected by the fire but were transported from the scene with the extinguishing water, and (iii) additives to the extinguishing water used for firefighting [2]. Pollution contributions from the first two sources vary widely and the chemical and particulate composition of the extinguishing water thus becomes unique to each individual fire. However, the contribution from the third source can likely be present at all fire locations.

Particle generation in fires is dependent on the burning material and combustion conditions where oxygen supply and temperature are important factors. Air supply and fire intensity can affect combustion rate and thus the compound formed from the burning material. For example, a well-ventilated fire with high combustion rate can generate a lower concentration of pollutants in the extinguishing water than an incomplete combustion of fire material [3]. Part of the particles contained in the fire extinguishing water comes from material, which has not undergone complete combustion or condensed substances after cooling of flue gas or formed by incomplete chemical reaction [4]. If the compound in the gas phase condenses when flue gas is cooled down, the enrichment of pollutants on the particles can be larger [5]. Usually, soot is present in a large amount of particles from a fire and can be formed under reducing conditions of incomplete combustion generating polycyclic aromatic hydrocarbons (PAHs) [4].

As particle formation in runoff water after any fire event depends on specific conditions and fire locations, it may be difficult to create a general understanding of their content [6]. However, there are a number of studies of particle formation in the air during the combustion of various materials and material mixtures (such as [7,8]). In these studies, a larger amount of particles were generated in the material, which burned for a longer time. Particle levels measured in flue gas showed a wide distribution of particles with the overall mass distribution in the range of 0.1–0.3 μm [7].

To understand how the transport of fire-born particles (FBPs) with water from firefighting operations affects contaminant transport requires understanding of the types of particles in the extinguishing water and their interaction with pollutants. Particle properties will affect their transport properties as well as the impact of particle interaction with the pollutants. Polycyclic aromatic hydrocarbons (PAHs) and metals are two groups of substances of environmental concern, which are commonly found in runoff water from fires at concentrations exceeding drinking water standards [1,9,10]. However, the characteristics of these particles when present in the extinguishing water, including properties affecting transport and interaction with other particles and contaminants are to date not well documented.

It is also critical to understand how the chemical signature of the water carrying the particles affects the transport. During fire-extinguishing operations the water chemistry can be strongly affected by the fire and the specific materials that are present and burning on the site. Upon infiltration into the ground, in contact with soil and mixing with soil and groundwater, this signature may start to change, affecting the transportability of FBPs. Water chemistry parameters such as pH and salt concentration are well known to affect particle transport in porous media (e.g., [11–16]). While water chemistry can vary strongly for different fires and extinguishing operations, it is important to understand how these parameters influence the particle mediated contaminant transport and ultimate fate of the fire-generated contamination.

A number of studies exist regarding fire-generated particles in air (e.g., [7,8]) but a thorough understanding of the properties of fire-born particles carried with the extinguishing water is still lacking. In order to understand and predict the subsurface transport and environmental impact of FBPs, it is necessary to understand their chemical and physical properties and their interaction with other constituents in the extinguishing water and soil system. The main objective of this study is to quantify the physical and chemical properties of the particles commonly found in discharge water from fire extinguishing operations and to predict the fate and transport of these particles in soil and groundwater. This case study is based on collection of original samples from several points related to the same fire in a warehouse facility in Uppsala, Sweden. A detailed characterization of the particles present in the extinguishing water was performed. Metal and PAH contaminants associated with the particles were quantified, and the properties related to the transportability of the particles in subsurface systems were measured including particle-size distribution, particle surface charge, and structures of the particles. Further, a series of column transport experiments in a controlled environment were performed using one of the collected firefighting water samples for varying water chemistry and physical conditions.

2. Materials and Methods

2.1. Sampling of Fire-Extinguishing Water

During autumn 2013, a fire occurred in a store and warehouse in Uppsala, Sweden, for tiles and miscellaneous tools. The part of the building that was most exposed by the fire was a shop, where tools were kept in plastic containers as well as a staff room with kitchen and wooden furniture. Upon extinguishing the fire, the fire department used approximately 50 m^3 of water, but no additives were used in the water. At the fire location, foam was observed in the runoff water in contact with the burned materials. The ground surface in and around the building consisted mainly of hardened surfaces, which resulted in pooled extinguishing water on these surfaces.

Water from the fire area was collected at three different points in the building near the most fire-prone locations. The sampling was done six days after the fire was extinguished. The collected water was deemed to be representative and relatively unchanged during this time period because it was pooled indoors under roof and on impermeable surfaces. Sampling site 1 was in the doorway to the room where the most intense fire occurred, sampling site 2 was on the floor outside this room and sampling site 3 was located in an area near site 2 where bags of cement powder were stored and the extinguishing water may have been affected by these materials. After collection, the samples were stored in glass and plastic bottles (depending on the planned analysis) until the analyses were carried out.

2.2. Characterization of Fire-Born Particles in the Extinguishing Water

The physical and chemical properties of the FBPs in sampled water were analyzed for each sampling site. All three original samples were used for measurement of pH and electrical conductivity (Table 1). The zeta potential of the FBPs was measured using a Zeta-Sizer (Malvern Instruments, Malvern, UK). Later, three sets of original fire extinguishing water samples from all three locations were prepared for the analysis of various metals and PAHs using Inductively Coupled Plasma/Mass spectroscopy (ICP/MS) and Gas Chromatography/Mass Spectroscopy (GC/MS) respectively at ALcontrol Laboratories, Linköping, Sweden. Samples from the second site were used for column transport experiments. This sample was filtered through an 11-μm membrane filter so that both particles smaller and greater than 11 μm could be analyzed for metals and PAH at ALcontrol Laboratories. The sample for filtered material of particles >11 μm was prepared by adding an equal amount of de-ionized (DI) water to the filtered material collected from the membrane filter. In addition to the chemical characterization of extinguishing water samples, two sets of samples (with particles <11 μm and <100 μm) were used for visualization and image analysis using Scanning Electron Microscopy (Leo 1550, Field Emission Scanning Electron Microscope, Jena, Germany).

Table 1. Characteristics of fire born particles (FBPs) in extinguishing water and column experiment parameters.

Experiment No.	FBP Size Range (μm)	Background pH (-)	Electrical Conductivity (mS/cm)	FBP Zeta Potential (mV)	Porous media Size Range (μm)	Porous media Average (μm)	Flow Rate (mL/min)	Porosity (-)
Sample 1	<100	8.64	4.06	−10.1	N.A.	N.A.	N.A.	N.A.
Sample 3	<100	7.72	3.02	−10.7	N.A.	N.A.	N.A.	N.A.
Sample 2								
1	<11	9.8	3.83	−5.76	250–400	325	1	0.33
2	<100	9.8	3.83	−5.76	250–400	325	1	0.33
3	<11	9.8	3.83	−5.76	400–500	450	1	0.33
4	<11	9.8	3.83	−5.76	177–250	213	1	0.34
5	<11	9.8	3.83	−5.76	125–177	151	1	0.34
6	<11	6.5	3.83	−4.06	250–400	325	1	0.33
7	<11	3	3.83	−3.11	250–400	325	1	0.33

N.A.: Not applicable.

2.3. Porous Media

Quartz sands (Sibelco Nordic, Baskarp, Sweden) with four different grain size distributions were used in this study. The sand was sieved to fine (0.125–0.177 mm), medium 1 (0.177–0.225 mm), medium 2 (0.25–0.4 mm), and coarse (0.4–0.5 mm) grain sizes. To remove all the impurities, the sand was washed using the same procedure as Mekonen *et al.* [15]. In brief, it was sequentially cleaned by 0.1 M hydrochloric acid (37%–38%) solution for 30 min, and then flushed and rinsed several times by DI water. The sand was then washed by 30% H_2O_2 and DI water with a volumetric ratio of 8:35 in the shaking flask for a total 40 min by stirring the sand every 10 min. Finally, the sand was washed by DI water and placed in the oven at 105 °C for 24 h. The rinsed and dried sand was labeled and stored in a clean plastic bottle.

2.4. Experimental Setup

Transport experiments were performed using a cylindrical glass column with 15 cm length and 2.5 cm diameter. A steel mesh filter (0.2 mm) was used on both sides of the glass column in order to stop the sand from entering into the tubing. The tubes connected to the column were flushed and filled with background solution to remove air bubbles before starting the experiment. A three-way valve was used to control the type of liquid flow (extinguishing water samples or background solution). The column was wet-packed in order to avoid any trapped air in the column. A high precision peristaltic pump (IPC8, Ismatec, Wertheim, Germany) was calibrated and used for injection of the samples at the chosen flow rates. The column effluents were collected by a fraction collector (CF-2, Spectrum Labs Inc., Breda, The Netherlands) and the concentration was measured by UV/VIS spectrophotometer (DR 5000, Hach Lange Ltd., Salford, UK) at a wavelength of 334 nm after a complete wavelength scanning. In addition, a number of tracer tests (using Brilliant Blue R, Sigma Aldrich, Stockholm, Sweden) were conducted separately to obtain the breakthrough curves with different collector grain sizes at the selected flow rates. The injected tracer concentration was 20 mg/L and the breakthrough concentrations were quantified at 300 nm wavelength using the spectrophotometer.

2.5. Column Experiments

Column experiments were performed to examine transport of the FBPs through a porous medium using sample number 2 only for different experimental condition listed in Table 1. First, the column was flushed with the corresponding aqueous solution at high flow rate for 10 pore volumes. Then, the flow rate was reduced to the experimental flow rate of 1 mL/min for another 2 pore volumes. Once the absorbance of the effluent sample had reached the background level (absorbance less than detection limit of the spectrophotometer), the experiment was performed. The water sample was first injected for 5.2 pore volumes (phase 1), and then the sand column was flushed with background solution for another 5.2 pore volumes (phase 2). Finally, DI water was injected for another 3 pore volumes (phase 3) in order to lower the ionic strength and study remobilization of particles related to deposition at secondary energy minima. Each column experiment was repeated twice for the experimental conditions shown in Table 1. Two sets of column experiments were conducted using water samples collected from site 2 after filtering them using 11 μm and 100 μm filter membranes, respectively. Three sets of column experiments were performed after adjusting the pH of water samples (filtered to particle sizes less than 11 μm) and background samples. Four sets of column experiments were also conducted at the original pH of the collected samples but changing the porous media grain sizes. The electrical conductivity of the samples was measured using a conductivity meter and the equivalent conductivity of the background solution during phase 2 was adjusted using NaCl salt solution.

3. Results and Discussion

3.1. Characterization of Fire-Born Particles

The concentrations of the 16 measured polyaromatic hydrocarbons from the three different sampling locations are shown in logarithmic scale in Figure 1. The figure indicates the highest concentration of all form of PAH in sample 1 and lowest in sample 3. The compounds from benso[a]anthracene and to the right in Figure 1 are considered to be of high molecular weight [17] and these, except benzo[ghi]perylene, are also considered carcinogenic (e.g., [18–22]). Low molecular weight PAHs may be found in the gas phase whereas the high molecular weight PAHs, which are also of greater environmental concern, have very low volatility and will generally be found in particles. All measured PAHs were found at all three sampling locations, and the relative abundance of the compounds was in relative agreement with previous studies (e.g., [23]) although, comparatively, somewhat lower concentrations of the low molecular weight PAHs were observed. This may be due to some volatilization of low molecular weight PAHs before sampling was performed. The differences in concentrations between the different sampling locations was likely related to the fact that the extinguishing water was found in different state of concentration at the different locations, and, particularly, large amounts of black soot was observed at location 1. However, the local thermo-physical characteristics of the fire and the burning materials should also have influenced the proportion of different compounds that were locally formed. The sum of carcinogenic PAHs was larger than 10 times the Swedish guideline value for groundwater at all sampling locations ($>10 \times 0.2 = 2$ μg/L), which (undiluted) would be considered a "very serious" state of contamination [24].

The concentrations of the 10 measured metals at the three different sampling locations are shown in logarithmic scale in Figure 2. The relative difference in metal concentrations between the different sampling locations is small as compared to the PAHs (Figure 1), which may indicate less influence of local factors on the amount of metal particles formed in the fire. Some of the metals (Cd, Cu, Zn, *etc.*) were highest in sample 1, but they don't follow any proper order. The metal concentrations are also in relative agreement with previous findings [23]. The concentrations of Pb, Cd, Cu and Zn were more than 10 times the Swedish water quality guideline values [24] for all samples. For As, Cr and Ni, the concentrations were above the guideline values, whereas for Ba, Co and V, Swedish guideline values have not been established.

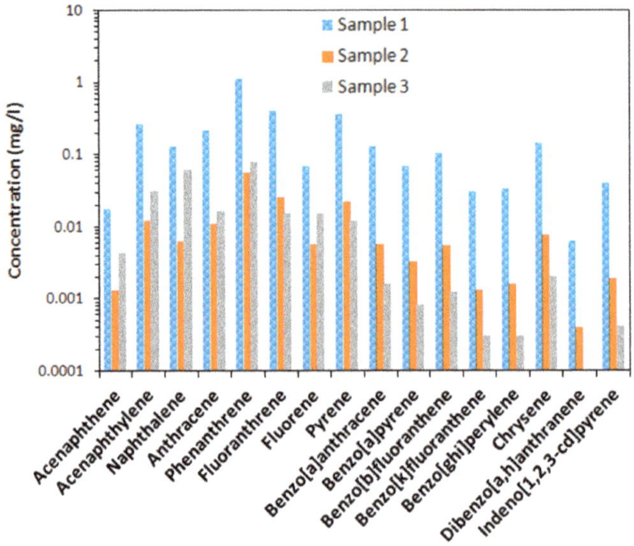

Figure 1. Concentration of PAHs in the collected fire extinguishing water samples from three different locations.

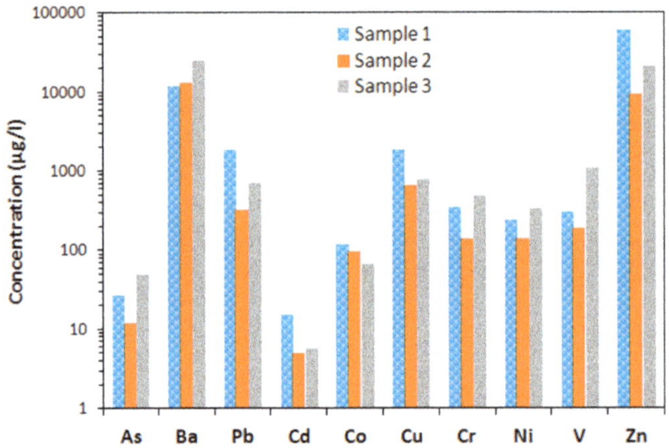

Figure 2. Concentration of metals in the collected fire extinguishing water samples from three different locations.

A critical factor for the mobility of particles in porous media is the size of the particles. It is therefore interesting to investigate how contaminants are distributed between different sizes of particles. This has also not previously been done for fire-generated particles in the extinguishing water. As shown in Figure 3, the concentrations of PAHs were typically at least one order of magnitude lower in the particle solution filtered to sizes <11 µm compared to the solution of sizes greater than 11 µm and less than 100 µm. The difference was more pronounced for the high molecular weight (and carcinogenic) PAHs, which indicates that these may have a larger tendency to form larger particles or aggregates as compared to the lower molecular weight PAHs. The summed concentration of carcinogenic PAHs was, however, still slightly larger than the Swedish groundwater quality guideline value of 0.2 µg/L even for the particles of size <11 µm. Concentrations in the unfiltered sample were higher than in the <100 µm filtrate, indicating that particles or aggregates >100 µm existed in the original sample.

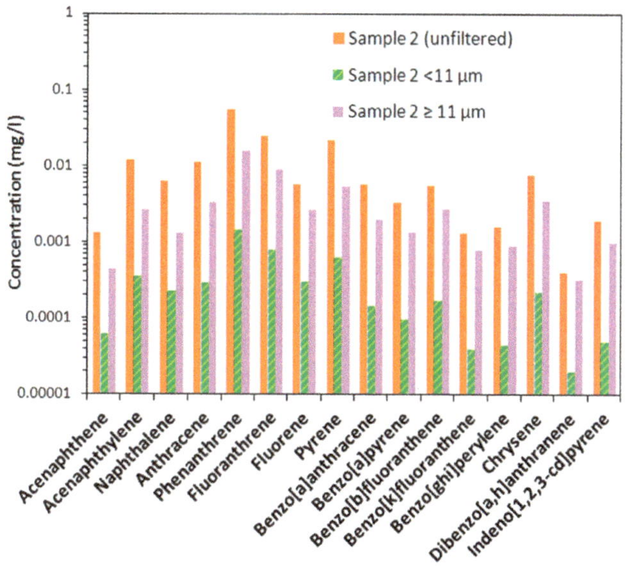

Figure 3. Concentration of PAHs in filtered and unfiltered samples at location 2.

As shown in Figure 4, six of the measured metals; As, Ba, Pb, Cd, Cr and Zn, had concentrations of at least one order of magnitude lower in the <11 µm filtrate compared to in particles between 11 µm and 100 µm. This indicates that these metals tend to be associated with larger particles, which may also be related to affinity of these metals to attach to larger soot particles. Four metals—Co, Cu, Ni and V—did not appear in significantly higher concentrations in the filtrate with larger particles. Co and Ni were even found in higher concentrations in the filtrate of smaller size particles (<11 µm). There appear to be a difference in characteristics between different metals, and the latter four may be less prone to aggregate. Contaminants found in smaller particles may also have larger mobility with water flowing through porous media, as will be further investigated in the following sections.

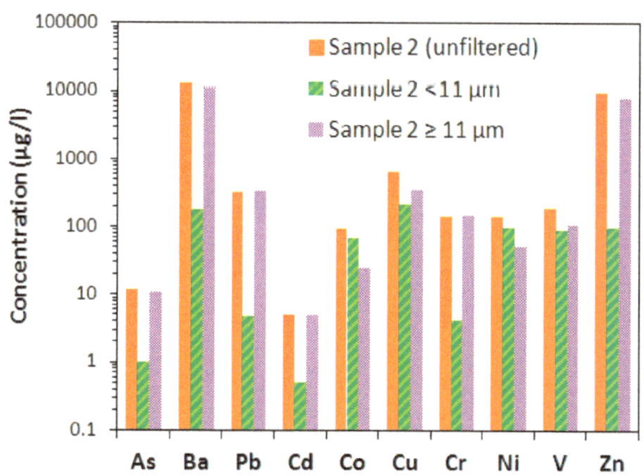

Figure 4. Concentration of metals in filtered and unfiltered samples at location 2.

3.2. Physical Factors for Transport of Fire-Born Particles in Porous Media

A set of column experiments with FBPs filtered to sizes <100 µm and <11 µm was conducted. As can be seen in Figure 5, the outflow concentrations from the column for samples filtered to less than 11 µm were close to their original (inflow) concentration. However, the outflow concentrations from the column for samples filtered to sizes less than 100 μm (note that this sample also contains particles <11 μm as opposed to in the chemical analyses) were much less than their original concentration. The steady state relative concentration of FBPs for finer sizes (<11 μm) was approximately 1, which means that the retention of these particles was negligible. It further indicates that attachment of FBPs did not occur at the natural chemical condition of the extinguishing water in combination with the selected physical condition (flow rate and porous media). However the results for FBPs <100 μm showed partial retention; at the same physical and chemical conditions, only 33% of the total particle concentration flowed through the column. The 77% retained particles in the column could not be mobilized even after changing the background solution to DI water (phase 3). It suggests that the particle larger than 11 μm were not deposited at the secondary energy minima locations. However, the result indicates that the selected porous media filtered out the larger particles in between 11 and 100 μm. Physical screening/straining was the likely retention mechanism in this scenario for FBPs, and other retention mechanisms should not have played any major role, as the remobilization of FBPs was negligible in phase 3 of the column experiment. Figure 6 shows the SEM images of FBPs for inflow and outflow samples from the column for both size ranges (<11 μm and <100 μm). The particles <11 μm were nearly spherical in shape and show a similar distribution in images of inflow and outflow solution (Figure 6a,b). This is in agreement with the column experiment results for this size range. However, the image of the inflow solution for particles <100 μm (Figure 6c) was quite different (fibrous structure) from that of the outflow solution (Figure 6d), where spherical shaped particles similar to the those in the sample filtered to sizes <11 μm can be observed. Thus, the larger fibrous kind of particles was filtered out during the flow through the porous media in the column. It can also be concluded that the FBPs <11 μm were highly mobile under the conditions tested here, posing a risk to reach the groundwater if they infiltrated into a subsurface system.

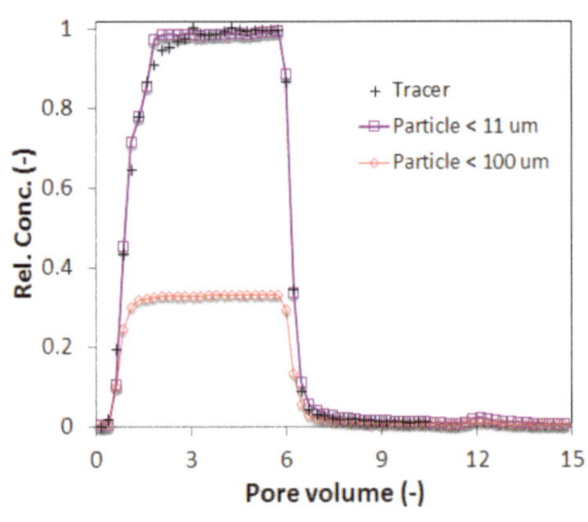

Figure 5. Representative FBP breakthrough curves from column experiments using extinguishing water collected at site 2 after filtration using two different particle sizes.

Column experiments for FBP transport through a wide range of porous media particle size distributions were conducted while keeping other physical and chemical conditions constant. Figure 7 shows the column outflow breakthrough curves for four porous media sizes with average sizes of 450 μm, 325 μm, 213 μm, and 151 μm, respectively. The results indicate that the FBPs are completely mobile in the three largest porous media grain sizes, while in the smallest grain size a small portion of approximately 2%–3% of the FBPs were retained. A small peak of remobilization could be observed in phase 3 in these experiments (Figure 7) but was not significant in terms of the total inflow mass of FBPs to the column. It can be concluded that even the average grain size of 151 μm (range 125–177 μm) was big enough to allow nearly full mobility of FBPs of sizes up to 11 μm. It can be noted that these FBPs were still more than one order of magnitude smaller than the sand grain size.

This further confirms that the retention mechanism for larger FBPs is likely be straining, and no other retention mechanisms were operative for the present experimental scenarios.

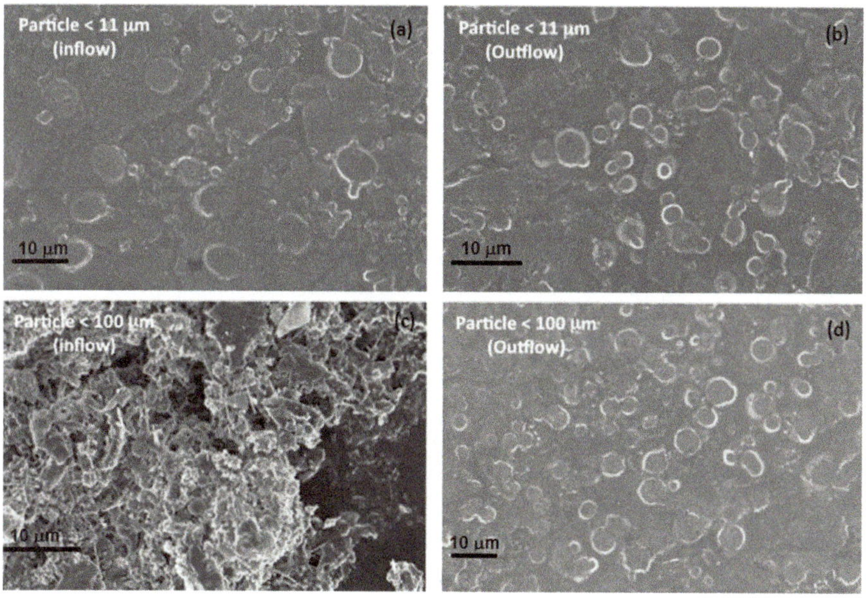

Figure 6. Scanning electron microscope images for fire-born particles from sampling site 2 for: (**a**) column inflow particle solution <11 μm; (**b**) column outflow particle solution <11 μm; (**c**) column inflow particle solution <100 μm; (**d**) column outflow particle solution <100 μm.

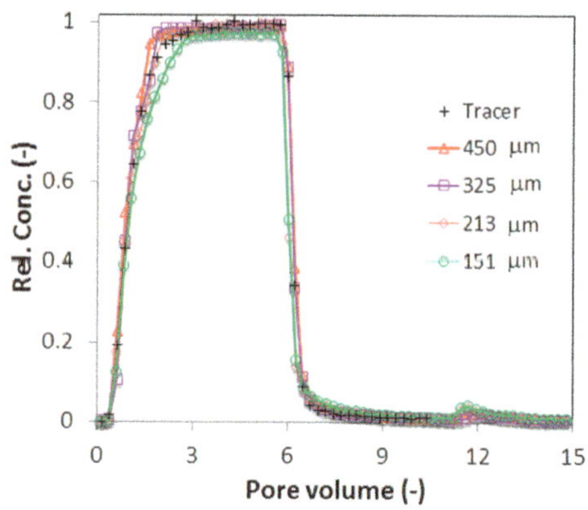

Figure 7. Representative FBP breakthrough curves from column experiments using fire extinguishing water collected at site 2 with four different porous media grain sizes.

3.3. Background Chemistry for Transport of Fire-Born Particles in Porous Media

The effect of background solution chemistry was also tested in this study. The ionic strength was not possible to vary in these fire extinguishing water samples because the salt concentration was very high (approximately 60 mM). The original pH of the fire extinguishing water was also high so the variation of solution chemistry is based on lowering the pH, as shown in Table 1. The effects of pH change on FBP outflow concentrations are shown in Figure 8. The results indicate that the relative concentration decreases with a decrease in solution pH. The relative concentration was close to 1.0, 0.92, and 0.88 for pH equal to 9.8, 6.5, and 3.0 respectively. An increase in retention of colloids/nanoparticles with a decrease in pH has been observed in several previous studies (such as, [15,16,25]). The Derjaguin-Landau-Verwey-Overbeek (DLVO) interaction forces between FBPs and sand surfaces might change due to a decrease in pH. The measurements presented in Table 1 show an increase in surface charges with a decrease in solution pH, which can favor particle retention at secondary energy minima. The surface charges are also in agreement with the absence of a secondary peak during phase 3 for pH 9.8 while small peaks were present for pH 6.5 and 3.0 (Figure 8). These results indicate a partial retention of FBPs at very low pH, however, at pH representative of natural soil and groundwater (pH 6.5), the FBPs had nearly full mobility.

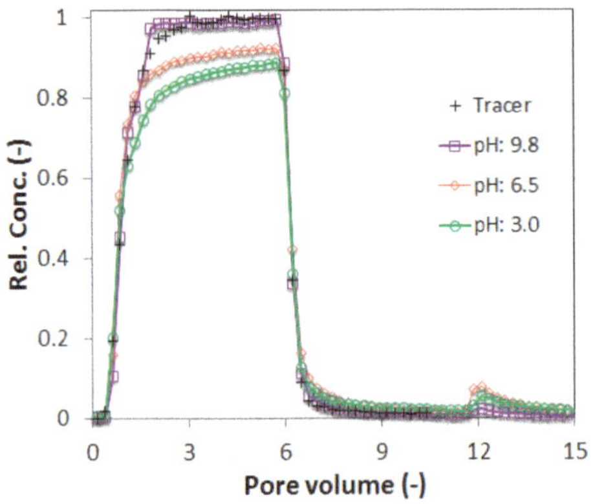

Figure 8. Representative FBP breakthrough curves from column experiments using fire-extinguishing water collected at site 2 adjusted to three different pH.

3.4. Mass Balance and Total Retention of Fire-Born Particles

An integration of total particle mass was done to calculate the ratio of outflow mass to inflow mass. Figure 9 shows the mass ratios for the different physical and chemical condition tested in this study. A high mass ratio indicates low retention of particles. The mass ratio was much higher for small size (<11 μm) FBPs than the larger (<100 μm) particles (Figure 9a). It can also be noted that the actual outflow mass was in the same order for both cases (data not shown here), which indicates that most of the filtered/strained FBPs were in the range of 11 to 100 μm.

The mass ratio for phases 1 and 2 are plotted in Figure 9b as a function of average grain size. Although the absolute differences in retention were relatively small, a clear trend of increase in outflow mass with increase in grain size could be observed (Figure 9b). Physical straining was the likely retention mechanism.

Figure 9c shows the effect of pH on the outflow to inflow mass ratio. There is a clear trend of increase in outflow mass of FBPs with increase in pH. As discussed in the previous section, deposition at secondary energy minima is a possible retention mechanism at low pH, which was also supported by the observed remobilization of FBPs in phase 3.

From this study, we attempted to identify the presence of harmful contaminants (metals and PAH) in different sizes of fire burned particles from the selected location then we tried to represent the extreme pH condition for possible mobilization of these contaminant attached fire particles in porous media in order to warn the agencies involved in fire fighting operations (to save the ground- or surface-water pollution).

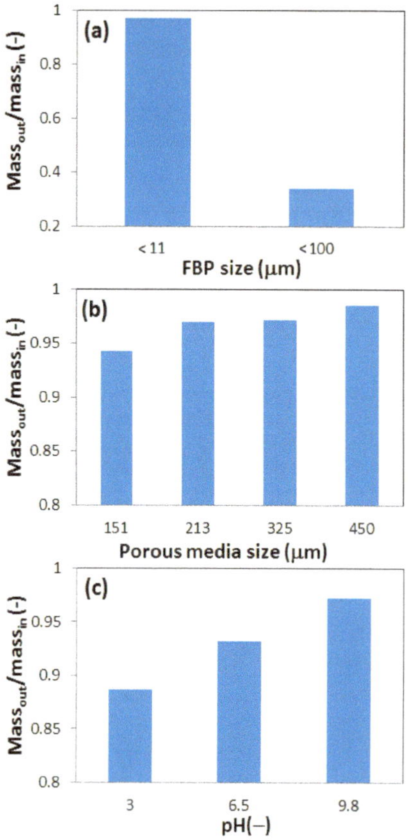

Figure 9. Ratio of outflow to inflow mass of FBPs in column experiments for: (**a**) FBPs filtered to sizes <11 μm and <100 μm, respectively; (**b**) different porous media grain sizes; (**c**) different pH of solution.

4. Conclusions

Detailed characterization of fire extinguishing water is important to understand both the contamination content and the potential subsurface mobility of the contaminants present. This can be fire and site specific. Depending on the type of fire, a number of contaminants such as metals and polyaromatic hydrocarbons can be present in the fire-generated particles. The transport and retention of fire-born particles (FBPs) upon infiltration to soil and groundwater systems are, however, also influenced by the physical and chemical conditions of the soil and the water carrying the particles. Some of the key characteristics that affect transport of fire-born particles in water-saturated soil and groundwater are their surface charge, size, and the solution chemistry of the water that carries them. Size affects physical straining while surface charge and solution chemistry influence the physicochemical mechanisms of particle and surface interactions. Sand grain size had only a small influence on the transport of smaller particles (<11 μm) through the column. For these particles, retention was only observed when the water chemistry was changed. The magnitudes of the negatively charged particle surface were less at low pH, which resulted less retention (*i.e.*, some particles being attracted to the sand grains leading to deposition at energy minima). The larger particles (>11 μm and <100 μm) had almost no mobility in the porous media, which might have filtered out by physical straining mechanism. The chemical analyses showed that the main part of the PAH content consists high molecular weight carcinogenic PAHs, which were found in the larger particles (>11 μm), while the total PAH content were predominant in the smaller particles (<11 μm). For metals, a similar behaviour was observed for As, Ba, Pb, Cd, Cr and Zn, while Co, Cu, Ni and V were almost equal for both finer

(<11 μm) and larger (>11 μm) particles. In combination with the results for particle mobility, this study also indicates that a large portion of the PAHs and several metals of high environmental concern were likely be filtered out to a large extent in the top soil upon infiltration to the ground. However, those contaminants associated with small particles (<11 μm) are likely to be highly mobile with the infiltrating water and may pose a threat of groundwater contamination. Although the main part of the total contaminant load was found in larger, immobile particles, the current results indicate that both PAHs and metals may be found at concentrations exceeding groundwater quality guideline values in the fraction of smaller mobile particles. Future research would be beneficial to further investigate a broader range of fires and local soil conditions.

Acknowledgments: The Swedish Civil Contingencies Agency (MSB) is highly acknowledged for funding the part of this research.

Author Contributions: Prabhakar Sharma has designed the main experimental work, monitored the implementation of the work in the laboratory, and prepared the manuscript. Fritjof Fagerlund planned and initiated this work after discussions with the Swedish Civil Contingencies Agency and also took part in the design of the experiments. Ulrika Iverfeldt and Amanda Eskebaek performed the sample preparation and conducted the flow through column experiments in the laboratory. The data analysis part was performed through discussion among all authors.

Conflicts of Interest: The authors declare no conflict of interest.

References

1. Myndigheten för Samhällsskydd Och Beredskap (MSB). *Bränder*; Myndigheten för Samhällsskydd Och Beredskap: Stockholm, Sweden, 2012.
2. Särdqvist, S. *Vatten Och Andra Släckmedel*; Räddningsverket: Karlstad, Sweden, 2006.
3. Norberg, P.; Lithner, D. *Rening Och Destruktion av Kontaminerat Släckvatten*; Myndigheten för Samhällsskydd Och Beredskap (MSB): Stockholm, Sweden, 2013.
4. Hertzberg, T. *Partiklar Från Bränder—Förstudie*; Sveriges Provnings och Forskningsinstitut (SP): Borås, Sweden, 2001.
5. Butler, K.M.; Mulholland, G.W. Generation and transport of smoke components. *Fire Technol.* **2004**, *40*, 149–176. [CrossRef]
6. Noiton, D.; Fowles, J.; Davies, H. *The Ecotoxicology of Fire-Water Runoff Part II—Analytical Results*; New Zealand Fire Service Commission: Wellington, New Zealand, 2001.
7. Hertzberg, T.; Blomqvist, P. Particles from fire—A screening of common materials found in buildings. *Fire Mater.* **2013**, *27*, 295–314. [CrossRef]
8. Blomqvist, P.; Simonson, M.; Stec, A.A.; Gylestam, D.; Karlsson, D. *Characterisation of Fire Generated Particles*; BRANDFORSK Project 700-0601 (SP Rapport 2010:01); Sveriges Provnings och Forskningsinstitut (SP): Borås, Sweden, 2010.
9. Leifeld, J.; Fenner, S.; Muller, M. Mobility of black carbon in drained peatland soil. *Biogeosciences* **2007**, *4*, 425–432. [CrossRef]
10. Diblasi, C.J.; Li, H.; Davis, A.P.; Ghosh, U. Removal and fate of pollycyclic aromatic hydrocarbon pollutants an urban stormwater bioretention facility. *Environ. Sci. Technol.* **2009**, *43*, 494–502. [CrossRef] [PubMed]
11. Elimelech, M.J.; Gregory, X.J.; Williams, R.A.; Gregory, J.; Xiaodong, J.; Williams, R.A. *Particle Deposition & Aggregation—Measurement, Modelling and Simulation*; Elsevier: Philadelphia, PA, USA, 1995.
12. Sharma, P.; Abdou, H.M.; Flury, M. Effect of the lower boundary condition and flotation on colloid mobilization in unsaturated sandy sediments. *Vadose Zone J.* **2008**, *7*, 930–940. [CrossRef]
13. Sharma, P.; Flury, M.; Zhou, J. Detachment of colloids from a solid surface by a moving air-water interface. *J. Colloid Interface Sci.* **2008**, *326*, 143–150. [CrossRef] [PubMed]
14. Sharma, P.; Flury, M.; Mattson, E.D. Studying colloid transport in porous media using a geocentrifuge. *Water Resour. Res.* **2008**, *44*. [CrossRef]
15. Mekonen, A.; Sharma, P.; Fagerlund, F. Transport and mobilization of multiwall carbon nanotubes in quartz sand under varying saturation. *Environ. Earth Sci.* **2014**, *71*, 3751–3760. [CrossRef]

16. Sharma, P.; Bao, D.; Fagerlund, F. Deposition and mobilization of functionalized multiwall carbon nanotubes in saturated porous media: Effect of grain size, flow velocity and solution chemistry. *Environ. Earth Sci.* **2014**, *72*, 3025–3035. [CrossRef]
17. Sterner, O. *Förgiftninger Och Miljöhot*; Studentlitteratur: Lund, Sweden, 2003.
18. Nisbet, I.C.; LaGoy, P.K. Toxic Equivalency Factors (TEFs) for Polycyclic Aromatic Hydrocarbons (PAHs). *Refulatory Toxicol. Pharmacol.* **1992**, *16*, 290–300. [CrossRef]
19. Ruokojärvi, P.; Aatamila, M.; Ruuskanen, J. Toxic chlorinated and polyaromatic hydrocarbons in simulated house fires. *Chemosphere* **2000**, *41*, 825–828. [CrossRef]
20. Larsson, I.; Lönnermark, A. *Utsläpp Från Bränder—Analyser av Brandgaser Och Släckvatten*; BRANDFORSK Projekt 707-021 (SP Rapport 2002:24); Sveriges Provnings och Forskningsinstitut (SP): Borås, Sweden, 2002.
21. Lönnermark, A.; Blomqvist, P. *Emissions from Fires in Electrical and Electronics Waste*; Sveriges Provnings Och Forskningsinstitut (SP): Borås, Sweden, 2005.
22. Lönnermark, A.; Blomqvist, P. *Emissions from Tyre Fires*; Sveriges Provnings Och Forskningsinstitut (SP): Borås, Sweden, 2005.
23. Blomqvist, P.; Lönnermark, A.; Simonson, M. *Miljöbelastning vid Bränder Och Andra Olyckor—Utvärdering av Provtagning Och Analyser*; Räddningsverket: Karlstad, Sweden, 2004.
24. Naturvårdsverket. *Metodik för Inventering av Förorenade Områden—Bedömningsgrunder för Miljökvalitet—Vägledning, Rapport 4918 för Insamling av Underlagsdata*; Naturvårdsverket: Värnamo, Sweden, 1999.
25. Sharma, P.; Fagerlund, F. Transport of surface-modified carbon nanotubes through soil and groundwater. *J. Vis. Exp.* **2015**. [CrossRef] [PubMed]

12

The Effects of Thermal Treatment and Steam Addition on Integrated CuO/CaO Chemical Looping Combustion for CO_2 Capture

Alvaro Recio [1], Sip Chen Liew [1,†], Dennis Lu [2,†], Ryad Rahman [3], Arturo Macchi [3,†] and Josephine M. Hill [1,*]

1 Department of Chemical and Petroleum Engineering, University of Calgary, 2500 University Drive NW, Calgary, AB T2N 1N4, Canada; arecdel@gmail.com (A.R.); scliew@ucalgary.ca (S.C.L.)
2 CanmetENERGY, Natural Resources Canada, 1 Haanel Drive, Ottawa, ON K1A 1M1, Canada; dennis.lu@canada.ca
3 Department of Chemical and Biological Engineering, University of Ottawa, 161 Louis Pasteur Street, Ottawa, ON K1N 6N5, Canada; ryad.rahman@gmail.com (R.R.); Arturo.Macchi@uottawa.ca (A.M.)
* Correspondence: jhill@ucalgary.ca
† These authors contributed equally to this work.

Academic Editor: Gustavo A. Fimbres Weihs

Abstract: The combination of Chemical Looping Combustion (CLC) with Calcium Looping (CaL) using integrated pellets is an alternative CO_2 capture process to the current amine-based sorbent processes, but the pellets lose sorption capacity over time. In this paper, the deactivation behavior of CaO, CuO and CuO/CaO integrated pellets used for multiple (16–20) cycles in a thermogravimetric analyzer was studied. The impact of thermal treatment and the presence of steam on the deactivation were also investigated. Nitrogen physisorption and scanning electron microscopy/energy-dispersive X-ray analysis were used to characterize the pellets. The analysis revealed significant migration of the copper to the surface of the composite pellets, which likely suppressed carbonation capacity by reducing the accessibility of the CaO. While thermal pre-treatment and steam addition enhanced the performance of the base CaO pellets, the former led to cracks in the pellets. In contrast, thermal pretreatment of the CuO/CaO composite pellets resulted in worse CLC and CaL performance.

Keywords: carbon capture; calcium oxide; copper oxide; calcium looping combustion; thermal treatment; steam carbonation

---

1. Introduction

Multiple solutions have been developed to economically capture carbon dioxide. Among these, absorption using amine-based sorbents remains the most extensively-employed commercial scale technology [1]. Due the inherent economics and energy penalty of amine based solvents, however, alternative technologies may be more suitable depending on the process situation [2]. For instance, CO_2 capture applications, such as flue gases from fossil fuel power plants (post-combustion) [3] and biomass gasifiers (pre-combustion) [4], with calcium looping (CaL) and chemical looping combustion (CLC) have been shown to be promising. One of the remaining challenges, however, is the deactivation (*i.e.*, loss of reactivity, Equation (1)) of CaO based pellets used in this process. CaL consists of two stages: the forward reaction, Equation (1) to capture CO_2 is named carbonation while the reverse reaction to release CO_2, is named calcination. The carbonation is exothermic, and characterized by an initial rapid rate, which is kinetically controlled, followed by an abrupt transition to a slow reaction rate, which is diffusion controlled [5,6]. The reverse endothermic calcination to regenerate the CaO sorbent can be carried out in an oxyfuel combustor at high temperatures (up to 900 °C) to produce a concentrated stream of CO_2 for compression and storage [7]. The integration of chemical looping combustion (CLC) within CaL was proposed by Abanades *et al.* [8], and Manovic and Anthony [3] to offset the energy penalty from oxyfuel combustion (air separation) units. The irreversibility of the reaction in Equation (1) is often ascribed to sorbent sintering or agglomeration [3,9–11], as inferred by pore size measurements and microscopy.

$$\begin{gathered} CaO_{(s)} + CO_{2(g)} \leftrightarrow CaCO_{3(s)} \\ \Delta H_{r,298K} = -178\text{kJ/mol} \end{gathered} \tag{1}$$

In CLC, metal oxides act as oxygen carriers and transfer the oxygen from the air to the fuel, thus avoiding direct contact and combustion [12]. CuO is a suitable oxygen carrier due to its high oxygen-carrying capacity (0.20 g O_2/g CuO), high reactivity, and relatively low cost compared to metals such as NiO and CoO [3]. Since the reduction of the metal oxide is an exothermic reaction, it can supply the necessary heat energy to release CO_2 from the sorbent via calcination, which in turn avoids the energy penalties inherent to technologies, such as oxyfuel combustion [13]. Once the combustion reaction Equation (2) is complete, the oxygen carrier is regenerated by exposure to air Equation (3).

$$\begin{gathered} CH_{4(g)} + 4CuO_{(s)} \leftrightarrow 4Cu_{(s)} + CO_{2(g)} + 2H_2O_{(g)} \\ \Delta H_{r,298K} = -179\text{ kJ/mol} \end{gathered} \tag{2}$$

$$\begin{gathered} O_{2(g)} + 2Cu_{(s)} \leftrightarrow 2CuO_{(s)} \\ \Delta H_{r,298K} = -312\text{ kJ/mol} \end{gathered} \tag{3}$$

The stream obtained after the combustion of the fuel contains mainly CO_2, possibly some unreacted alkanes from the fuel (e.g., natural gas), and steam that can be easily separated by condensation to produce a stream of CO_2, which can be subsequently compressed and transported for storage [14]. The viability of various options of using CaL/CLC for post combustion CO_2 capture has been investigated via process simulation [15]: Namely, integrated CaL/CLC using single CuO/CaO pellets (Figure 1a), and a dual-loop process with decoupled CuO and CaO pellets (Figure 1b). The simulation results showed that both configurations of CaL/CLC process had similar energy penalty reductions up to 40% as compared to just CaL process from the elimination of an air separation unit, but at the cost of increased solids circulation and process complexity. The dual-loop process required 10% to 20% less solids circulation to achieve similar energy efficiency as compared to the single-loop process. It is possible to offset the difference in the solids circulation between the two configurations if the carbonation capacity of the combined CaL/CLC is substantially improved which was part of the objective of this study.

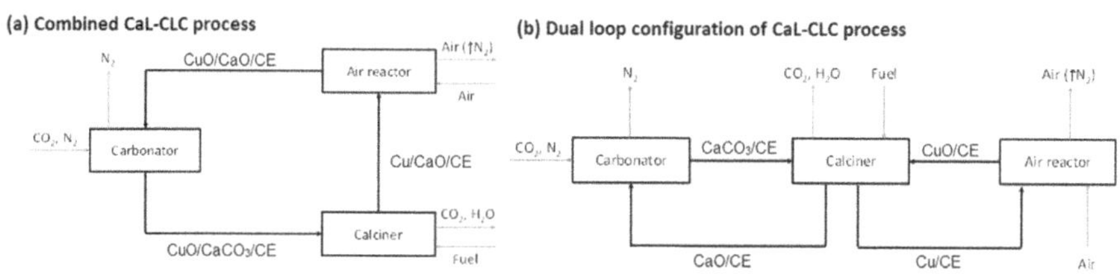

Figure 1. Schematic of the CaL-CLC process in a (**a**) combined configuration; or (**b**) dual loop configuration. Black and gray arrows represent flow of solids and gases, respectively.

Several researchers have proposed combining CaL and CLC functions in a single composite pellet but all reported decreased carbonation performance relative to pure CaO sorbents [3,8,16]. Different strategies have been proposed to improve the performance. For example, Qin *et al.* [17] studied various heat pre-treatment procedures with CaO/CuO pellet mixing to overcome the loss of performance and found that certain thermal treatments improved the multi-cycle CaL stability.

Steam, which is present in post combustion flue gas, has also been demonstrated to have a positive effect on the stability of pure CaO pellets to capture CO_2 over multiple cycles [17–24]. Several theories have been proposed to explain how steam improves the CaL process. Manovic *et al.* [19] suggested that the presence of steam promotes carbonation by enhancing the solid-state diffusion of CO_2 through the $CaCO_3$ layer forming over the CaO. Dou *et al.* [24] suggested that steam promotes the formation of $Ca(OH)_2$ that will react with CO_2 to form $Ca(HCO_3)_2$ and increase the amount of carbonation. In contrast, Yang and Xiao [22] reported that the performance of CaO was improved even without the formation of $Ca(OH)_2$. Arias *et al.* [25] reported that the kinetic-controlled part of the carbonation reaction was not affected by the presence of steam, but the diffusion-controlled part was enhanced.

In an effort to better understand the deactivation and promotion processes, both individual CuO and CaO pellets, as well as integrated CuO/CaO pellets, were characterized after multiple CaL and/or CLC cycles. The impact of the two reported methods to reduce deactivation—thermal pretreatment and exposure to steam during carbonation—on the characteristics of the pellets has also been investigated. The CLC and/or CaL performance was measured with a thermogravimetric analyzer (TGA) using carbonation at 650 °C, calcination/reduction at 875 °C, and oxidation at 900 °C. The pellets were characterized with physisorption to determine surface area and pore volume, and scanning electron microscopy with field emission imaging to determine the distribution of components. The rates of deactivation were fit to an exponential decay model.

2. Experimental Section

2.1. Materials

Pellets containing CaO, CuO or a combination of CaO and CuO, and held together with a calcium aluminate-based cement (CE) binder were used in simulated dual loop (CaO/CE and CuO/CE) or single loop (CuO/CaO/CE) configurations (Figure 1). The CaO in the pellets originated from powdered (< 45 μm) Cadomin limestone (Cadomin, AB, Canada), while the CuO powder (Sigma-Aldrich Corporation, St. Louis, MO, USA) consisted of particles less than 5 μm in size. The calcium aluminate binder was a commercially available cement with the majority of particles (> 80%) of sizes less than 45 μm (CA-14, 71% Al_2O_3, and 28% CaO, Almatis Inc., Rotterdam, The Netherlands). The components were mixed together and pressed into pellets with sizes of 250–650 μm and the compositions given in Table 1. The compositions were chosen based on previous studies [3].

Table 1. Composition, surface area and pore volume of the three types of pellets.

Pellets	Composition (wt %)			Surface Area	Pore Volume
	Cu	Ca	Cement	(m^2/g) *	(cm^3/g) *
CuO/CaO/CE	50	40	10	15 ± 1.1	0.078 ± 0.006
CaO/CE	-	90	10	20 ± 0.6	0.11 ± 0.008
CuO/CE	50	-	50	34 ± 3.1	0.052 ± 0.004

* Average of three measurements plus or minus the standard error.

2.2. Characterization

Physisorption analysis with N_2 adsorption was carried out at −196 °C using a Tristar 3000 adsorption instrument (Micromeritics Instrument Corporation, Norcross, GA, USA). Prior to analysis, the received samples were degassed under vacuum at 300 °C for 3 h. The reported values are the averages of measurements on three different samples.

Scanning electron microscopy (SEM) images were obtained using a Quanta 250 FEG field emission scanning electron microscope (FEI Company, Hillsboro, OR, USA). This instrument also had energy-dispersive X-ray analysis capabilities for element detection and mapping. The samples were not sputter coated prior to imaging analysis because surface charging was minimal.

2.3. *CaL/CLC Cyclic Reactivity Measurement*

The performance of the pellets was studied using two different thermogravimetric analyzer (TGA) instruments. CaL/CLC experiments without steam were performed on a Cahn Thermax 500 instrument (Thermo Fisher Scientific Inc., Rockford, IL, USA) at the University of Calgary; while experiments with steam were performed using a Cahn 1000 TGA at Canmet Energy National Laboratory, Ottawa. Sample sizes of approximately 25–30 mg were used for each experiment, and the gas flow rates were 300 mL/min.

According to their composition, the pellets were cycled as shown in Figure 2. Each cycle of CaL and CLC (for 50% CuO/CaO/CE samples) consisted of the following steps: (1) oxidation in flowing air (Praxair Inc., Danbury, CT, USA) at 900 °C for 10 min; (2) carbonation in flowing of CO_2 (20%, balance N_2, Praxair) at 650 °C to simulate a coal-fired flue gas environment for 15 min; (3) calcination in flowing N_2 (Praxair, 99.999%) while heating from 650 °C to 875 °C; and (4) reduction in flowing CH_4 (Praxair, 99.7%) at 875 °C for 10 min. The standard heating and cooling rates were 15 °C/min and 10 °C/min, respectively. Experiments with the CaO/CE samples required only calcination and carbonation steps while those with the CuO/CE samples required only oxidation and reduction steps, all at the conditions given for the integrated pellets (as shown in Figure 2b,c). A maximum of 16 cycles was performed with the CuO/CaO/CE samples while 20 cycles were performed with the CaO/CE and CuO/CE samples according to the limits of the TGA software (Thermal Analyst, Version 1.3.2.2). Typical TGA data for each of the samples are given in Figure 2. Each reported experiment was performed three times to check the reproducibility, and determine standard errors.

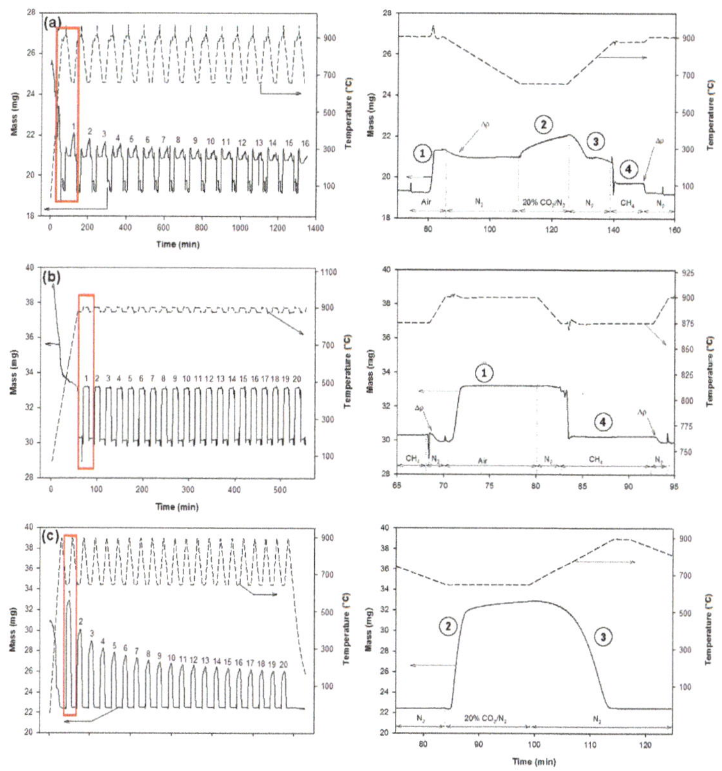

Figure 2. Mass (left axis) and temperature (right axis) profiles recorded by TGA during multiple CaL and/or CLC cycles for (**a**) CaO/CuO/CE pellets; (**b**) CuO/CE pellets; and (**c**) CaO/CE pellets. The graphs in the right column are magnifications of the highlighted sections of each respective graph. ① oxidation; ② carbonation; ③ calcination; ④ reduction. Δρ are differences of buoyancies due to either the change of temperature or change of gas.

The recorded mass change during carbonation was attributed only to $CaCO_3$ formation as given in Equation (1), such that the CO_2 capacity and conversion for carbonation, were calculated as follows:

$$CO_2\ \text{Capacity}\left(\frac{gCO_2}{gCaO}\right) = \frac{m_{carb} - m_{oxid}}{m_{oxid}} \tag{4}$$

$$\%\text{carbonation} = \frac{m_{carb} - m_{oxid}}{m_{oxid}y_{CaO}\left(\frac{MW_{CaCO_3}}{MW_{CaO}} - 1\right)} \cdot 100 \tag{5}$$

where m_{carb} is the mass of carbonated pellets, m_{oxid} is the mass of oxidized pellets, y_{CaO} is the mass fraction of CaO in the pellets, $MW_{CaCO_3}$is the molecular weight of $CaCO_3$ and MW_{CaO} is the molecular weight of CaO.

To test the effects of thermal pretreatment, some pellets were heated in an ex-situ reactor at rates of 1.5 °C/min or 7.5 °C/min to 900 °C in either flowing N_2 or air (50 mL/min) and then held for 2 h. After 2 h, the samples were cooled in flowing N_2 (50 mL/min). In these cases, the heating rate and atmosphere have been appended to the sample names. For example, CaO/CE-1.5N_2 refers to CaO/CE pellets heated at 1.5 °C/min in N_2.

3. Results and Discussion

The surface areas and pore volumes of the pellets are small (Table 1). The CuO/CE pellets had the highest surface area of 34 m^2/g, while the CaO/CE pellets had the highest pore volume of 0.11 cm^3/g. These results are consistent with those reported previously [26]. The higher proportion of cement, which contains high surface area alumina [18] in the CuO/CE pellets is consistent with their higher surface area.

3.1. *Performance of Pellets over Multiple CaL and/or CLC Cycles*

Figure 3 compares the performance of the CuO/CaO/CE, CaO/CE, and CuO/CE pellets. The CaL performance is shown based on both the CO_2 capacity Equation (4), which is the amount of CO_2 captured per theoretical amount of CaO ($gCO_2/gCaO$, Figure 3a), and the percent carbonation, Equation (5), which is the actual amount of CO_2 divided by the theoretical amount of CO_2 capacity based on the carbonizable/available CaO (Figure 3b); while, the CLC performance is shown in terms of the O_2 capacity ($gO_2/gCuO$, Figure 3c) and % oxidation (Figure 3d).

For both CuO/CaO/CE and CuO/CE pellets, the oxidation capacity remained essentially constant –0.22 $gO_2/gCuO$ (Figure 3c) and ~90% oxidation (Figure 3d)—over multiple cycles, suggesting its suitability as an oxygen carrier under the conditions of the tests. Using similar pellets, Rahman *et al.* [4] observed a ~15% loss in oxygen conversion after 15 cycles with reduction with syngas (10% H_2 and 20% CO) in pre-combustion CO_2 capture. In contrast, the capacity for CO_2 (carbonation) decreased with increasing cycle number (Figure 3a,b) and the CO_2 capacity was significantly lower for the CuO/CaO/CE pellets than for the CaO/CE pellets. More specifically, the initial CO_2 capacity of the CuO/CaO/CE pellets was approximately 25% of that of the CaO/CE pellets. Given that the capacity has been normalized to the amount of CaO in each type of pellets, this result indicates that the presence of CuO is hindering the accessibility of the CaO. The carbonation capacity of the CuO/CaO/CE pellets decreased exponentially, losing more than 50% capacity after 6 cycles and more than 70% capacity after 16 cycles. Similarly, the carbonation capacity of the CaO/CE pellets decreased exponentially with a loss of 66% after 20 cycles. The loss in capacity for CaO/CE was similar to that of the CuO/CaO/CE pellets comparing at 16 cycles (~64%). Fitting the carbonation data to a two-parameter exponential decay equation gave decay constants of 0.5 and 0.2 for the CaO/CE and CuO/CaO/CE pellets, respectively. Thus, the decay rate of the CaO/CE pellets was approximately 2.5 times higher than that of the CuO/CaO/CE pellets. The initial CO_2 capacity of the former pellets, however, was 4 times higher than that of the latter pellets, and so there was more capacity to lose.

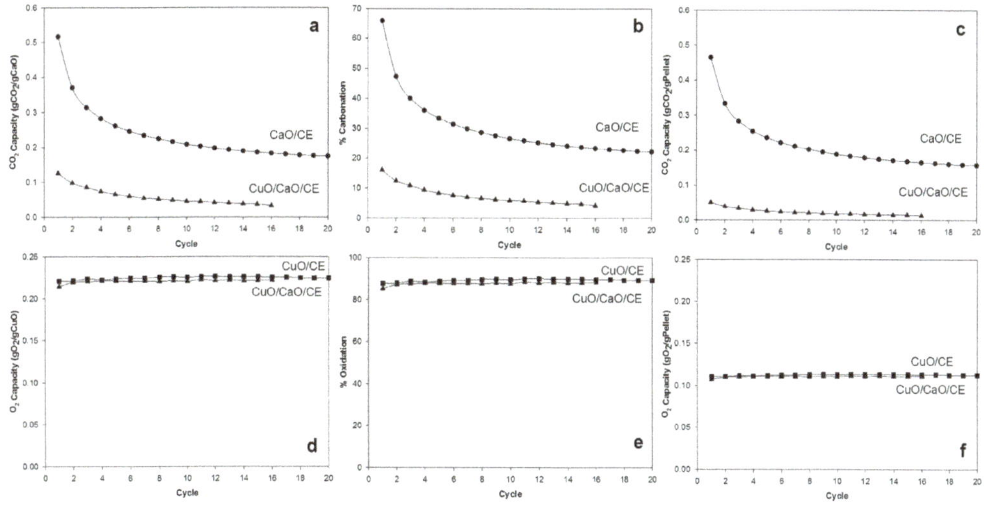

Figure 3. Performance of CaO/CuO/CE, CuO/CE, and CaO/CE pellets over multiple cycles in terms of (**a**) CO_2 capacity per gram of CaO; (**b**) % Carbonation; (**c**) CO_2 capacity per gram of pellets; (**d**) O_2 capacity per gram of CuO; (**e**) % Oxidation; and (**f**) O_2 capacity per gram of pellets.

A decrease in the CO_2 carrying capacity of the pellets with cycling has been observed previously, and sintering of Cu has been suggested as the main cause for the deactivation [3,9–11]. To verify that Cu is sintering, the pellets were examined with SEM/EDX as shown in Figure 4. The images shown are typical of all pellets without pretreatment. The "wrapping effect" of CuO over CaO with cycling is evident in these images, which shows the migration of Cu to the surface of the pellets with increased cycling. The melting points of Cu and CuO are 1084 °C and 1336 °C, respectively [16]. Although these melting points are several hundreds of degrees above the calcination/reduction (875 °C) and oxidation (900 °C) temperatures, the metal and metal oxide phases will soften and be mobile at the Tamman temperature of Cu and CuO (405 °C and 527 °C, respectively). Furthermore, the presence of $CaCO_3$ has been shown to lower the melting points of other minerals including Cu and CuO [16].

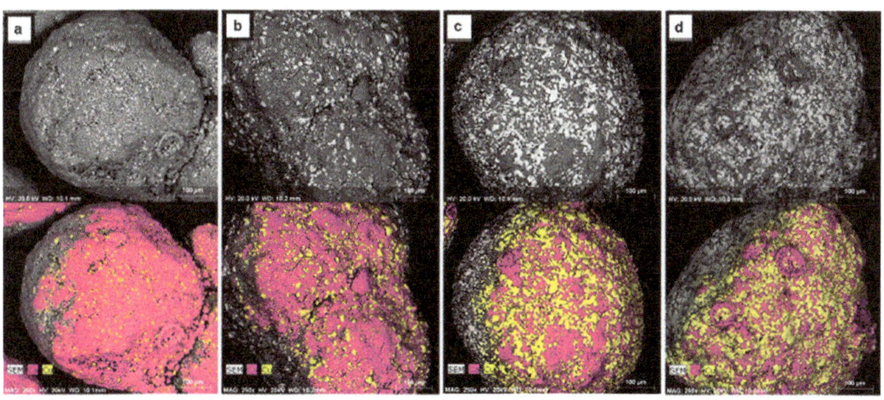

Figure 4. SEM micrographs with corresponding EDX maps of CaO/CuO pellets after (**a**) 0; (**b**) 5; (**c**) 10; and (**d**) 15 cycles of CaL and CLC. The elements Ca and Cu are represented in purple and yellow, respectively.

3.2. Effect of Thermal Pre-Treatments on CO_2 Capacity

The CO_2 capacity was higher in the first cycle for the untreated CaO/CE pellets than for CaO/CE pellets that were thermally treated (Figure 5). After three cycles, however, all thermally treated pellets had higher capacities, suggesting that the thermal treatment stabilized the pellets. At the end of 20 cycles, the capacity of the untreated CaO/CE pellets was 0.17 $gCO_2/gCaO$ which is less than 0.22 $gCO_2/gCaO$ for the pellets treated at 1.5 °C/min in either air or N_2 and 0.23 $gCO_2/gCaO$ for the pellets heated at the higher rate of 7.5 °C/min in N_2. The corresponding percent decreases in capacity over 20 cycles for these pellets were 65%, 53%, 51% and 46%, respectively. Since there were no noticeable differences in the results based on the use of air or N_2 as a medium during the thermal pretreatment, it can be concluded that oxygen does not exert any significant effect on the state of the pellets.

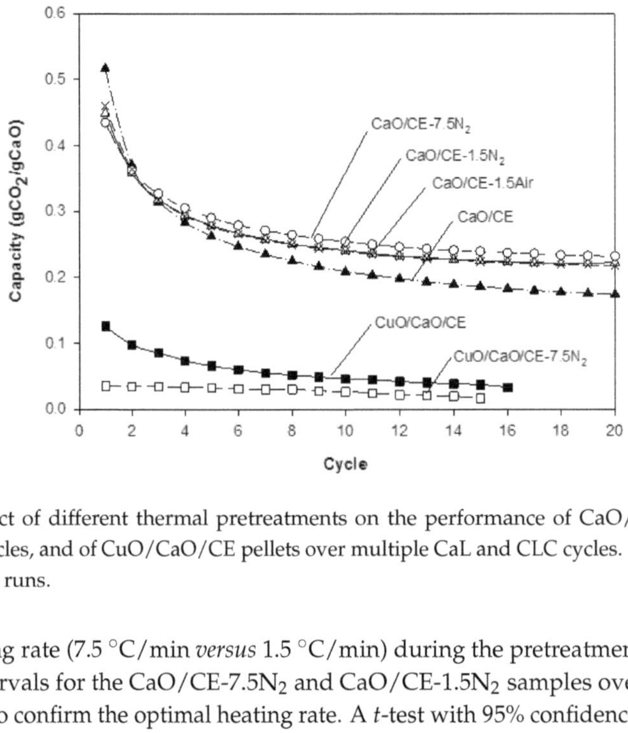

Figure 5. Impact of different thermal pretreatments on the performance of CaO/CE pellets over multiple CaL cycles, and of CuO/CaO/CE pellets over multiple CaL and CLC cycles. Each curve is the average of three runs.

A faster heating rate (7.5 °C/min *versus* 1.5 °C/min) during the pretreatment may be beneficial; the confidence intervals for the CaO/CE-7.5N_2 and CaO/CE-1.5N_2 samples overlap and so further testing is required to confirm the optimal heating rate. A *t*-test with 95% confidence interval confirmed a significant improvement of CaO/CE-7.5N_2 over CaO/CE at the 20th cycle. Other authors have used different temperatures and holding times [10,27] than those in the study herein (*i.e.*, 900 °C and 2 h) but also saw an improvement in the stability of the CaO after thermal treatment. SEM micrographs of the CaO/CE pellets as received, after heating at 7.5 °C/min but before use, and after heating at 7.5 °C/min and 20 cycles are shown in Figure 6. There is a distribution of particles sizes in the as-received pellets (Figure 6a). The pellets heat-treated at 7.5 °C/min in flowing N_2 had evidence of fractures /fissures on most pellets (Figure 6b). Although these fractures will be beneficial for diffusion, the particles will be less mechanically stable within the process. The CaO/CE-7.5N_2 pellets in Figure 6c (after 20 cycles) were obtained at the end of a cycle (*i.e.*, after carbonation) and so these pellets are larger than those in Figure 6b (at 0 cycles).

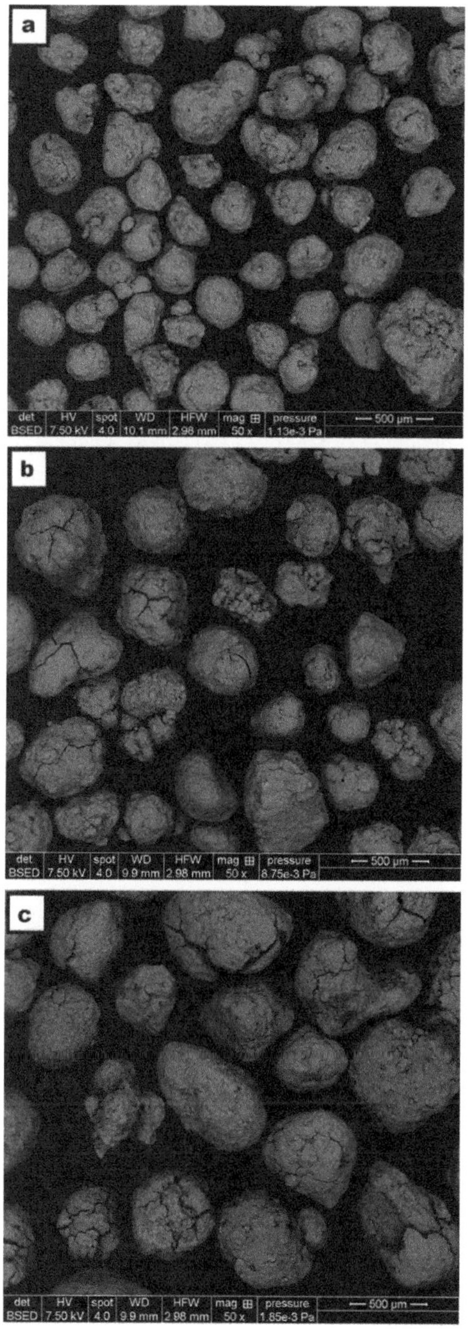

Figure 6. SEM micrographs of (**a**) CaO/CE as received; (**b**) CaO/CE-7.5N_2 at 0 cycle; and (**c**) CaO/CE-7.5N_2 at 20 cycles (carbonated).

For the CuO/CaO/CE pellets, a thermal pretreatment with a heating rate of 7.5 °C/min in flowing N_2 was tested because this treatment gave the best results for the CaO/CE pellets. For the integrated pellets, the thermal treatment resulted in more stable performance but the CO_2 capacity was less than that of the untreated CuO/CaO/CE pellets (Figure 5). The additional heat treatment may have enhanced the sintering of the Cu. When applied before mixing with the CaO, heat treatment has been shown to improve the carbonation stability over 40 cycles [17]. Specifically, Qin *et al.* [17] used a heat pretreatment (800–1000 °C, 0.5–8 h) with the copper precursor (copper acetate) and observed a 50% increase in the carbonation capacity of their CuO/CaO/MgO pellets after 40 cycles. The authors suggested that this approach allowed the majority of the Cu/CuO components to form stable clusters that would not coat the CaO active surfaces during carbonation. Their wet impregnation method likely produced much smaller (nanometer-sized) Cu particles than the dry mixing of micron-sized CaO and CuO particles used herein.

3.3. Effect of Steam Addition on CO_2 Capacity

The addition of steam enhanced the performance of both the CaO/CE (Figure 7a) and CuO/CaO/CE pellets (Figure 7b). The capacity of the CaO/CE pellets after 15 cycles was 0.142 $gCO_2/gCaO$ without steam, 0.169 $gCO_2/gCaO$ with 10% steam, and 0.237 $gCO_2/gCaO$ with 30% steam. The results without steam and with 10% steam are not significantly different (*t*-test) even at the 15th cycle. With a higher level of steam (30%), there was a significant improvement in the performance of the CaO/CE pellets. At 30% steam, however, the carbonation conversion of CaO/CE pellets fluctuated between cycles. Five, instead of three, experiments were performed to confirm that this behavior was reproducible. The averages of the minimum and maximum capacities for each cycle in the five runs are plotted in Figure 7a. The inset graph shows representative data from one of the five runs. Even with the fluctuations, the capacity with 30% steam was always higher than with no steam. The reason for the fluctuations is not known.

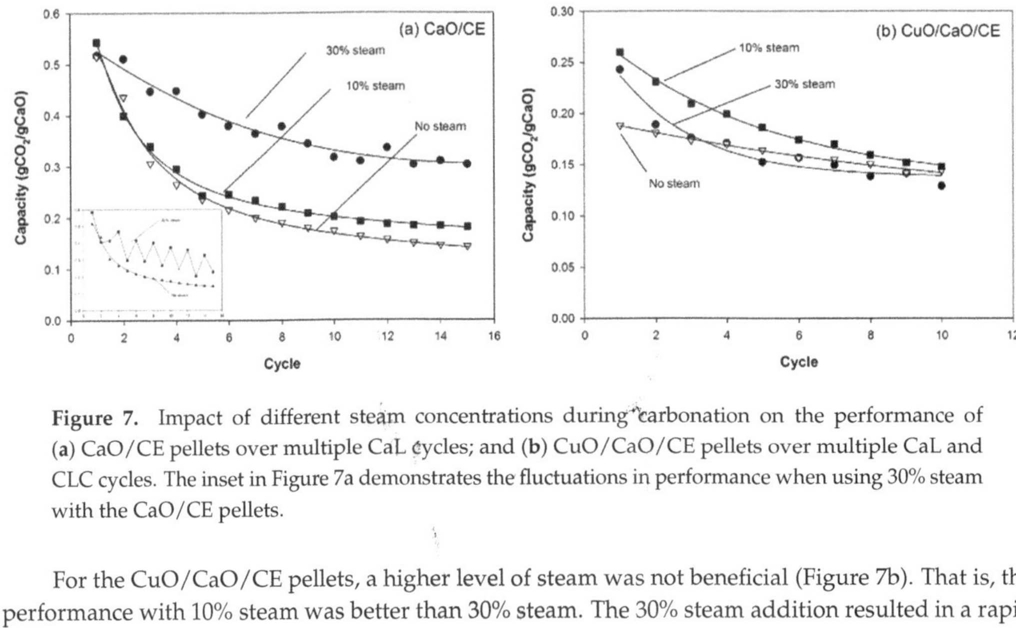

Figure 7. Impact of different steam concentrations during carbonation on the performance of (**a**) CaO/CE pellets over multiple CaL cycles; and (**b**) CuO/CaO/CE pellets over multiple CaL and CLC cycles. The inset in Figure 7a demonstrates the fluctuations in performance when using 30% steam with the CaO/CE pellets.

For the CuO/CaO/CE pellets, a higher level of steam was not beneficial (Figure 7b). That is, the performance with 10% steam was better than 30% steam. The 30% steam addition resulted in a rapid loss in capacity and after 4 cycles the CO_2 capacities were approximately 0.170 $gCO_2/gCaO$ with 30% steam or without steam. A fluctuation pattern similar to the experiment with the CaO/CE pellets and 30% steam, but with a smaller amplitude, was observed.

Figure 8 compares the carbonation of CuO/CaO/CE pellets in the first cycle with and without steam over 20 min. During carbonation, it has been suggested that there is a transition from kinetic to diffusion control, which is evident in the change in slope at approximately 1 min. The rates of carbonation are higher in the presence of steam, which supports the theory that steam enhances solid-state diffusion [19]. Unlike typical CaO based pellets which exhibited higher carbonation conversion with increasing steam concentration [19,25], the carbonation conversion of the CuO/CaO/CE pellets studied herein was higher with 10% steam than 30% steam (Figures 7b and 8). This result could be due to side reactions catalyzed by the presence of Cu, which inhibit the carbonation of CaO.

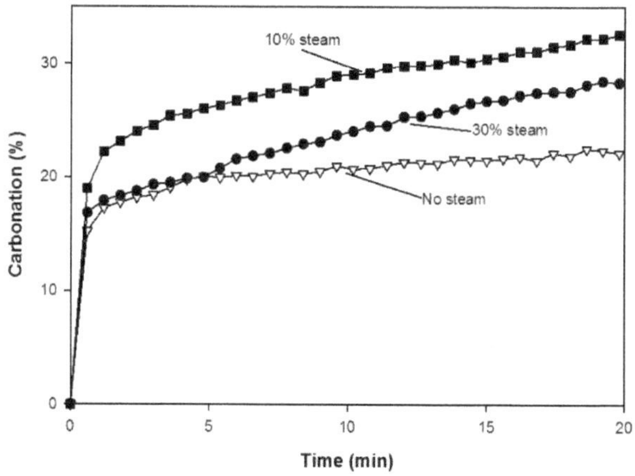

Figure 8. Comparison of carbonation conversion (%) over one cycle for CuO/CaO/CE pellets with 0, 10% and 30% steam.

4. Conclusions

In this work, the performance of integrated CuO/CaO/CE pellets was evaluated against CaO/CE and CuO/CE pellets in a multi-cycle CaL and CLC process. The oxidation capacity of the integrated pellets remained constant and on par with the CuO pellets after 16 cycles. However, the carbonation capacity was significantly decreased with the addition of CuO to the pellets. SEM/EDX images showed an increasing surface enrichment of Cu on the integrated pellets with cycle number. Thermal pretreatments with either air or N_2 stabilized the performance of the CaO/CE pellets but also caused fractures. Pretreatment with heating at 7.5 °C/min also stabilized the CuO/CaO/CE pellets but reduced the CO_2 capacity and so was not beneficial. The addition of steam improved the performance of both CaO/CE and CuO/CaO/CE pellets. For the CaO/CE pellets, the initial CO_2 capacity was unchanged by the addition of steam but there was significantly less loss in capacity with 30% steam, while only a minor improvement with 10% steam. In contrast, the initial CO_2 capacity was enhanced for the CuO/CaO/CE pellets in the presence of steam but after 10 cycles the capacities were comparable with or without steam. The performance of the integrated pellets was better in 10% steam than 30% steam.

Acknowledgments: The authors acknowledge funding from Carbon Management Canada (Project A346).

Author Contributions: All authors collaborated on the design of the experiments for this study as part of a collaborative project entitled "Pre- and post-combustion CO_2 capture using novel composite CaO/CuO sorbents". Alvaro Recio and Sip Chen Liew carried out the TGA experiments and characterization of the pellets; Dennis Lu, Ryad Rahman, and Arturo Macchi assisted Alvaro Recio with the experiments in the presence of steam; Ryad Rahman prepared the pellets; Alvaro Recio, Sip Chen Liew and Josephine M. Hill prepared the manuscript with input and feedback from Dennis Lu, Ryad Rahman, and Arturo Macchi.

Conflicts of Interest: The authors declare no conflict of interest.

References

1. Li, L.; Zhao, N.; Wei, W.; Sun, Y. A review of research progress on CO_2 capture, storage, and utilization in Chinese academy of sciences. *Fuel* **2013**, *108*, 112–130. [CrossRef]
2. Adanez, J.; Abad, A.; Garcia-Labiano, F.; Gayan, P.; de Diego, L.F. Progress in chemical-looping combustion and reforming technologies. *Prog. Energy Combust. Sci.* **2012**, *38*, 215–282. [CrossRef]

3. Manovic, V.; Anthony, E.J. Integration of calcium and chemical looping combustion using composite CaO/CuO-based materials. *Environ. Sci. Technol.* **2011**, *45*, 10750–10756. [CrossRef] [PubMed]
4. Rahman, R.A.; Mehrani, P.; Lu, D.Y.; Anthony, E.J.; Macchi, A. Investigating the use of CaO/CuO sorbents for *in situ* CO_2 capture in a biomass gasifier. *Energy Fuels* **2015**, *29*, 3808–3819. [CrossRef]
5. Butler, J.W.; Lim, C.J.; Grace, J.R. CO_2 capture capacity of CaO in long series of pressure swing sorption cycles. *Chem. Eng. Res. Design* **2011**, *89*, 1794–1804. [CrossRef]
6. Rout, K.R.; Fermoso, J.; Chen, D.; Jakobsen, H.A. Kinetic rate of uptake of a synthetic Ca-based sorbent: Experimental data and numerical simulations. *Fuel* **2014**, *120*, 53–65. [CrossRef]
7. Blamey, J.; Anthony, E.J.; Wang, J.; Fennell, P.S. The calcium looping cycle for large-scale CO_2 capture. *Prog. Energy Combust. Sci.* **2010**, *36*, 260–279. [CrossRef]
8. Abanades, J.C.; Murillo, R.; Fernandez, J.R.; Grasa, G.; Martínez, I. New CO_2 capture process for hydrogen production combining Ca and Cu chemical loops. *Environ. Sci. Technol.* **2010**, *44*, 6901–6904. [CrossRef] [PubMed]
9. Kazi, S.S.; Aranda, A.; Meyer, J.; Mastin, J. High performance CaO-based sorbents for pre- and post-combustion CO_2 capture at high temperature. *Energy Procedia* **2014**, *63*, 2207–2215. [CrossRef]
10. Manovic, V.; Anthony, E.J. Lime-based sorbents for high-temperature CO_2 capture—A review of sorbent modification methods. *Int. J. Environ. Res. Public Health* **2010**, *7*, 3129–3140. [CrossRef] [PubMed]
11. Manovic, V.; Wu, Y.; He, I.; Anthony, E.J. Core-in-shell CaO/CuO-based composite for CO_2 capture. *Ind. Eng. Chem. Res.* **2011**, *50*, 12384–12391. [CrossRef]
12. Mattisson, T.; Järdnäs, A.; Lyngfelt, A. Reactivity of some metal oxides supported on alumina with alternating methane and oxygen application for chemical-looping combustion. *Energy Fuels* **2003**, *17*, 643–651. [CrossRef]
13. García-Labiano, F.; de Diego, L.F.; Adánez, J.; Abad, A.; Gayán, P. Temperature variations in the oxygen carrier particles during their reduction and oxidation in a chemical-looping combustion system. *Chem. Eng. Sci.* **2005**, *60*, 851–862. [CrossRef]
14. Leion, H.; Lyngfelt, A.; Mattisson, T. Solid fuels in chemical-looping combustion using a NiO-based oxygen carrier. *Chem. Eng. Res. Design* **2009**, *87*, 1543–1550. [CrossRef]
15. Duhoux, B. Combined Calcium Looping and Chemical Looping Combustion Process Simulation Applied to CO_2 Capture. M.Sc. Thesis, University of Ottawa, Ottawa, NO, Canada, 2015.
16. Qin, C.; Yin, J.; Liu, W.; An, H.; Feng, B. Behavior of CaO/CuO based composite in a combined calcium and copper chemical looping process. *Ind. Eng. Chem. Res.* **2012**, *51*, 12274–12281. [CrossRef]
17. Qin, C.; Yin, J.; Luo, C.; An, H.; Liu, W.; Feng, B. Enhancing the performance of CaO/CuO based composite for CO_2 capture in a combined Ca-Cu chemical looping process. *Chem. Eng. J.* **2013**, *228*, 75–86. [CrossRef]
18. Manovic, V.; Fennell, P.S.; Al-Jeboori, M.J.; Anthony, E.J. Steam-enhanced calcium looping cycles with calcium aluminate pellets doped with bromides. *Ind. Eng. Chem. Res.* **2013**, *52*, 7677–7683. [CrossRef]
19. Manovic, V.; Anthony, E.J. Carbonation of CaO-based sorbents enhanced by steam addition. *Ind. Eng. Chem. Res.* **2010**, *49*, 9105–9110. [CrossRef]
20. Champagne, S.; Lu, D.Y.; Macchi, A.; Symonds, R.T.; Anthony, E.J. Influence of steam injection during calcination on the reactivity of CaO-based sorbent for carbon capture. *Ind. Eng. Chem. Res.* **2013**, *52*, 2241–2246. [CrossRef]
21. Dobner, S.; Sterns, L.; Graff, R.A.; Squires, A.M. Cyclic calcination and recarbonation of calcined dolomite. *Ind. Eng. Chem. Process Design Dev.* **1977**, *16*, 479–486. [CrossRef]
22. Yang, S.; Xiao, Y. Steam catalysis in cao carbonation under low steam partial pressure. *Ind. Eng. Chem. Res.* **2008**, *47*, 4043–4048. [CrossRef]
23. Symonds, R.T.; Lu, D.Y.; Macchi, A.; Hughes, R.W.; Anthony, E.J. CO_2 capture from syngas via cyclic carbonation/calcination for a naturally occurring limestone: Modelling and bench-scale testing. *Chem. Eng. Sci.* **2009**, *64*, 3536–3543. [CrossRef]
24. Dou, B.; Song, Y.; Liu, Y.; Feng, C. High temperature CO_2 capture using calcium oxide sorbent in a fixed-bed reactor. *J. Hazard. Mater.* **2010**, *183*, 759–765. [CrossRef] [PubMed]
25. Arias, B.; Grasa, G.; Abanades, J.C.; Manovic, V.; Anthony, E.J. The effect of steam on the fast carbonation reaction rates of CaO. *Ind. Eng. Chem. Res.* **2012**, *51*, 2478–2482. [CrossRef]

26. Xu, L.; Sithambaram, S.; Zhang, Y.; Chen, C.-H.; Jin, L.; Joesten, R.; Suib, S.L. Novel urchin-like CuO synthesized by a facile reflux method with efficient olefin epoxidation catalytic performance. *Chem. Mater.* **2009**, *21*, 1253–1259. [CrossRef]
27. Valverde, J.M.; Sanchez-Jimenez, P.E.; Perez-Maqueda, L.A. High and stable CO_2 capture capacity of natural limestone at Ca-looping conditions by heat pretreatment and recarbonation synergy. *Fuel* **2014**, *123*, 79–85. [CrossRef]

13

Development of a CMOS Route for Electron Pumps to Be Used in Quantum Metrology

Sylvain Barraud [1], Romain Lavieville [1], Louis Hutin [1], Heorhii Bohuslavskyi [1], Maud Vinet [1], Andrea Corna [2], Paul Clapera [2], Marc Sanquer [2] and Xavier Jehl [2,*]

[1] Laboratoire d'électronique des technologies de l'information (LETI), Université Grenoble Alpes and CEA, Minatec Campus, F-38054 Grenoble, France; sylvain.barraud@cea.fr (S.B.); r.lavieville@gmail.com (R.L.); louis.hutin@cea.fr (L.H.); heorhii.bohuslavskyi@cea.fr (H.B.); maud.vinet@cea.fr (M.V.)

[2] Institut Nanosciences et Cryogénie (INAC), Université Grenoble Alpes and CEA, F-38054 Grenoble, France; andrea.corna@cea.fr (A.C.); paul.clapera@gmail.com (P.C.); marc.sanquer@cea.fr (M.S.)

* Correspondence: xavier.jehl@cea.fr

Academic Editor: Alexander V. Sergienko

Abstract: The definition of the ampere will change in the next few years. This electrical base unit of the S.I. will be redefined by fixing the value of the charge quantum, *i.e.*, the electron charge e. As a result electron pumps will become the natural device for the *mise en pratique* of this new ampere. In the last years semiconductor electron pumps have emerged as the most advanced systems, both in terms of speed and precision. Another figure of merit for a metrological device would be its ability to be predictible and shared. For that reason a mature fabrication process would certainly be an advantage. In this article we present electron pumps made within a CMOS (Complementary Metal Oxide Semiconductor) research facility on 300 mm silicon-on-insulator wafers, using advanced microelectronics tools and processes. We give an overview of the whole integration scheme and emphasize the fabrication steps which differ from the normal CMOS route.

Keywords: electron pumps; electrical quantum metrology; microelectronics; MOSFETs

---

1. Introduction

In this article we present a CMOS route for the fabrication of single electron pumps using advanced multi-gate Fully-Depleted Silicon on Insulator (FDSOI) technology. The motivation arises from the upcoming redefinition of the only electrical base unit in the S.I., the ampere, expected in 2018. Following the trend initiated by atomic physics with the definition of the second, the aim is to use a quantum effect instead of the electro-mechanical definition of the ampere which is still in use today although it cannot be realized with sufficient accuracy. Instead, National Metrology Institutes (NMIs) across the world use daily the quantum Hall and Josephson effects to realize an ohm and a volt with very high reproducibility, although these are not the S.I. volt and ohm. These two quantum effects rely on measured constants (more precisely the von Klitzing constant $R_{K-90} = 25812.807\,\Omega$ and the Josephson constant $K_{J-90} = 483597.9 \times 10^9$ Hz·V^{-1}) which in theory depend only on the two fundamental constants e and h. Just like fixing the speed of light allowed to define the meter, fixing e will allow to redefine the ampere as a number of electrical charges per unit of time. While research in NMIs is very active to determine the numerical values of e and h with the highest precision, there will still be a need after the redefinition for the *mise en pratique* (ie realization) of the new, quantum definition of the ampere. Electron pumps are the natural devices to achieve this goal. These quantized current sources deliver a dc current I given by $I = nef$, with n an integer, and f a frequency used to force the transfer of exactly n charges during each cycle. These quantized currents should be robust (insensitive) to changes of driving parameters (bias voltages, amplitude of the ac drive for instance) over a finite and large enough range. Considering the value of e, the relation $I = ef$ means a current of approximately 0.16 pA/MHz, which implies to work at relatively high frequencies to generate a current high enough to be exploited. Currently the goal is to reach at least 100 pA (roughly 600 MHz) with an absolute precision better than 1 ppm. Recently NMIs and academic laboratories have focused on electron pumps made with GaAs-based semiconducting nanostructures operated in the non-adiabatic regime [1–4]. Silicon has emerged also as an interesting alternative [5,7,8]. All devices make use of the Coulomb blockade phenomenon which allows to control single electrons by an electric field. A consequence however is the necessity to work at cryogenic temperatures, typically below 0.5 K for the most advanced metrological experiments. Although this is not a problem for a niche application such as quantum electrical metrology, it can set a limit for other applications of pumps. Indeed in microelectronics having a well controlled and compact current source could be of great interest for adiabatic computation for instance. Compared to individual transistors, pumps can be regarded as active circuit components since suitably applied gate voltages generate well defined currents irrespective of bias voltages, rather than simply switching them. However this would be relevant only at higher temperatures, at least above 2 or 3 K, where cryogen-free machines based on pulse-tubes can be used.

At CEA-Grenoble we developped a CMOS route for the fabrication of electron pumps [5] using fully depleted SOI technology on 300 mm wafers. Here we review the main fabrication steps and challenges.

2. Samples Requirements: 65 nm Gate Pitch

Our building block for electron pumps is a small metallic island isolated by two field-effect transistor channels acting as tunable barriers. The total capacitance C of the island defines the charging energy $E_c = e^2/2C$ which corresponds to the electrostatic Coulomb repulsion to compensate to add an extra charge onto the island. This energy becomes relevant provided that the temperature T is small enough (thermal energy $k_B T \ll E_c$, with k_B the Boltzmann constant) and if the charge on the island is a good quantum number, which happens in practice when the barriers resistance are large enough, at least larger than the resistance quantum $\frac{h}{e^2} \approx 25.8\,\mathrm{k\Omega}$. The basic design to realize this system is a nanowire transistor with two gates in series, as presented in [5]. In principle this device can be realized with a full-CMOS process, using only optical (deep UV, 193 nm wavelength) lithography. With the scanner currently used at CEA-Leti this implies a gate pitch of 190 nm, as illustrated in Figure 1. While recent results have shown that this distance can allow for electron pumping at temperatures below 100 mK (Clapera *et al.*, unpublished [6]), it is crucial to obtain a smaller spacing, hence a smaller island having a larger charging energy. For that reason and despite its inherently low throughput for mainstream CMOS production we have developped a custom electron-beam lithography step for the gate level.

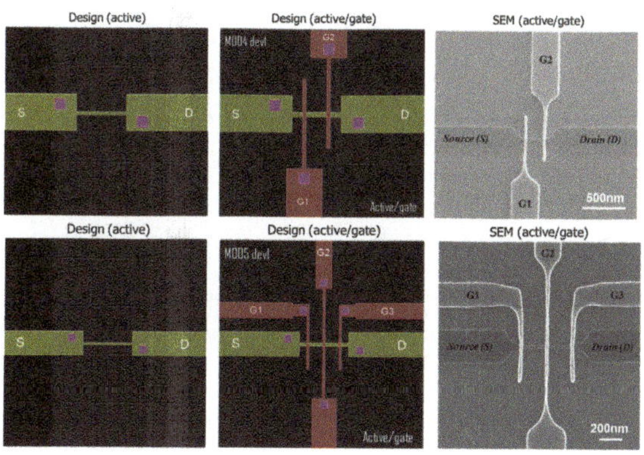

Figure 1. Two-gate and three-gate devices realized with deep-UV lithography. **Left**: mask design with the nanowire (yellow) and its Source (S) and Drain (D) extremities, and the two and three gates in red; **Right**: Scanning electron micrograph of actual devices, with a gate pitch of 190 nm.

3. Electron Pumps Layout and Processing with FDSOI Technology

3.1. Active Area Patterning

[110]-oriented Nanowire (NW) field-effect-transistors (N- and P-MOS) with high-k/metal gate stack were fabricated on (100) SOI wafers with BOX (buried oxide) thickness of 145 nm. The silicon film thickness is approximately 13 nm. The silicon layer is patterned to create NWs by the mesa isolation technique, *i.e.*, each transistor is isolated from each other with the SiO_2 buried oxide of the SOI substrate. NW patterns are defined by deep-UV (193nm) optical lithography and followed by a resist trimming process in order to shrink the dimensions of nanowires. Nanowire width down to 10 nm can be achieved after active patterning [9]. In Figures 2–4, we show the active area of various structures, highlighted in green. Undoped or channel doping can be used.

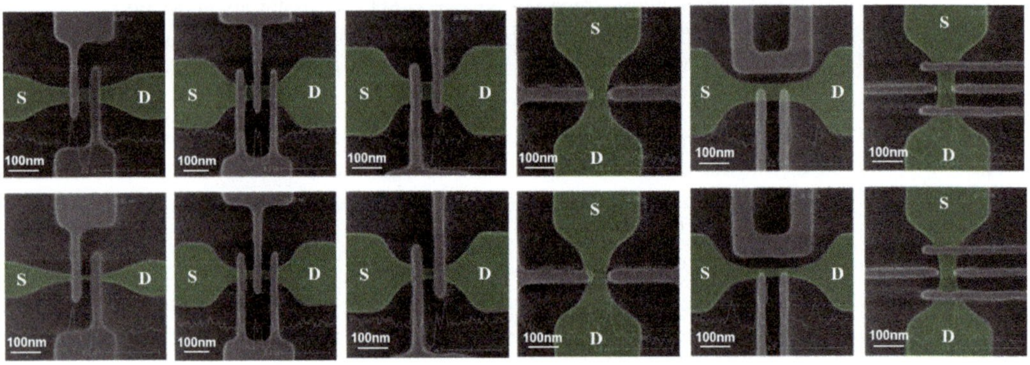

Figure 2. Examples of samples after gate patterning. Active (resp. gate) area is represented in green (grey) color. Multi-gate devices are realized with a 65 nm gate pitch.

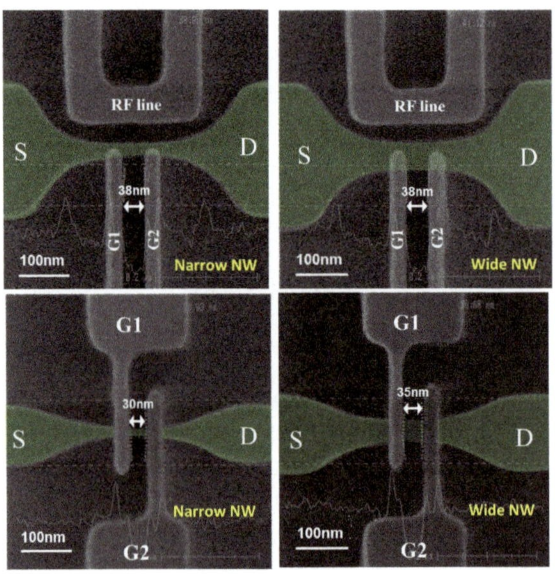

Figure 3. Examples of two-gate devices after gate patterning for narrow (15 nm) and wide (40 nm) nanowire width. Active (resp. gate) area is represented in green (grey) color. **Top**: Scanning electron micrograph of two-gate device. The two gates partially overlap the active area and a RF line is located in the vicinity of the Si islands; **Bottom**: Scanning electron micrograph of two-gate device used for electron pumps. The minimal spacing between gates is 30 nm.

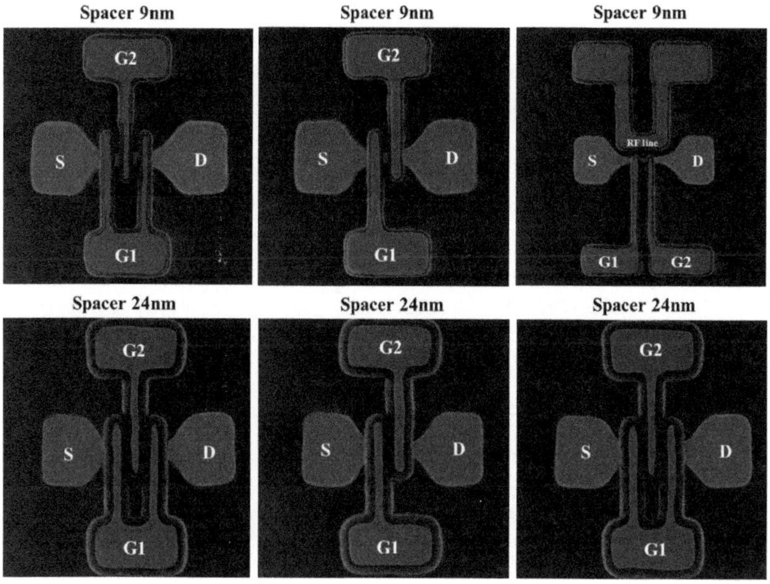

Figure 4. Examples of samples after spacer patterning. Multi-gate devices are realized with a 65 nm gate pitch. Active (resp. gate) area is represented in green (grey) color. **Top**: a thin spacer (9 nm) is used in order to create a silicide area between the two gates; **Bottom**: a thick spacer (24 nm) is used in order to avoid salicidation between the two gates.

3.2. *Gate Patterning with a 65 nm Pitch*

After the active area patterning a high-k/metal gate is deposited. The gate stack consists in a SiO_2 oxide of 2.5 nm, 2 nm CVD HfO_2 (high-k dielectric), 5 nm TiN (metal gate) made by atomic layer deposition and Poly-Silicon (50 nm) layers. The gate is wrapped around the channel like an omega-gate configuration. As for the active patterning, the same photo-resist trimming is used to achieve gate lengths down to 20 nm. Note that the 2.5 nm of SiO_2 is markedly larger than the usual value used for microelectronics applications, because we infer a large impact of remote Coulomb scattering in the gate stack. Hence pushing this interface further away could be an important point for low temperature applications. The benefits of e-beam gates is illustrated in Figures 2 and 3. First, for two gates in series the spacing can be as low as 35 nm, compared to 170 nm in Figure 5. Another advantage is the relatively straightforward integration of more complex designs such as 3 or 4-gate pumps or pumps with an RF antenna in the vicinity of the island. Note however that a nearby detector based on another active area was not implemented in this study. This would require advanced patterning for the active area layer [10] which is possible in principle but has not been used in this work.

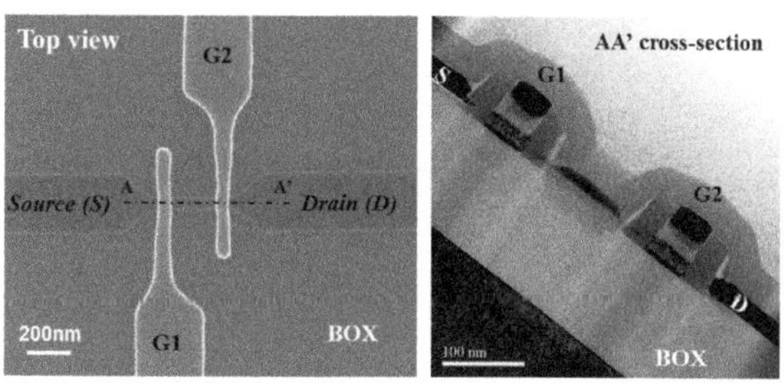

Figure 5. Two-gate device realized with deep-UV lithography. **Left**: Scanning electron micrograph of an actual device, with a gate pitch of 190 nm; **Right**: Transmission electron micrograph of two-gate device realized with wide spacer (40 nm).

Afterwards a nitride spacer is formed on the sidewalls of the gate. Spacers are self-aligned since they consist of a nitride layer anisotropicaly etched, thus remaining only where a significant step height is present, *i.e.*, at this stage all around the nearly 60 nm high gate stack. Spacer of thicknesses 9, 15 or 30 nm have been implemented (Figure 6). The thinner spacer allows creating a silicide area between the two gates in series, similar to the devices of "type C" in [11]. Then, low access resistances are realized by epitaxial silicon growth on the source and drain (T_{Si} =18 nm). A second offset spacer consisting of a tetraethyl orthosilicate (TEOS) liner and a nitride layer was fabricated prior to source/drain implantations, activation spike anneal and silicidation (NiPtSi silicide), in order to obtain low contact resistances.

Finally, tungsten contact and standard Cu back-end-of-the-line process flows were used, as illustrated in Figure 7. As a result and in contrast with all other electron pumps reported so far, the devices are encapsulated with an industrial process. This could become important when exchanging and comparing samples between NMIs becomes a reality. It also ensures straightforward compatibility with circuit design tools, in order to build more complex circuits on-chip with electron pumps [12].

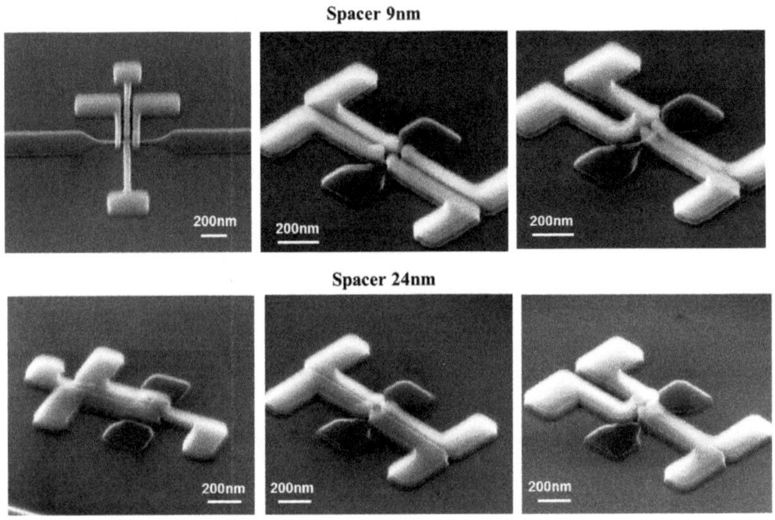

Figure 6. Examples of samples after Si source-drain epitaxy. Multi-gate devices are realized with a 65 nm gate pitch. **Top**: a thin spacer (9 nm) is used in order to create a silicide area between the two gates; **Bottom**: a thick spacer (24 nm) is used in order to avoid salicidation between the two gates.

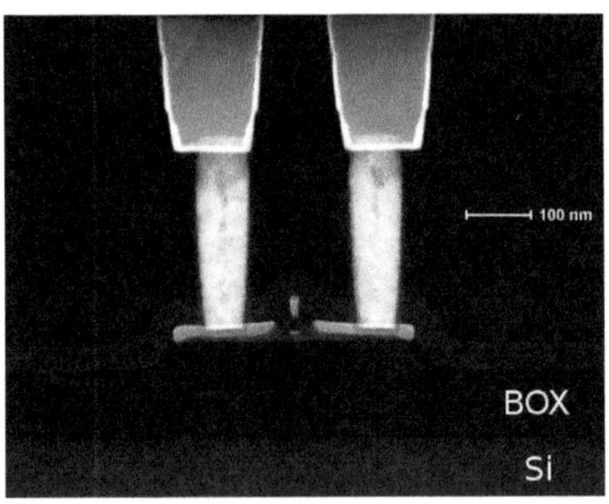

Figure 7. Transmission electron micrograph of single-gate device after back-end process. The device is at the center, and only the Source and Drain vertical vias and copper lines are visible. Tungsten contact and standard Cu back-end-of-the-line process are used, as usual for advanced CMOS transistors.

4. Electrical Characterizations at 300 K

The simple design of electron pumps relying on two NW FETs in series allows testing at 300 K, as illustrated in Figure 8. For this measurement a very small source-drain bias is applied (here 50 μV), in order to not saturate the high-gain (V/I of the order of 2×10^8) transimpedance amplifier used for subsequent measurements at low temperature. The maximum current measured (10 nA) thus corresponds to a resistance of 5 kΩ. The device shown on the left has a width of 50 nm and two gates of length 50 nm. For each backgate voltage a set of three curves is shown to illustrate the behaviour of each gate independently (green and blue curves, obtained by fixing one gate at +0.8 V while sweeping the other one) and the global behaviour ($V_{g1} = V_{g2}$, red curves). A reasonable electrostatic control is obtained despite the two-gate layout which is not optimal: a slope of 90 mV/decade is obtained in the sub-threshold region. Moreover the backgate has a strong effect on the curves, allowing to tune the access resistances to the central island [13]. This additional knob offered by SOI technology is important for operation at low temperature, in particular to get away from impurities like donor states which induce a perturbation on the current maps [5]. On the right of Figure 8 we show the characteristics of a device with thinner silicon layer (T_{Si} = 11 nm), narrower nanowire (W = 15 nm), 40 nm long gates and 30 nm spacing between them. It results in a nearly optimal slope of 68 mV/decade. The devices developped with our approach are now under systematic measurements to assess their metrological capabilities, in collaboration with NMIs. It is important to emphasize that the aggressive pitch (65 nm) we reached with our electron beam lithography gate level is now accessible with the most advanced industrial process [14] (14 nm node).

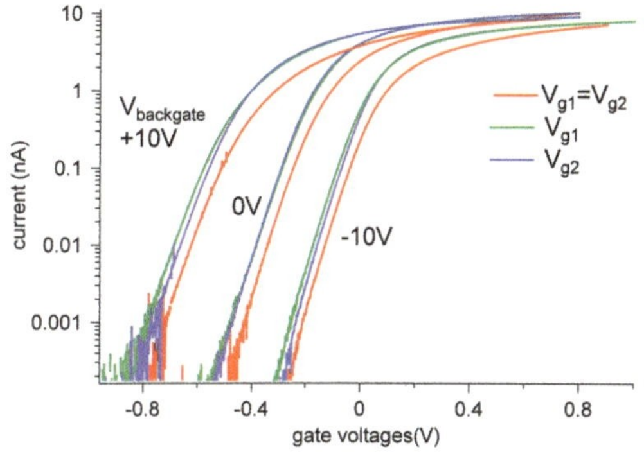

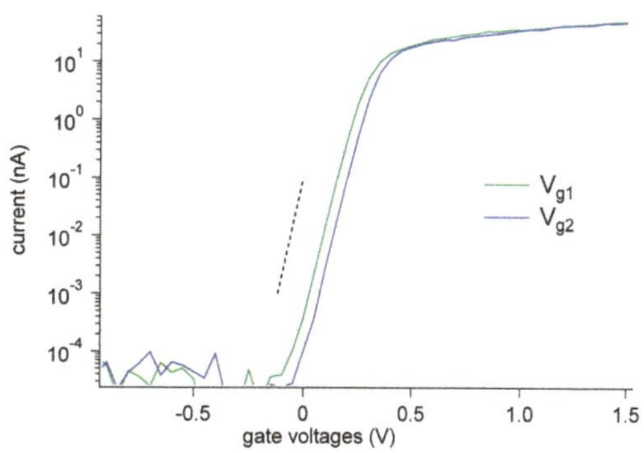

Figure 8. Left: Drain-Source current versus gate voltages at 300 K and at three values of the backgate voltage in a two-gate electron pump. For the red curves the two gates are swept together. For the blue and green curves one gate is swept while the other one is kept at +0.8 V. The sub-threshold slope measured here is ≈ 90 mV/decade. The degradation compared to the ideal value of 60 mV/decade at 300 K is mostly due to the thick silicon layer (T_{Si} = 20 nm); **Right**: Same data (except for the two gates at the same voltage, not shown) at zero backgate voltage in a device with very good electrostatic control (slope of 68 mV/decade) thanks to a much thinner silicon layer (T_{Si} = 11 nm). The black dashed line shows the ideal slope of 60 mV/decade.

5. Conclusions

There is a need for robust and practical electron pumps to be shared and compared by NMIs for the quantum metrology of the forthcoming new ampere. In contrast with all other pumps realized in academic or small-scale fabrication facilities, we use a nearly industrial process to design and fabricate electron pumps and circuits. In this article we gave an overview of the most relevant process steps, including the custom gate level defined by electron beam lithography instead of the standard deep UV optical process used in production. This allows for a more aggressive pitch, of the order of 65 nm.

We have shown a very good electrostatic control of the devices even for two-gate layouts which are the building blocks of electron pumps.

Acknowledgments: This work is partially supported by the EU through the FP7 ICT collaborative project SiAM (Project No. 610637). We also acknowledge support from the Fondation Nanosciences in Grenoble, through the Ph.D. and excellence chair programs.

Author Contributions: Sylvain Barraud, Romain Lavieville, Louis Hutin, Heorhii Bohuslavskyi and Maud Vinet designed and fabricated the devices, measured by Andrea Corna, Paul Clapera, Xavier Jehl and Marc Sanquer. All the authors participated in writing the manuscript.

Conflicts of Interest: The authors declare no conflict of interest.

References

1. Blumenthal, M.D.; Kaestner, B.; Li, L.; Giblin, S.; Janssen, T.J.B.M.; Pepper, M.; Anderson, D.; Jones, G.; Ritchie, D.A. Gigahertz quantized charge pumping. *Nat. Phys.* **2007**, *3*, 343–347.
2. Giblin, S.; Kataoka, M.; Fletcher, J.; See, P.; Janssen, T.; Griffiths, J.; Jones, G.; Farrer, I.; Ritchie, D. Towards a quantum representation of the ampere using single electron pumps. *Nat. Commun.* **2012**, *3*, 930.
3. Fricke, L.; Wulf, M.; Kaestner, B.; Hohls, F.; Mirovsky, P.; Mackrodt, B.; Dolata, R.; Weimann, T.; Pierz, K.; Siegner, U.; *et al.* Self-Referenced Single-Electron Quantized Current Source. *Phys. Rev. Lett.* **2014**, *112*, 226803.
4. Kaestner, B.; Kashcheyevs, V. Non-adiabatic quantized charge pumping with tunable-barrier quantum dots: A review of current progress. *Rep. Prog. Phys.* **2015**, *78*, 103901.
5. Jehl, X.; Voisin, B.; Charron, T.; Clapera, P.; Ray, S.; Roche, B.; Sanquer, M.; Djordjevic, S.; Devoille, L.; Wacquez, R.; *et al.* Hybrid Metal-Semiconductor Electron Pump for Quantum Metrology. *Phys. Rev. X* **2013**, *3*, 021012.
6. Clapera, P. Design and operation of CMOS-compatible electron pumps fabricated with optical lithography. *IEEE Electron Device Lett.* **2016**, submitted.
7. Rossi, A.; Tanttu, T.; Tan, K.Y.; Iisakka, I.; Zhao, R.; Chan, K.W.; Tettamanzi, G.C.; Rogge, S.; Dzurak, A.S.; *et al.* An Accurate Single-Electron Pump Based on a Highly Tunable Silicon Quantum Dot. *Nano Lett.* **2014**, *14*, 3405–3411.
8. Yamahata, G.; Karasawa, T.; Fujiwara, A. Gigahertz single-hole transfer in Si tunable-barrier pumps. *Appl. Phys. Lett.* **2015**, *106*, 023112.
9. Barraud, S.; Coquand, R.; Casse, M.; Koyama, M.; Hartmann, J.; Maffini-Alvaro, V.; Comboroure, C.; Vizioz, C.; Aussenac, F.; Faynot, O.; *et al.* Performance of Omega-Shaped-Gate Silicon Nanowire MOSFET With Diameter Down to 8 nm. *IEEE Electron. Device Lett.* **2012**, *33*, 1526–1528.
10. Gaben, L.; Barraud, S.; Pimenta-Barros, P.; Morand, Y.; Pradelles, J.; Samson, M.P.; Previtali, B.; Besson, P.; Allain, F.; Monfray, S.; *et al.* Omega-Gate Nanowire Transistors Realized by Sidewall Image Transfer Patterning: 35 nm channel pitch and opportunities for stacked-Nanowires architectures. In Proceedings of the International Conference on Solid-State Devices and Materials, Sapporo, Japan, 27–30 September 2015.
11. Pierre, M.; Roche, B.; Wacquez, R.; Jehl, X.; Sanquer, M.; Vinet, M. Intrinsic and doped coupled quantum dots created by local modulation of implantation in a silicon nanowire. *J. Appl. Phys.* **2011**, *109*, 084346.
12. Clapera, P.; Ray, S.; Jehl, X.; Sanquer, M.; Valentian, A.; Barraud, S. Design and Cryogenic Operation of a Hybrid Quantum-CMOS Circuit. *Phys. Rev. Appl.* **2015**, *4*, 044009.
13. Roche, B.; Voisin, B.; Jehl, X.; Wacquez, R.; Sanquer, M.; Vinet, M.; Deshpande, V.; Previtali, B. A tunable, dual mode field-effect or single electron transistor. *Appl. Phys. Lett.* **2012**, *100*, 032107.
14. Natarajan, S.; Agostinelli, M.; Akbar, S.; Bost, M.; Bowonder, A.; Chikarmane, V.; Chouksey, S.; Dasgupta, A.; Fischer, K.; Fu, Q.; *et al.* A 14 nm logic technology featuring 2nd-generation FinFET, air-gapped interconnects, self-aligned double patterning and a 0.0588 μm^2 SRAM cell size. In Proceedings of the IEEE International Electron Devices Meeting (IEDM), San Francisco, CA , USA, 15–17 December 2014.

14

Experimental Tests on Steel Plate-to-Plate Splices Bonded by C-FRPS Laminas with and without Wrapping

Mario D'Aniello *, Francesco Portioli and Raffaele Landolfo

Department of Structures for Engineering and Architecture, University of Naples "Federico II", Naples 80134, Italy; fportiol@unina.it (F.P.); landolfo@unina.it (R.L.)
* Correspondence: mdaniel@unina.it

Academic Editors: Luciano Feo and Francesco Ascione

Abstract: The results of an experimental investigation carried out on steel splices bonded by (Carbon-Fiber–Reinforced Polymers) C-FRPs are presented in this paper. The main aim of the study is to examine the influence of different parameters on the type of failure and on the ductility of splices. Different configurations of the specimens were considered, including butt and lapped joints using different arrangements for end anchorage of the bonded C-FRP laminas, such as (i) external bonding; and (ii) anchored jacketing with C-FRP sheets transversally wrapped to the longitudinal axis of the joints. The results in terms of failure modes and response curves are described and discussed, highlighting the potentiality of these types of bonded connections for metal structures. In particular, experimental results showed that (i) the failure modes exhibited by both butt and lapped wrapped splices were substantially similar; (ii) the wrapped anchoring is beneficial in order to achieve large deformations prior to failure, thus allowing a satisfactory ductility, even though a more timely installation process is necessary.

Keywords: steel; C-FRPs; bonded splices; double strap test; lap shear test

---

1. Introduction

Historic metallic structures represent a significant portion of the architectural and cultural heritage in most European countries. The majority of these constructions is still in service and is exposed to service actions that are larger than was expected [1]. The reliability and the structural performance of this category of constructions are also affected by age-related and environmentally-induced deterioration and the low mechanical performance of the constituting materials. Outcomes from the European research project "Sustainable Bridges" [2] showed that aged steels do not usually fulfill the requirements of EN 10025 [3] for standardized materials. Thus, there is an urgent need to verify the compliance of historic steel structures with current standards and to assess their residual lifetime.

The traditional strengthening method for steel structures is to cut and replace steel plates or to attach additional plates to existing members. However, these techniques do not preserve the structure from deterioration, since steel plates are prone to corrosion and fatigue [4].

The use of composite materials is a very promising and increasingly economic method for strengthening steel structures [4–8]. In particular, the application of externally bonded Carbon-Fiber–Reinforced Polymer (C-FRP) plates [4–6,9]. This technique allows the increase of both the strength and buckling capacity of steel members and also the fatigue-life extension [5,6,10] and crack-patching [11,12].

The effectiveness and reliability of this strengthening/repairing method is mainly guaranteed by the anchorage of the fibers, which typically fails before the plates achieve their tensile capacity.

Therefore, the debonding of FRP from the steel elements must first be addressed in order to obtain a fully composite action. The first critical aspect is to clarify the bond-slip characteristics of FRP-plated steel members, which are mainly influenced by the bonding length and the adhesive layer behavior.

In particular, the adhesive is the weakest link in C-FRP-to-steel bonded joints [4,11,12]. To analyze the adhesive characteristics and to investigate the type of failure of CFRP-bonded joints, many authors have performed a series of experimental tests.

Zhao *et al.* [4] discussed the main research activity on FRP-strengthened steel structures, and illustrated the progress on the bond between the steel and FRP, the strengthening of steel hollow section members, and the fatigue crack propagation in the FRP-steel system.

Xia *et al.* [13] studied the behavior of the epoxy layer by performing some pull-out tests in which the FRP-to-steel interface is subjected to direct shear. A bond-slip model was also performed, and the main adhesive characteristics were discussed.

Yong-xin *et al.* [14] performed some bonding tests on CFRP-strengthened steel elements varying the type of adhesive, and then they studied the durability of the chosen type of adhesive. An aging model of the bond behavior of C-FRP-to-steel in a man-made seawater environment has been also proposed.

Nowadays, existing standard [15,16] and purely empirical procedures [4–14,17] mislead the derivation of the bond-slip characteristics because of the difficulty in measuring the strain distributions and capacities in the small region where debonding occurs.

The present paper describes the results of an experimental study aimed at investigating the influence of different bonding lengths on both the type of failure mode of the joint and the ductility of the failure process.

This work is part of an extensive experimental activity devoted to determining the main characteristics of CFRP-strengthened steel elements, in order to design and test C-FRP strengthening on riveted connections, which represent the main typology of joint for historic metal structures [1].

The paper is organized as follows: in Section 2 the experimental program is illustrated, in Section 3 the main experimental results are described, and in Section 4 the main conclusions are drafted.

2. Experimental Program

2.1. Investigated Parameters and Experimental Program

A total of 30 bonding tests were prepared, as summarized in the program matrix reported in Table 1. The investigated parameters are explained as follows:

(1) Bonding length: This parameter may noticeably affect the failure load [18]. Therefore, different strip lengths were examined in order to verify the effectiveness of the strengthening in accordance with Italian code provisions [15], in particular 150 mm, 200 mm (value recommended by the Italian code [15]) and 250 mm.

(2) Wrapping sheet: The effects of the anchored jacketing by wrapping composite sheets were analysed considering, alternatively, 100 mm and 150 mm widths of wrapping sheets.

(3) Load eccentricity: Both butt and lapped plates were considered in order to analyse the effects of secondary bending moments induced by load eccentricity on adhesive deformation and strength.

The geometries of the investigated connections are shown in Table 2. Specimens were labeled as C-B-W, where:

C is the splice configuration (e.g., B: Butt; U: Unsymmetrical lapped);

B is the bonding length (e.g., 150, 200, 250 mm);

W is the wrapping depth (e.g., 100, 150 mm);

Three nominally identical samples were built up for every type of specimen.

Table 1. Experimental program matrix.

Specimen Tag	Butt	Unsymmetrical	B (mm)	W (mm)	Test No.
		Unwrapped			
B150	√		150		3 (a, b, c)
B200	√		200		3 (a, b, c)
B250	√		250		3 (a, b, c)
U-B150		√	150		3 (a, b, c)
U-B200		√	200		3 (a, b, c)
U-B250		√	250		3 (a, b, c)
		Wrapped			
B150W100	√		150	100	3 (a, b, c)
B150W150	√		150	150	3 (a, b, c)
U-B150W100		√	150	100	3 (a, b, c)
U-B150W150		√	150	150	3 (a, b, c)
					Total test 30

Note: B: bonding length; W: wrapping depth.

Table 2. Geometry of specimens.

	Butt	Lapped
Unwrapped	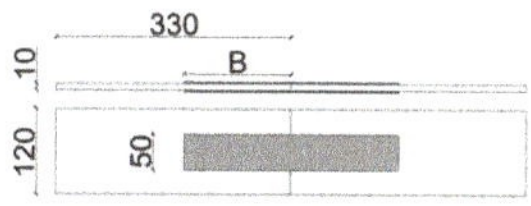	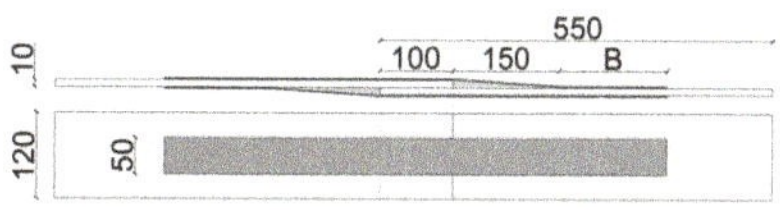
Wrapped	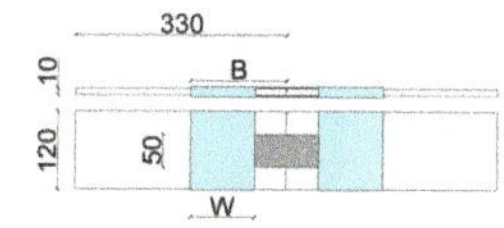	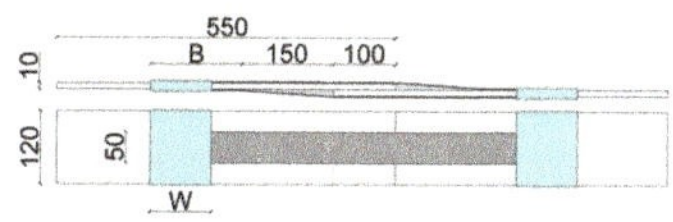

2.2. Experimental Set-Up and Monitored Parameters

The experimental setup is shown in Figure 1a. In particular, a universal electro-mechanical Zwick/Roell testing machine (see Figure 1b) was used to carry out tests. The specimens were loaded in tension under displacement control until failure, *i.e.*, after the load decreased.

The maximum load reached and the types of failure modes were observed for each test. The relative in-plane displacement of tested specimens was measured by means of a couple of LVDTs (Linear Variable Differential Transformer), which were characterized by a displacement range of ±150 mm and positioned on both ends of the C-FRP lamina (Figure 1c,d). The displacement rate was fixed at 0.1 mm/s and an acquisition frequency of 10 Hz was assumed [1].

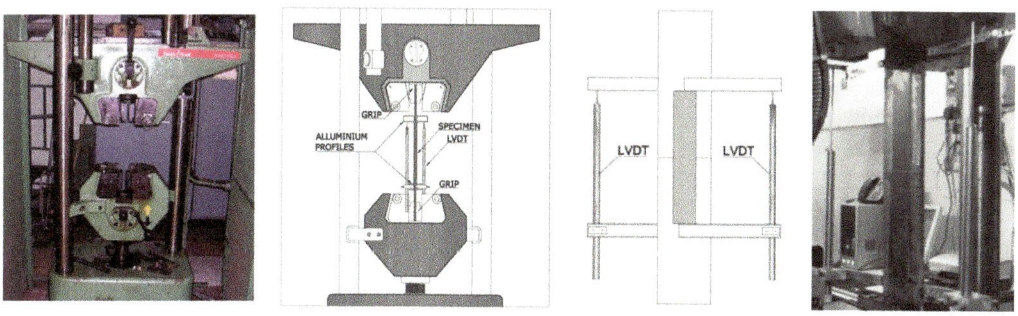

(a) (b) (c) (d)

Figure 1. Test setup (**a**); the testing machine (**b**); the layout of LVDTs (**c**,**d**).

The parameters that have been monitored and post-processed for the tests are illustrated in Figure 2a for C-FRP-bonded splice joints without wrapping and in Figure 2b for the wrapped specimens, where the symbols are explained as follows:

- $\delta = (\delta_{LVDT1} + \delta_{LVDT2})/2$: Average displacement (being δ_{LVDTi} the displacement recorded by the ith LVDT);
- F_y: Conventional elastic strength determined according to European recommendations for testing steel elements developed by European Convention for Constructional Steelwork (ECCS), *i.e.*, ECCS-45 [19]. In particular, this approach requires the following steps: (1) evaluating the tangent line at the origin of the force-displacement, it gives a slope equal to α_y (see Figure 2b); (2) locating a tangent line tilt of $\alpha_y/10$ on the experimental response (see Figure 2b); (3) the intersection of the two tangents defines the level of F_y;
- F_u: Ultimate strength, which is the maximum recorded average load;
- δ_y: Slip corresponding to F_y;
- δ_u: Displacement corresponding to a load equal to F_y on the post-peak branch of the response curve
- $\mu = \delta_u/\delta_y$: Displacement capacity.

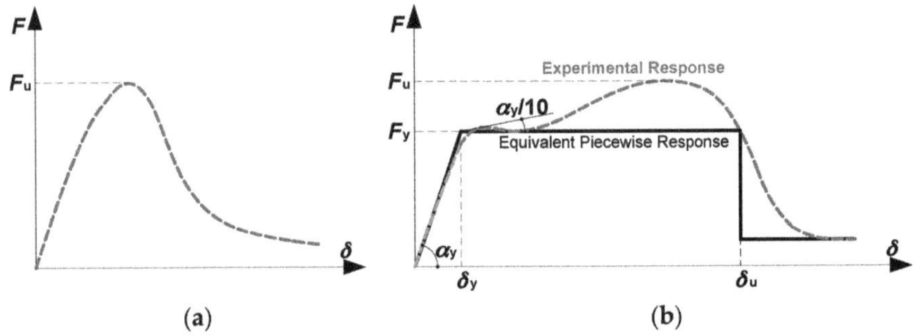

Figure 2. Monitored mechanical parameters [20]: without wrapping (**a**); with wrapping (**b**).

In particular, it should be noted that the displacement δ_y corresponding to F_y and the displacement capacity μ were monitored to provide an estimate of the equivalent ductility of the splices. These parameters were solely extracted for specimens with wrapping because these splices were able to provide a large displacement capacity. Conversely, as shown hereinafter, the specimens without wrapping showed a very brittle behavior.

3. Experimental Results

3.1. *Unwrapped Specimens*

Test results in terms of both failure modes and experimental response curves are summarized in Figures 3–5. Three basic types of failure modes were observed: (a) adhesive layer failure (Figure 3a,b); (b) C-FRP and adhesive interface failure (Figure 3c,d); (c) C-FRP delamination, namely the separation of some carbon fibers from the resin matrix (Figure 3e,f). In some cases a mixed mechanism made of adhesive layer failure and C-FRP delamination occurred. A similar performance (namely failure mode and force-displacement response) was observed for both butt and lapped specimens without wrapping.

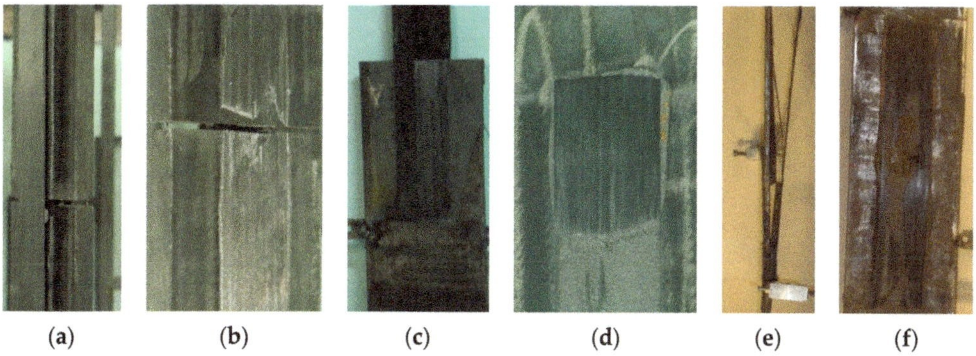

Figure 3. Failure modes of C-FRP bonded splice joints without wrapping [20].

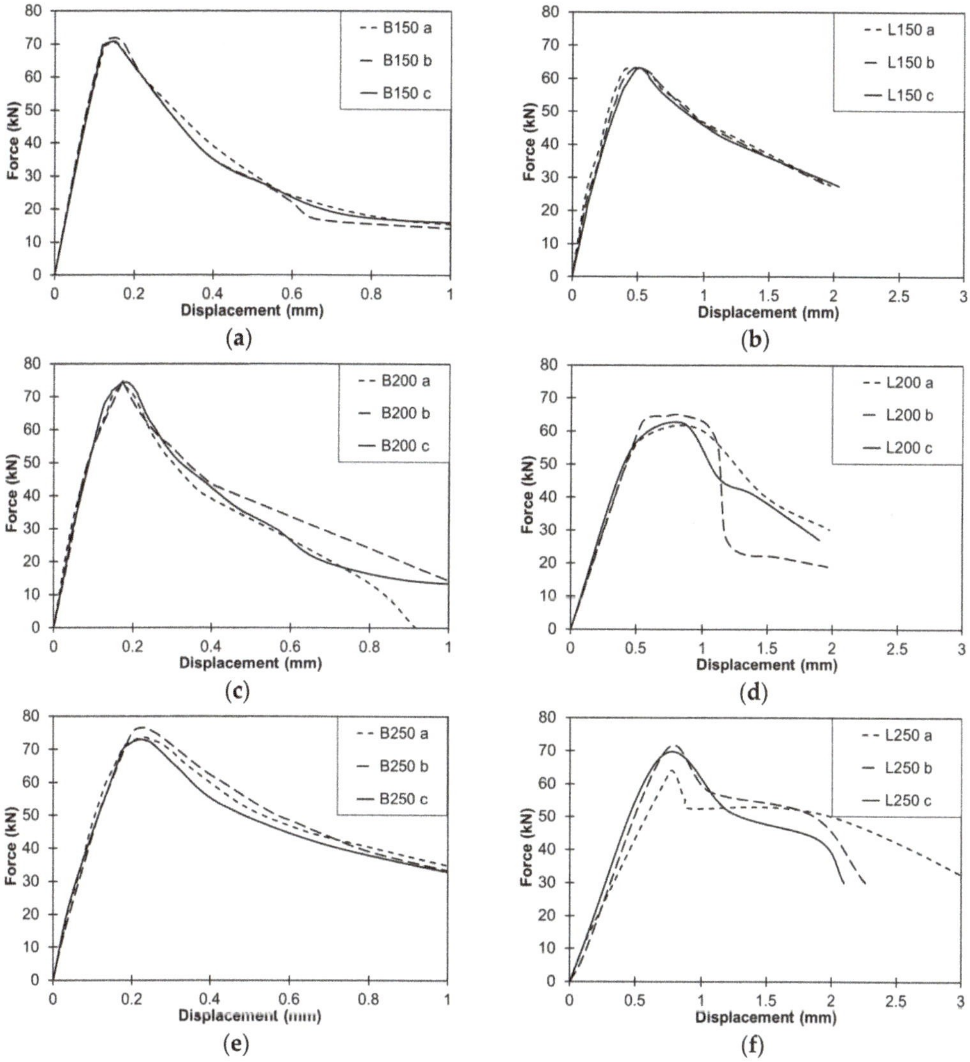

Figure 4. Experimental response curves of C-FRP bonded splice joints without wrapping [20].

Figure 5. Failure modes of C-FRP bonded splice joints with wrapping [20].

Tests showed that the debonding started at the gap area between the steel plates. With the strength of steel plates larger than the strength of the composite strips, we also consider the type of adhesive layer and the stiffness of composite laminas. In the tests presented in this paper, the steel plates behave elastically and the axial stiffness of the laminas is lesser than those of the steel plates. These aspects might have induced the brittle response of splices, which can be observed in the force-displacement response curves depicted in Figure 4. From these plots it can be noted that all butt splices showed an initial stiffness larger than the rigidity exhibited by lapped configurations. This result is ascribable to the different lengths of the two sets of specimens, with the lapped splices being longer than the butt specimens in order to allow gripping the plates in the test machine. In addition, the presence of secondary bending effects was another source of deformability.

Table 3 summarizes the geometrical and mechanical parameters monitored during each test. In particular, Table 4 reports the measured thickness of the adhesive layer, considering that this parameter could have considerable effect on the bond behavior. In the present study it was found that in all cases with a thicker adhesive layer, the failure mode was mainly characterized by C-FRP delamination, thus being consistent with the results in the literature [21,22].

Table 3. Specimen details and test results for C-FRP-bonded splice joints without wrapping [20].

Specimen	t_a * (mm)			F_u (kN)				Failure Mechanism
	Mean	SD	CV	Test Value	Mean	SD	CV	
B150 a	0.49	0.09	0.18	70.37				Delamination
B150 b	0.56	0.08	0.14	71.86	71.00	0.77	0.01	Delamination
B150 c	0.54	0.11	0.21	70.76				Delamination
B200 a	0.41	0.08	0.19	73.22				C-FRP to adhesive interface + Delamination
B200 b	0.40	0.09	0.22	74.53	73.78	0.67	0.01	Adhesive/Delamination
B200 c	0.41	0.10	0.24	73.60				Adhesive/Delamination
B250 a	0.38	0.07	0.18	76.59				Adhesive/Delamination
B250 b	0.39	0.09	0.23	73.01	74.71	1.80	0.02	Adhesive/Delamination
B250 c	0.41	0.09	0.22	74.53				C-FRP to adhesive interface
L150 a	0.38	0.09	0.24	63.03				Adhesive/Delamination
L150 b	0.36	0.08	0.22	63.18	63.08	0.09	0.00	Adhesive/Delamination
L150 c	0.39	0.11	0.28	63.01				Adhesive/Delamination
L200 a	0.40	0.08	0.20	60.61				C-FRP to adhesive interface
L200 b	0.38	0.09	0.24	63.89	63.15	2.26	0.04	C-FRP to adhesive interface
L200 c	0.37	0.10	0.27	64.94				C-FRP to adhesive interface + Adhesive/Delamination
L250 a	0.41	0.07	0.17	63.89				C-FRP to adhesive interface
L250 b	0.39	0.09	0.23	69.91	67.18	3.05	0.05	Adhesive/Delamination
L250 c	0.39	0.09	0.23	67.73				Adhesive/Delamination

* t_a is the thickness of the adhesive layer measured at 10 sections equally spaced along the axis of the splice.
SD = Standard Deviation, CV = Coefficient of Variation.

Table 4. Specimen details and test results for C-FRP-bonded splice joints with wrapping [20].

Specimen	t_a * (mm)			F_y (kN)				F_u (kN)				$\mu = \delta_u/\delta_y$				$\frac{F_{y,wrapped}}{F_{u,simple\ bonded}}$	$\frac{F_{u,wrapped}}{F_{u,simple\ bonded}}$
	Mean	SD	CV	Test Value	Mean	SD	CV	Test Value	Mean	SD	CV	Test Value	Mean	SD	CV	(-)	(-)
B150 W100 a	0.38	0.09	0.23	76.60				84.49				6.12					
B150 W100 b	0.35	0.10	0.28	78.36	77.32	1.15	0.01	79.36	81.62	2.62	0.03	1.29	3.83	2.42	0.63	1.09	1.15
B150 W100 c	0.37	0.11	0.30	77.00				81.01				4.07					
B150 W150 a	0.39	0.08	0.20	78.20				91.96				6.38					
B150 W150 b	0.38	0.10	0.26	76.34	77.78	1.32	0.02	94.75	93.92	1.70	0.02	4.26	5.68	1.23	0.22	1.10	1.32
B150 W150 c	0.36	0.10	0.28	78.80				95.05				6.41					
L150 W100 a	0.37	0.10	0.27	80.08				85.60				4.64					
L150 W100 b	0.41	0.09	0.21	76.02	78.29	1.33	0.03	83.90	83.97	1.59	0.02	5.42	4.92	0.43	0.09	1.24	1.33
L150 W100 c	0.36	0.12	0.34	78.78				82.42				4.71					
L150 W150 a	0.38	0.09	0.23	78.76				104.95				8.26					
L150 W150 b	0.40	0.10	0.25	78.20	78.85	1.71	0.01	107.00	107.85	3.41	0.03	6.25	7.61	1.18	0.15	1.25	1.71
L150 W150 c	0.39	0.11	0.28	79.60				111.60				8.33					

* t_a is the thickness of the adhesive layer measured at 10 sections equally spaced along the axis of the splice. SD = Standard Deviation, CV = Coefficient of Variation.

As far as the shear strength (F_u), for both butt and lapped splices with 150 mm of bonding length, no appreciable differences can be recognized; the only effect of the lapped configuration was a mean reduction in the peak strength of about 11.2%. Also, for the longer bonding length a similar strength reduction was observed (e.g., 14.4% for 200 mm and 10.1% for 250 mm). Furthermore, longer lapped joints showed a post-peak response different from those experienced by the butt specimens. Indeed, comparing Figure 4c,e to Figure 4d,f, it is possible to observe a post-peak displacement capacity without a sharp softening for the lapped splices. This feature was due to the non-contemporary failure of the bonded C-FRP strips at both sides of the specimen. Such a response depended on the presence of the secondary bending moment due to the load eccentricity, which led to an increase in the applied force on one side of the cover bonded strip and a reduction on the opposite side. Experimental tests showed that the ultimate loads did not vary with the bonded length of the splice in the range of 150 mm–250 mm. This result implies that the effective bonding length is smaller than the length of the tested splices. Indeed, once the bond strength is achieved due to the full development of the bonding law, increasing the bonded length beyond this threshold does not provide a strength increase, although some improvement in terms of ductility can be recognized [20].

3.2. Wrapped Specimens

The experimental tests showed that the overall failure modes were very similar for both butt and lapped wrapped specimens. In particular, the failure modes were characterized by a sequence of nonlinear events. The former which experimentally observed was the adhesive failure, causing the initiation of slipping between the C-FRP strips and the steel elements (see Figure 5a,b). Once this phenomenon occurred, the interactions between the wrapped jacketing and the laminas allowed transferring the applied load from the steel plates to the C-FRP laminas by friction/adhesion. Increasing the applied displacements of the sliding actions of the C-FRP laminas within the wrapped jacket caused bending of the fibers, constituting the wrapped fabric in the transverse direction of the jacket (Figure 5c). During this behavioral phase, some specimens showed a secondary failure mode characterized by the unwrapping of the fabrics (Figure 5d), which impaired the joint capacity which dropped sharply. For all specimens the failure occurred when the jacketed anchorage was not able to restrain the C-FRP slip, thus resulting in large in-plane displacements due to the separation of the steel plates (see Figure 5e for both lapped and butt splices).

The experimental evidence indicated that the presence of wrapped jacketing at the anchorage allowed an improvement of performance, as is recognizable comparing Figure 6 to Figure 4. Indeed, all wrapped specimens exhibited a relatively large displacement capacity, which is substantially different from the brittle response of simply bonded splices. Indeed, the anchorage by wrapping allowed us to achieve higher C-FRP deformations prior to failure, delaying/preventing premature delamination. This result confirms the advantages of using wrapped fabrics to anchor bonded C-FRP strips, which has been previously demonstrated for applications on reinforced concrete structures [20]. The force-displacement response curves of all wrapped specimens were characterized by a saw-teeth profile. In this regard, it is interesting to highlight that during the tests, sudden slips with cracks in the resin of the wrapped anchorage were observed at each peak of the saw-teeth profile of the response curves. This feature can be explained by considering that the fibers of the fabric restrain the laminas and each sudden drop of resistance corresponds to the bonding failure of each one.

Table 4 summarizes the measured values of the conventional elastic and ultimate forces, and the displacement capacity $\mu = \delta_u/\delta_y$ is reported per test. As it can be noted, all wrapped specimens showed an ultimate strength larger than that experienced by the corresponding bonded specimens. The beneficial effect of the wrapped anchorage was largely observed for lapped splices with a jacketing width equal to 150 mm, which reached a mean ultimate strength larger than 70% of the strength of the unwrapped specimens. It should be noted that all specimens with a jacketing width equal to 150 mm showed an increase of ultimate strength significantly larger than those developed by specimens with 100 mm of jacketing width, as it can be easily observed comparing Figure 6a,b to Figure 6c,d, respectively.

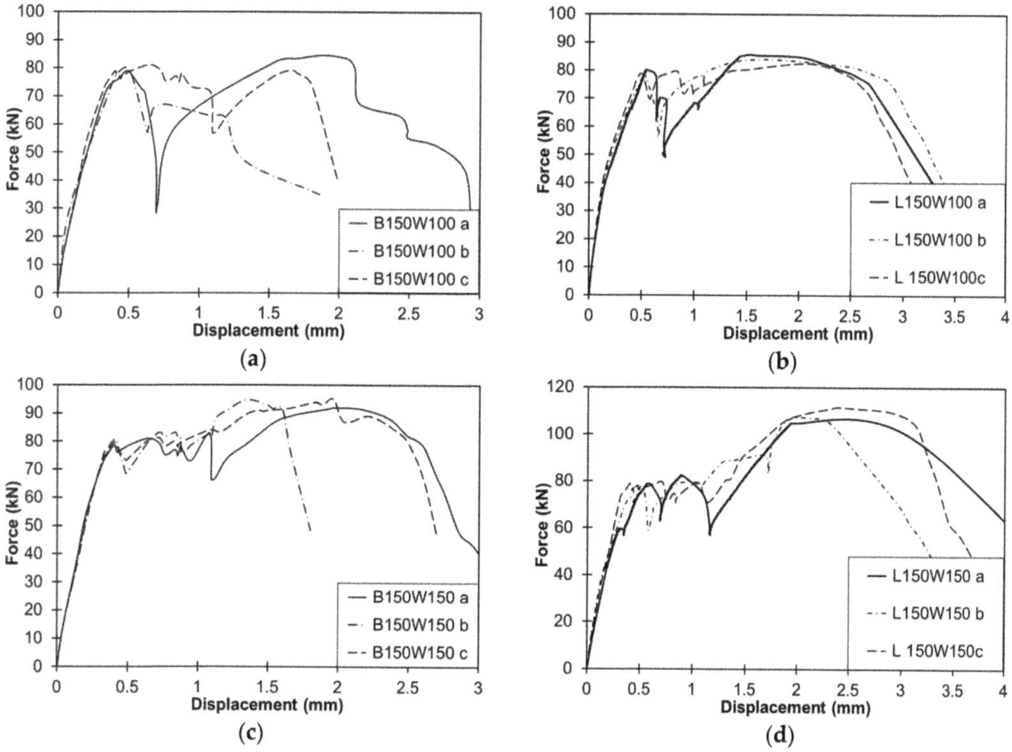

Figure 6. Experimental response curves of C-FRP bonded splice joints with wrapping [20].

4. Conclusions

The tests were carried out on both butt and lapped joint configurations using different arrangements for the end anchorage of the bonded composites, such as (i) external bonding and (ii) anchored jacketing with C-FRP sheets transversally wrapped to the longitudinal axis of the joints.

The results and discussions presented allow us to draft the following remarks:

- The failure modes exhibited by both butt and lapped bonded splices were similar, characterized by a non-ductile response with a clear post-peak softening behaviour. The lapped specimens experienced a peak strength lesser than those of butt splices of about 11.2%.
- Tests on unwrapped bonded splices showed that the ultimate bonding strength is practically insensitive to the bonding length of the strips in the range of 150 mm–250 mm. Hence, the limit of 200 mm proposed by the Italian code [15] is conservative.
- Tests of wrapped splices highlighted the beneficial role of anchoring externally bonded FRP laminas in order to achieve larger deformations prior to failure, even though a more timely installation process is necessary.
- The anchorage with wrapped C-FRP fabrics modified the post-elastic response of the splices by providing a pseudo-ductile behaviour, thus allowing a non-negligible ductility.
- Similarly to the bonded specimens, the failure modes exhibited by both butt and lapped wrapped splices were substantially similar. As a consequence, it was not possible to observe a strict dependence of the failure mechanism on the splice geometry.

Acknowledgments: The authors are sincerely grateful to BASF-Italia spa and, in particular, Arch. Raffaele Hassler for providing all composite materials and assistance in the application of FRPs. They also wish to thank RFI (Rete Ferroviaria Italiana) and in particular: Eng. Antonio D'Aniello, former Director of the Department of Naples at RFI for his courtesy, cooperation and assistance throughout this research. Finally, authors would like to thank Eng. Giuseppe La Manna Ambrosino and Arch. Renata Marmo, for their help in the preparation of the specimens.

Author Contributions: All Authors equally contributed to the research described and discussed by this paper. All authors have read and approved the final manuscript.

Conflicts of Interest: The authors declare no conflict of interest.

References

1. D'Aniello, M.; Portioli, F.; Fiorino, L.; Landolfo, R. Experimental investigation on shear behaviour of riveted connections in steel structures. *Eng. Struct.* **2011**, *33*, 516–531. [CrossRef]
2. Sustainable bridges—European Research Project under the EU 6th Framework Programme, 2006. Available online: http://www.sustainablebridges.net/ (accessed on 4 February 2016).
3. British Standards Institution (BSI). *Hot Rolled Products of Non-Alloy Structural Steels. Technical Delivery Conditions for Non-Alloy Structural Steels*; BS EN 10025-2:2004; BSI: London, UK, 2004.
4. Zhao, X.; Zhang, L. State-of-the-art review on FRP strengthened steel structures. *Eng. Struct.* **2007**, *29*, 1808–1823. [CrossRef]
5. Hollaway, L.C.; Cadei, J. Progress in the technique of upgrading metallic structures with advanced polymer composites. *Prog. Struct. Eng. Mater.* **2002**, *4*, 131–148. [CrossRef]
6. Hollaway, L.C. Polymers, Fibres, Composites and the Civil Engineering Environment: A Personal Experience. *Adv. Struct. Eng.* **2010**, *13*, 927–960. [CrossRef]
7. Song, K.; Zhang, Y.; Meng, J.; Green, E.C.; Tajaddod, N.; Li, H.; Minus, M.L. Structural Polymer-Based Carbon Nanotube Composite Fibers: Understanding the Processing–Structure–Performance Relationship. *Materials* **2013**, *6*, 2543–2577. [CrossRef]
8. Guades, E.; Aravinthan, T.; Islam, M.; Manalo, A. A review on the driving performance of FRP composite piles. *Compos. Struct.* **2012**, *94*, 1932–1942. [CrossRef]
9. Deng, J.; Lee, M.M.K. Adhesive bonding in steel beams strengthened with CFRP. *Proc. Inst. Civ. Eng. Struct. Build.* **2009**, *162*, 241–249. [CrossRef]
10. Stratford, T.J.; Chen, J.F. Designing for tapers and defects in FRP-strengthened metallic structures. In Proceedings of the Designing for Tapers and Defects in Frp-Strengthened Metallic Structures, BBFS2005, Hong Kong, China, 7–9 December 2005.
11. Kim, Y.J.; Brunell, G. Interaction between CFRP-repair and initial damage of wide flange steel beams subjected to three-point bending. *Compos. Struct.* **2011**, *93*, 1986–1996. [CrossRef]
12. Bassetti, A.; Nussbaumer, A.; Hirt, M.A. Crack repair and fatigue life extension of riveted bridge members using composite materials. In Proceedings of the Crack Repair and Fatigue Life Extension of Riveted Bridge Members Using Composite Materials, ESE-IABSE-FIB, Sharm El Sheikh, Egypt, 26–30 March 2000.
13. Xia, S.H.; Teng, J.G. Behaviour of FRP-to-steel bonded joints. In Proceedings of the International Symposium on Bond Behaviour of FRP in Structures, BBFS2005, Hong Kong, China, 7–9 December 2005.
14. Teng, J.G.; Yu, T.; Fernando, D. Strengthening of steel structures with fiber reinforced polymer composites. *J. Constr. Steel Res.* **2012**, *78*, 131–143. [CrossRef]
15. Istruzioni per la Progettazione, l'Esecuzione ed il Controllo di Interventi di Consolidamento Statico Mediante l'utilizzo di Compositi Fibrorinforzati. Available online: http://www.cnr.it/documenti/norme/IstruzioniCNR_DT200_2004.pdf (accessed on 4 February 2016).
16. Cadei, J.M.C.; Stratford, T.J.; Hollaway, L.C.; Duckett, W.G. *Strengthening Metallic Structures Using Externally Bonded Fibre-Reinforced Polymers*; Publication C595, Construction Industry Research and Information Association (CIRIA): London, UK, 2004.
17. Yang, Y.; Yue, Q.; Peng, F. Experimental research on bond behaviour of CFRP to steel. In Proceedings of the Experimental Study on Bond Behaviour between UHM CFRP Laminate and Steel, BBFS2005, Hong Kong, China, 7–9 December 2005.
18. Mancusi, G.; Ascione, F. Performance at collapse of adhesive bonding. *Compos. Struct.* **2013**, *96*, 256–261. [CrossRef]

19. ECCS-45. *Recommended Testing Procedure for Assessing the Behaviour of Steel Elements under Cyclic Loads*; European Convention for Constructional Steelwork: Brussels, Belgium, 1986.
20. D'Aniello, M.; Portioli, F.; Landolfo, R. Lap shear tests on hot-driven steel riveted connections strengthened by means of C-FRPs. *Compos. Part B Eng.* **2014**, *59*, 140–152. [CrossRef]
21. Fawzia, S.; Zhao, X.L.; Al-Mahaidi, R. Bond-slip models for double strap joints strengthened by CFRP. *Compos. Struct.* **2009**, *92*, 2137–2145. [CrossRef]
22. Da Silva Lucas, F.M.; Rodriguesa, T.N.S.S.; Figueiredoa, M.A.V.; de Mouraa, M.F.S.F.; Chousal, J.A.G. Effect of adhesive type and thickness on the lap shear strength. *J. Adhes.* **2006**, *82*, 1091–1115. [CrossRef]

15

A Preliminary Assessment of the Initial Compression Power Requirement in CO_2 Pipeline "Carbon Capture and Storage (CCS) Technologies"

Abdussalam El-Suleiman [1,*], Nnamdi B. Anosike [2] and Pericles Pilidis [3]

1 Department of Mechanical Engineering Technology, Federal Polytechnic, Nasarawa 962101, Nigeria
2 Department of Mechanical Engineering, Nnamdi Azikiwe University, Awka 420107, Nigeria; nb.anosike@unizik.edu.ng
3 Centre for Propulsion Engineering, Cranfield University, Cranfield MK43 0AL, UK; p.pilidis@cranfield.ac.uk
* Correspondence: a.elsuleiman@gmail.com

Academic Editor: Manoj Gupta

Abstract: CO_2 captured from fossil-fueled power generation plants is said to be economically transported via pipelines over long distances. The CO_2 must be compressed to pipeline specifications using compressors and pumps that are driven by gas turbine (GT) or other prime movers. This paper presents the evaluation of actual work transfer or required prime power by modeling the governing equations of compression using the Peng–Robinson equation of state (PR-EOS). A computer code was developed to carry out the modeling and subsequent simulation of the compression power requirement. The simulation of prime mover power was carried out for different technology (head per stage) of the compressor ranging from 10-staged compression to double stage compression. The results show that the current technology of the centrifugal compressor could require as much as 23MW of prime mover power to compress 1.5 million tonnes per year of CO_2—a projected equivalent CO_2 released from a 530MW combined cycle gas turbine (CCGT) power generation plant.

Keywords: CO_2; pipeline; prime mover; CO_2 compression; combined cycle gas turbine (CCGT)

1. Introduction

Sustainability of fossil fuel-fired power generation plants require the removal of carbon in the form of carbon dioxide (CO_2) from either the fuel before combustion or flue gas after combustion. This becomes necessary because CO_2 emission, which is a consequence of combustion or oxidation of fossil fuel, has been identified as a main greenhouse gas [1,2]. Once the CO_2 is removed, it is usually compressed and transported to the point of storage away from the atmosphere. The point of CO_2 storage is usually far from the source of emission; hence, transportation over long distance is highly envisaged. As a result, the CO_2 is envisaged to be transported via pipelines, which primarily have gained wide acceptance for long-distance transportation [2–5]. Prior to transport, equipment driven by gas turbines is used to compress the captured CO_2 to a dense-phase state, which is said to be the most efficient and economical state for CO_2 pipeline transport [3,4]. The required power for compression or compression duty using appropriately sized GT is addressed in this paper.

It is worth mentioning that the quantification of compression duty has been conducted by many researchers, especially in areas that deal with cost prediction models of CO_2 pipeline transport [6]. In such studies, the required energy for compression is either treated as an efficiency penalty on the overall power generation plant's efficiency or assumed to be supplied from the grid [7–10]. The power consumption of the compression process is said to be about 40% of the auxiliary load and could constitute 7%–12% of the power plants output [11,12]. This energy consumption is huge in economic terms, and the need for a separate source of energy for compression becomes desirable. Similarly, the requirement for booster stations in remote locations along a trunk pipeline where the cost of transporting electricity becomes prohibitive necessitates a suitable prime mover. Thus, GTs being extensively used in pipelines to compress gas at suitable points becomes a prime mover of choice for this duty. While considerable operational experience has been gathered in natural gas transmission, the same is not true for CO_2 transmission, especially since the properties and behavior of the gases are quite different. The distinguishing feature of this work hinges on the close coupling of the compression system with the GT prime mover, with a view of assessing the CO_2 compression power requirement. Analysis of this kind becomes imperative to aid the techno-economic assessment of deploying this prime mover in a CO_2 pipeline transport.

2. The Pipeline Transmission of CO_2

The economics for transporting large volumes of CO_2 over long distance in the gaseous phase is quite unfavorable [13,14]. This is because such pure gas phase transportation will be restricted to operate below 30–50 bar at which the densities and capacities will be too small. However, the pure gas phase will fit a system where the CO_2 is initially transported on gathering lines from low CO_2 producers for connection to larger trunk pipelines [13]. Low-pressure liquid phase is not desirable either except for transportation in vessels, tanks, and barges. The liquid phase for pipeline transportation comes with problems of frost formation in cold climates and the need for pipeline material suitable for cryogenic application. Furthermore, unlike natural gas, the critical point of CO_2 is near the potential compressor (and pipeline) operating point such that slight changes in ambient temperature or soil temperature for a buried pipeline can result in two-phase flow [3].

Abrupt pressure drops occasioned by two-phase flows and the detrimental effect of impurities must be avoided during CO_2 transmission in pipelines. Hence, the CO_2 is transported in the dense-phase state especially for onshore pipelines, as has been earlier highlighted [3,4]. In addition, the dense-phase transport gives an advantage for high delivery pressure at injection to sink [2]. The pipeline operating pressure in this state is between 10 and 15 MPa [5,7,15]. In general though, the CO_2 can be transported at temperature and pressure ranges of 12–44 °C and 85–150 bar, respectively [5,16]. The upper temperature limit is fixed by the compressor-station discharge temperature, while the lower limit is fixed by winter ground temperature for buried pipelines. The lower pressure limit is dictated by the phase behavior of CO_2, especially with impurities present and the need to maintain supercritical conditions, while the upper pressure limit is mostly due to economic concerns [5].

Highly recommended within this pipeline operating conditions is the "API 5L X-70" line pipe specification in the US Code of Federal Regulations—CFR (the CFR regulates the design, construction and operation of CO_2 pipeline transport in the USA) [8,17]. Similarly, the ASME/ANSI class 900# flanges are quite suitable since they can tolerate an operating pressure of about 153 bar (15.3 MPa) at a temperature of about 311.15 K (38 °C) [3,16].

3. Methodology

This work was carried out by adapting the TERA (Techno-Economic and Environmental Risk Analysis) framework developed in Cranfield University, UK [18,19]. The core of the TERA is the performance module, which is a detailed and rigorous thermodynamic representation of the component parameters and power plant achieved by simulating the design, off-design, and degraded performance of the power plant. The economic, environmental, and risk modules are built around the performance module and integrated with an optimizer to compare power plants optimized for a particular duty such as fuel consumption and operating cost. Applying this to the GT-driven transmission of CO_2 in pipelines, the adapted framework consists of four sets of FORTRAN-coded integrated modules, *i.e.*, a compression module, an engine performance module, an emission module, and an economic module. The compression module addressed in this paper involved code development to simulate the centrifugal compressor power requirement using the Peng–Robinson equation of state (PR-EOS), said to be among the most precise and proper for engineering application [4,20]. The CO_2 is assumed to be compressed from atmospheric pressure (1 bar) to the pipeline intake pressure of 10 MPa (100 bars) using a centrifugal compressor. A maximum temperature of 40 °C was maintained at the suction end of the centrifugal compressor. Considering the limitation imposed by maximum head achievable and allowable temperature in the compressor, multistage compression with intercooling was simulated. The CO_2 was assumed to be pure and the required CO_2 properties of density and viscosity were calculated using the code developed by McCollum and Ogden [9].

4. Physical Properties of Pure CO_2

Since the captured CO_2 will be compressed to a dense-phase state, an insight into the phase behavior of pure CO_2 depicted by Figure 1 is worthwhile. CO_2 in its pure state is colorless, odorless, and non-inflammable at ambient temperature and pressure [5].

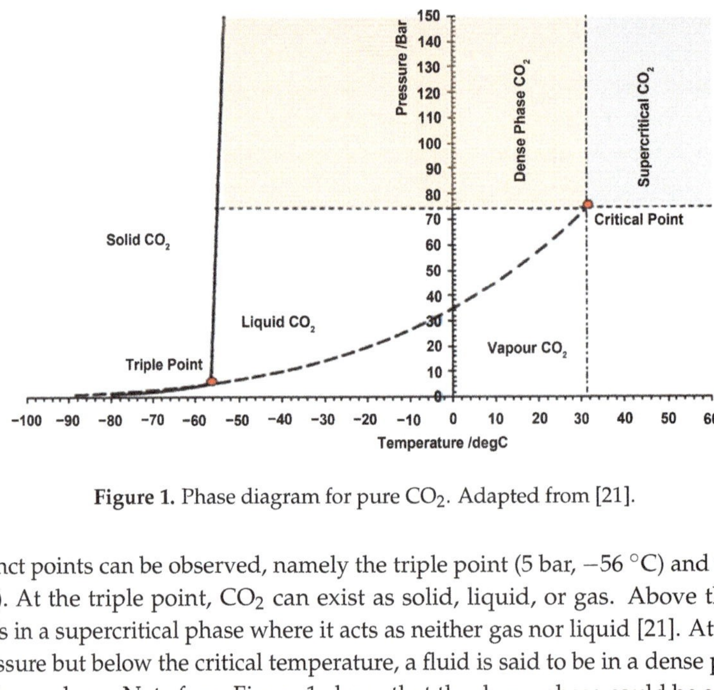

Figure 1. Phase diagram for pure CO_2. Adapted from [21].

Two distinct points can be observed, namely the triple point (5 bar, −56 °C) and the critical point (74 bar, 31 °C). At the triple point, CO_2 can exist as solid, liquid, or gas. Above the critical point, the pure CO_2 is in a supercritical phase where it acts as neither gas nor liquid [21]. At pressures above the critical pressure but below the critical temperature, a fluid is said to be in a dense phase; otherwise, it is in a non-dense phase. Note from Figure 1 above that the dense phase could be achieved without passing the liquid area. The dense phase is a very peculiar state in that, on one hand, it is not a liquid because it has a viscosity similar to that of a gas while, on the other hand, its density is closer to that of a liquid and has similar flow conditions as that of a liquid [13].

5. Modeling the Compressor Power (Head) Requirement

Equations for evaluating the head, work or power of a centrifugal compressor in units of Nm/kg, Nm or Nm/sec respectively have been presented in the literature. These equations are primarily used to evaluate the adiabatic head (H_{ad}) (Equation (1)) and, with slight modification, the polytropic head (H_{poly}) (Equation (2)) as expressed below [22–24].

$$H_{ad} = T_s Z_{ave} \frac{R_o}{mw} \frac{\gamma}{\gamma - 1} \left[\left(\frac{P_d}{P_s} \right)^{\frac{\gamma-1}{\gamma}} - 1 \right] \text{(KJ/kg), and} \tag{1}$$

$$H_{poly} = T_s Z_{ave} \frac{R_o}{mw} \frac{n}{n - 1} \left[\left(\frac{P_d}{P_s} \right)^{\frac{n-1}{n}} - 1 \right] \text{(KJ/kg)}, \tag{2}$$

where

$$\gamma = \frac{C_p}{C_v} = \frac{C_p}{C_p - R}\text{, and} \tag{3}$$

$$\frac{n}{n-1} = \frac{\gamma}{\gamma - 1}\eta_P\,. \tag{4}$$

T_s is the suction temperature (K); Z_{ave} is the average compressibility factor; R is the specific gas constant (KJ/kg-K); R_o is the universal gas constant (KJ/Kmol-K); mw is the molecular weight (kg/Kmol); γ is ratio of specific heats; P_d and P_s are pressures at discharge and suction, respectively (MPa); C_p and C_v are specific heats at constant pressure and volume respectively (KJ/kg-K); n is the polytropic index; and η_P is the polytropic efficiency.

Considering the "polytropic head", the required compression power, or gas power (P_{poly}) as it is also called, is defined by Equation (5) as

$$P_{poly} = \frac{m\,H_{poly}}{\eta_p}\ (KW)\,. \tag{5}$$

Further, the shaft or prime mover power (P_{shaft}) is given by Equation (6) as

$$P_{shaft} = \frac{m\,H_{poly}}{\eta_p}\,\frac{1}{\eta_{mech}}\ (KW)\,. \tag{6}$$

Mechanical efficiency (η_{mech}) of about 87% was assumed in the simulation; m is the CO_2 mass flow rate (kg/s).

In actual applications, the compression process is rarely polytropic or adiabatic. Thus, to achieve an estimate of the actual compression head, modifications such as the "Schultz polytropic head correction factor", the Mollier diagrams, or the use of equations of state have been readily explored [23–25].

In light of the above, the modeling of the centrifugal compressor power presented here was achieved using the PR-EOS [20]. Further reading about other equations of states, the Mollier diagram and the Schultz polytropic factor can be found in [23,25].

Equations (1) and (2) can be re-cast in terms of the enthalpy difference between the suction and discharge states determined by two of the following given properties of the compressed gas: pressure, temperature, and entropy. Hence, the actual compressor head, H_{actual} can be represented as [23,26]

$$H_{actual} = \Delta H = H_d - H_s\ (KJ/kg)\,. \tag{7}$$

This is further defined as

$$H_{actual} = \Delta H = H_d - H_s = \Delta H^*_{T_s} + \int_{T_s}^{T_d} C_p dT - \Delta H^*_{T_d}\ (KJ/kg)\,, \tag{8}$$

where ΔH is change in enthalpy (KJ/kg); H_d and H_s are enthalpies at discharge and suction respectively (KJ/kg); and dT is change in temperature (K). The term ΔH^* is called the enthalpy departure function defined by the PR-EOS as

$$\Delta H^* = R_O T\,(Z-1) + \frac{T\frac{da}{dT} - a}{2\sqrt{2}b}\ln\left(\frac{Z + 2.414B}{Z - 0.414B}\right)\,. \tag{9}$$

The terms of this equation are fully defined in [20]:

$$C_p = C_p\ (T)\,. \tag{10}$$

The mathematical expression for Equation (10) specifically for CO_2 is found in [26]. T is temperature (K).

6. Modeling the PR-EOS for Compressibility Factor (Z)

To fully solve Equation (8) for the actual head, the CO_2 compressibility factor (Z) needs to be determined by using the PR-EOS to evaluate ΔH^*. From [20],

$$Z^3 - (1 - B)\,Z^2 + \left(A - 3B^2 - 2B\right) Z - \left(AB - B^2 - B^3\right) = 0, \tag{11}$$

where

$$A = \frac{aP}{R_o{}^2 T^2}, \tag{12}$$

$$B = \frac{bP}{R_o T}, \text{ and} \tag{13}$$

$$Z = \frac{Pv}{R_o T}. \tag{14}$$

The terms A and B are fully defined in [20].

7. The Simulation of the Required Compression Power

The power required for compression is evaluated for the CO_2 flow rate that will meet both available GT driver power and pipeline compressor actual suction capacity. Consideration was given to the range of CO_2 flow rates in million tonnes per year (MTPY), which are representative of the kilograms of CO_2 produced per kilowatt hour from conventional (subcritical), supercritical, and advanced steam coal power generation plants as well as a CCGT power plant. A typical CCGT power generation plant will give a CO_2 equivalence of 1.5 MTPY [2,27]. A maximum of 25 MW driver power is considered while the compressor is expected to operate at best efficiency over the chosen CO_2 flow range. An actual design CO_2 flow rate of 1.5 MTPY (~100,000 m^3/h) at design speed, which is within the operational limit specified for pipeline compressor (PCH) as shown in Figure 2, is maintained. Inlet flows above and up to twice this capacity will be assumed driven by the same driver in a parallel flow arrangement. The simulation entails multistage compression power evaluation as well as consideration of pressure drop due to intercooling. All simulated results are assumed to be at a steady state.

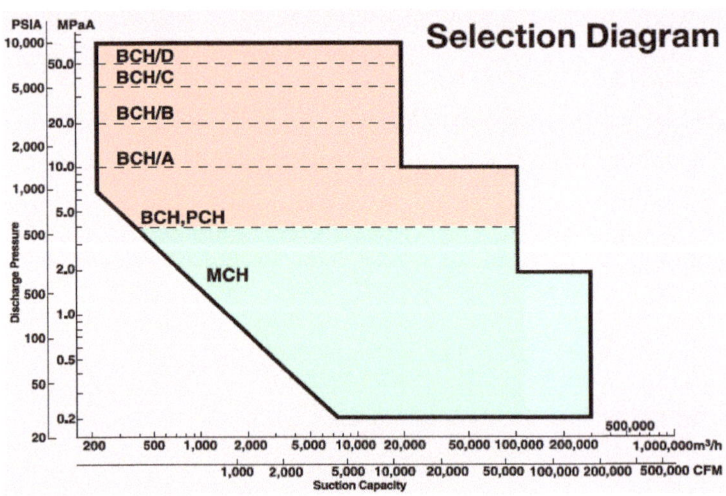

Figure 2. Centrifugal compressor selection chart [Courtesy Hitachi Plant Technologies].

A constant compressor polytropic efficiency value of 80% per stage was assumed for the simulation considering existing technology [24]. However, it should be noted that the compressor efficiency value decreases in successive stages for multi-stage compression due to fouling, speed matching, and mechanical constraints. A polytropic efficiency value of 90% was also used in the simulation to evaluate the effect of change in efficiency on the compressor head. A 10% increment on simulated power is applied to meet the API 617 minimum power margin specification for GT driver selection.

8. A Validation of the Simulated Compressibility Factor (Z)

The compressibility factor is crucial to the evaluation of required compression power. Therefore, the simulated CO_2 compressibility values from the code developed using the PR-EOS was validated against the values obtained using Aspen HYSYS—a commercial pipeline simulation software with a built-in PR-EOS library. The simulated values show a good agreement with Aspen-HYSYS, as is evident from the plot obtained in Figure 3.

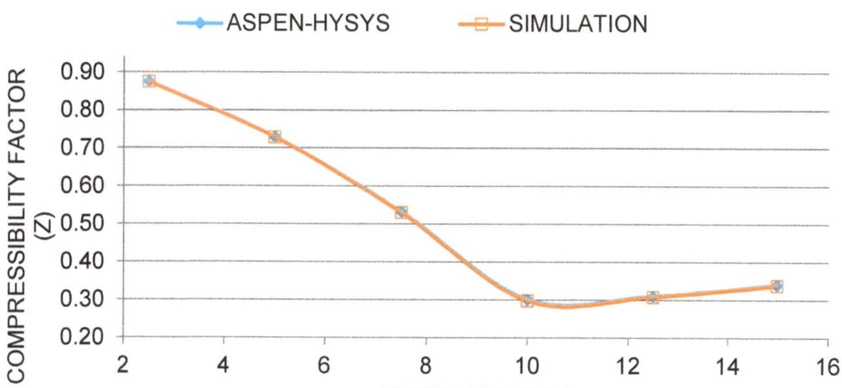

Figure 3. Comparison between simulated compressibility factors with the Peng–Robinson equation of state (PR-EOS) property table from Aspen-HYSYS for pure CO_2 at 40 °C.

However, the PR-EOS like all other equations of state has limits of applicability. Thus, the code will give useful results from zero pressure and temperature to pressures and temperatures of about 30 MPa and 589.29 K, respectively. It is worth mentioning that the maximum pressure required for the CO_2 pipeline is 15 MPa, which is within the EOS applicability range.

9. Results and Discussion

Figures 4–6 depict the GT power required for 10, 8, and 6 compressor stages of CO_2 compression, respectively, within the operating conditions earlier highlighted. The process requirement is to give a constant discharge pressure which is accomplished by varying the speed of the GT driver. In so doing, the compressor efficiently develops a constant head by lowering its speed at flows below the rated flow. As a result, it can be observed from the graphs that, as the CO_2 mass flow rate increases, the power requirement increases.

Further deducible from the plots is the steady increase in compression power requirement as the intercooler pressure drop increases. This is necessary to ensure the required discharge pressure is attained during the compression process.

At the rated flow of 1.5 MTPY, the power required considering a 5% intercooler pressure drop is 22.98 MW, 23.03 MW, and 23.41 MW for the 10, 8, and 6 compression stages, respectively. The values indicate about a 2% increase in power requirement as the number of stages reduces from 10 to 6 (compression ratios varying from 1.58 to 2.26). The rise in compressor head or otherwise rise in stage compression ratio with fewer stages accounts for this development.

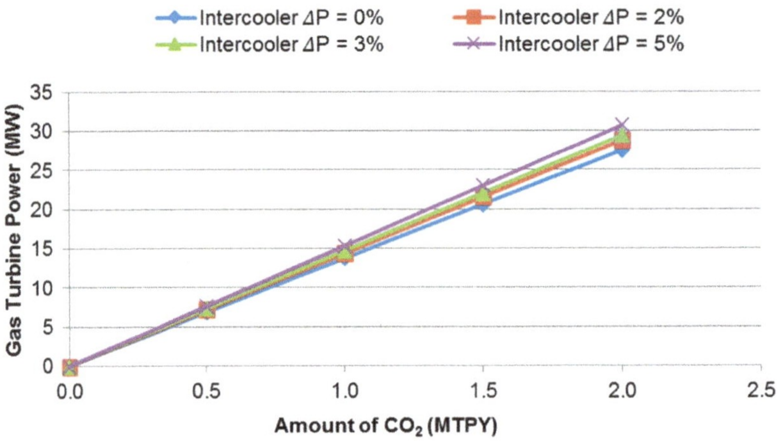

Figure 4. Gas turbine power requirement for a 10-stage CO_2 compression.

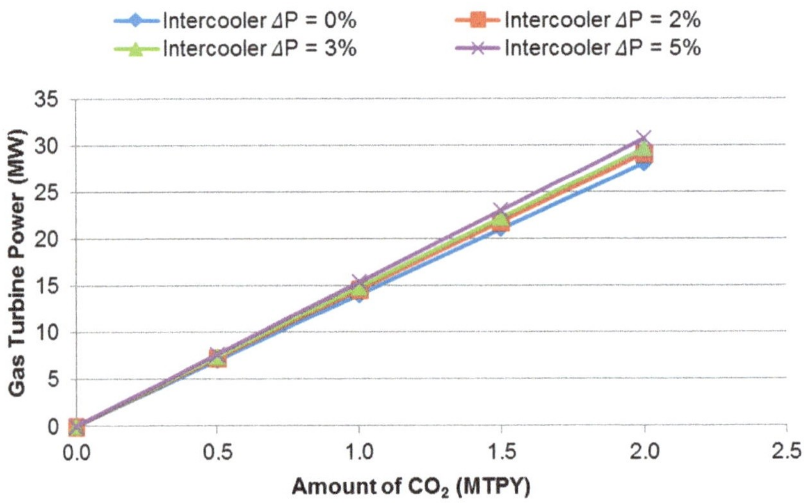

Figure 5. Gas turbine power requirement for a 8-stage CO_2 compression.

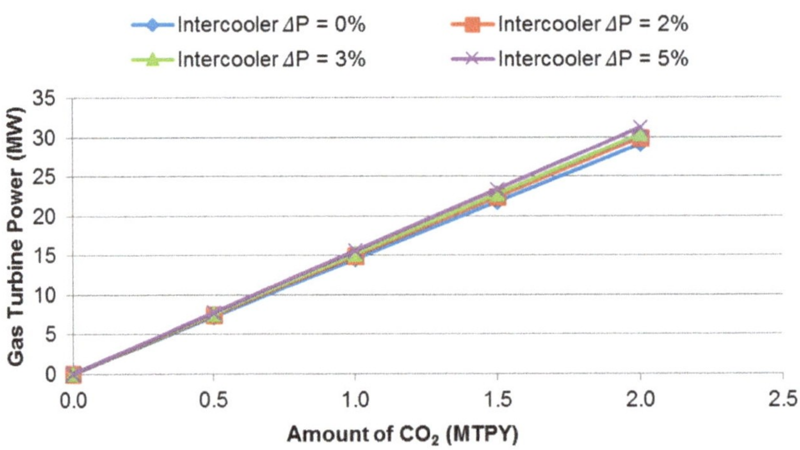

Figure 6. Gas turbine power requirement for a 6-stage CO_2 compression.

From the operational point of view, the increase in required power means an increase in energy cost. However, this is a trade-off for the compactness and comparative light weight of centrifugal compressors with fewer numbers of stages, which is advantageous during installation and scheduled maintenance. Another merit of the increase compressor head from the GT driver performance point of view is the ability to operate near design speed at a rated compressor operating point, thus minimizing part load operation. This is so considering the fact that higher head per stage is achieved at a higher compressor speed.

10. Conclusions

This paper has established the estimated power demand by the GT prime mover to compress typical CO_2 flows within the limits of assumptions considered. The results have shown that about 23 MW of power will be required for the initial compression of the CO_2 to pipeline operating pressure. This amount of power corresponds to 1.5 MTPY of CO_2—a projected equivalent CO_2 released from a 530-MW combined cycle gas turbine (CCGT) power generation plant. Future work will look at the off-design performance of a selected GT prime mover as well as cost implications.

Acknowledgments: The overseas study fellowship of the Federal Government of Nigeria is highly appreciated. Many thanks to Pericles Pilidis of the Centre for Propulsion Engineering, Cranfield University, UK for his candid support and supervision. The corresponding author is also grateful to Rahmatullah Yaqoob for proofreading this paper.

Author Contributions: Abdussalam El-Suleiman designed, run and analyzed the codes for the research; Nnamdi B. Anosike validated the codes; while Pericles Pilidis supervised and justified the results obtained. The paper was written by Abdussalam El-Suleiman. All authors have read and approved the final manuscript.

Conflicts of Interest: The authors declare no conflict of interest.

References

1. Safdarnejad, S.M.; Hedengren, J.D.; Baxter, L.L. Dynamic Optimization of a Hybrid System of Energy-Storing Cryogenic Carbon Capture and a Baseline Power Generation Unit. *Appl. Energy* **2016**, *172*, 66–79. [CrossRef]
2. IPCC. *Special Report on Carbon Dioxide Capture and Storage*; Prepared by Working Group III of the Intergovernmental Panel on Climate Change; Cambridge University Press: Cambridge, UK; New York, NY, USA, 2005.
3. Kurz, R.; Ohanian, S.; Brun, K. Compressors in High Pressure Pipeline Applications. In Proceedings of the ASME Turbo Expo, Glasgow, UK, 14–18 June 2010; pp. 631–640.
4. Li, H.; Yan, J. Impacts of Impurities in CO_2-Fluids on CO_2 Transport Process. In Proceedings of the ASME Turbo Expo 2006: Power for Land, Sea, and Air, Barcelona, Spain, 8–11 May 2006; pp. 367–375.
5. Joana, S.; Joris, M.; Evangelos, T. *Technical and Economic Characteristics of a CO_2 Transmission Pipeline Infrastructure*; EUR 24731 EN; European Commission Joint Research Centre, Institute for Energy: Petten, The Netherlands, 2011.
6. Knoope, M.M.J.; Ramirez, A.; Faaij, A.P.C. A State-of-the-Art Review of Techno-Economic Models Predicting the Costs of CO_2 Pipeline Transport. *Int. J. Greenh. Gas Control* **2013**, *16*, 241–270. [CrossRef]
7. Zhang, Z.X.; Wang, G.X.; Massarotto, P.; Rudolph, V. Optimization of Pipeline Transport for CO_2 Sequestration. *Energy Convers. Manag.* **2006**, *47*, 702–715. [CrossRef]
8. International Energy Agency Greenhouse Gas Programme (IEA GHG). *CO_2 Pipeline Infrastructure: An Analysis of Global Challenges and Opportunities*; Element Energy Ltd.: Cambridge, UK, 2010.
9. McCollum, D.L.; Ogden, J.M. *Techno-Economic Models for Carbon Dioxide Compression, Transport, and Storage & Correlations for Estimating Carbon Dioxide Density and Viscosity*; Institute of Transportation Studies, University of California: Davis, CA, USA, 2006.
10. Chandel, M.K.; Pratson, L.F.; Williams, E. Potential Economies of Scale in CO_2 Transport through Use of a Trunk Pipeline. *Energy Convers. Manag.* **2010**, *51*, 2825–2834. [CrossRef]
11. Justin, Z.; Sara, T. CO_2 Capture and Sequestration Options: Impact on Turbo-Machinery Design. In Proceedings of the ASME Turbo Expo, Berlin, Germany, 9–13 June 2008; pp. 945–963.

12. Cohen, S.M.; Rochelle, G.T.; Webber, M.E. Optimizing Post-Combustion CO_2 Capture in Response to Volatile Electricity Prices. *Int. J. Greenh. Gas Control* **2012**, *8*, 180–195. [CrossRef]
13. Skovholt, O. CO_2 Transportation System. *Energy Convers. Manag.* **1993**, *34*, 1095–1103. [CrossRef]
14. Koornneef, J.; Ramírez, A.; Turkenburg, W.; Faaij, A. The Environmental Impact and Risk Assessment of CO_2 Capture, Transport and Storage—An Evaluation of the Knowledge Base. *Prog. Energy Combust. Sci.* **2012**, *38*, 62–86. [CrossRef]
15. Safdarnejad, S.M.; Kennington, L.; Baxter, L.L.; Hedengren, J.D. Investigating the Impact of Cryogenic Carbon Capture on Power Plant Performance. In Proceedings of the American Control Conference, Palmer House Hilton, Chicago, IL, USA, 1–3 July 2015; pp. 5016–5021.
16. Mohitpour, M.; Golshan, H.; Murray, A. *Pipeline Design and Construction: A Practical Approach*, 2nd ed.; ASME Press: New York, NY, USA, 2003.
17. McCoy, S.T.; Rubin, E.S. An Engineering-Economic Model of Pipeline Transport of CO_2 with Application to Carbon Capture and Storage. *Int. J. Greenh. Gas Control* **2008**, *2*, 219–229. [CrossRef]
18. Ogaji, S.; Pilidis, P.; Sethi, V. Power Plant Selection in a Carbon Constrained World the TERA (Technoeconomic Environmental Risk Analysis). *J. Aerosp. Power* **2009**, *24*, 896–902.
19. MacCapani, M.; Khan, R.S.R.; Burgmann, P.J.; Lorenzo, G.D.; Ogaji, S.O.T.; Pilidis, P.; Bennett, I. A TERA Based Comparison of Heavy Duty Engines and their Artificial Design Variants for Liquefied Natural Gas Service. *J. Eng. Gas Turbines Power* **2014**, *136*. [CrossRef]
20. Peng, D.-Y.; Robinson, D.B. A New Two—Constant Equation of State. *Ind. Eng. Chem. Fundam.* **1976**, *15*, 59–64. [CrossRef]
21. Maroto-Valer, M.M. *Developments and Innovation in Carbon Dioxide (CO_2) Capture and Storage Technology*; CRC Press: Boca Raton, FL, USA; Oxford, UK, 2010.
22. Kurz, R. *The Physics of Centrifugal Compressor Performance*; Pipeline Simulation Interest Group: Palm Springs, CA, USA, 2004.
23. Paresh, G. *Performance Evaluation of Pumps and Compressors*, 1st ed.; Lulu.com: Raleigh, NC, USA, 2008.
24. Boyce, M.P. *Centrifugal Compressors: A Basic Guide*; PennWell: Tulsa, OK, USA, 2003.
25. Mokhatab, S.; Poe, W.A.; Speight, J.G. Natural Gas Compression. In *Handbook of Natural Gas Transmission and Processing*; Gulf Professional Pub.: Burlington, MA, USA, 2006; pp. 295–322.
26. Walsh, P.P.; Fletcher, P. *Gas Turbine Performance*, 2nd ed.; Blackwell Science: Oxford, UK, 2004.
27. Coal Industry Advisory Board (CIAB). *Power Generation from Coal: Measuring and Reporting Efficiency Performance and CO_2 Emissions*; OECD/IEA: Paris, France, 2010.

16

Drilling of Copper Using a Dual-Pulse Femtosecond Laser

Chung-Wei Cheng [1,*] and Jinn-Kuen Chen [2]

1 Department of Mechanical Engineering, National Chiao Tung University, No. 1001, Ta Hsueh Road, Hsinchu 30010, Taiwan

2 Department of Mechanical and Aerospace Engineering, University of Missouri, Columbia, MO 65211, USA; chenjnk@missouri.edu

* Correspondence: weicheng@nctu.edu.tw

Academic Editors: Chia-Lung Kuo, Jin-Chen Hsu, Chao-Ching Ho and Yuan-Jen Chang

Abstract: The drilling of copper using a dual-pulse femtosecond laser with wavelength of 800 nm, pulse duration of 120 fs and a variable pulse separation time (0.1–150 ps) is investigated theoretically. A one-dimensional two-temperature model with temperature-dependent material properties is considered, including dynamic optical properties and the thermal-physical properties. Rapid phase change and phase explosion models are incorporated to simulate the material ablation process. Numerical results show that under the same total laser fluence of 4 J/cm^2, a dual-pulse femtosecond laser with a pulse separation time of 30–150 ps can increase the ablation depth, compared to the single pulse. The optimum pulse separation time is 85 ps. It is also demonstrated that a dual pulse with a suitable pulse separation time for different laser fluences can enhance the ablation rate by about 1.6 times.

Keywords: femtosecond laser; dual pulse; two-temperature model

1. Introduction

Ultrashort (pico- or femto-second) lasers have been successfully demonstrated in drilling, cutting, surface structural modification and internal modification of transparent material because of their minimal heat-affected zone [1] and high peak power intensity [2]. Many theoretical and experimental works on ultrashort laser-material interactions have been reported. In the theoretical studies on femtosecond laser ablation of metal, most works have focused on two-temperature model (TTM) [1,3–5] or hydrodynamic model [6]. Experiments with single- or multi-shot ablation or the drilling of metal under different laser parameters have also been presented, e.g., laser fluence [7–9] and pulse duration [10–12]. For example, the ablation rate of iron can be optimized by variation of the single pulse duration in the range of 0.3–3.6 ps [12]. In our previous study, both numerical simulation and experiments was performed to investigate the femtosecond laser ablation of copper by a single-shot or multi-pulses [13,14].

Recently, numerous experiments on the enhancement of metal drilling using a dual-pulse nanosecond laser have been presented [15–17]. Nanosecond laser pulses separated by several tens of nanoseconds are shown that they can enhance the material removal rates while minimizing the heat-affected zone [16]. Experiments on the ultrashort dual-pulse laser drilling of copper using different wavelengths (800 nm and 532 nm) have also been investigated [18–20]. For a pulse separation time longer than 10 ps, the resulting plasma was found to shield the target surface and consequently decrease the ablation rate. On the other hand, the enhanced grooving of the steel when using a dual-pulse femtosecond laser was reported recently [21]. However, the optimum pulse delay time for ablation enhancement remains unclear, especially for the pulse separation time in the order of picoseconds.

In this study, the thermal ablation of copper by means of a dual-pulse femtosecond laser is investigated using a computational model, including a TTM with dynamic optical properties, two phase change models for melting and evaporation, and a phase explosion criterion for ejection of the mixture of metastable liquid droplets and vapor. The numerical results of the thermal response and the ablation depth generated by the dual-pulse femtosecond laser under different separation times and laser fluences are presented and discussed herein.

2. Modeling

In Figure 1, a copper foil is normally irradiated by a dual-pulse femtosecond laser on the front surface and the laser beam is propagated along the *z*-axis. The dual pulse is Gaussian in time with a full width at half maximum of t_p. The pulse separation time is t_{sep}. For simplicity without losing accuracy, the problem is approximated to be one dimensional since laser spot sizes are much larger than the thermally affected depth. The 1D TTM for electron and lattice temperature in a copper foil is expressed as follows [22,23]:

$$C_e \frac{\partial T_e}{\partial t} = \frac{\partial}{\partial z}(k_e \frac{\partial T_e}{\partial z}) - G(T_e - T_l) + S(z,t) \tag{1}$$

$$C_l \frac{\partial T_l}{\partial t} = \frac{\partial}{\partial z}(k_l \frac{\partial T_l}{\partial z}) + G(T_e - T_l) \tag{2}$$

where t is the time, T the temperature, C the heat capacity, k the thermal conductivity, G the electron-phonon coupling factor, and $S(z, t)$ the laser heat density. The subscripts e and l denote the electron and lattice, respectively. The k_l is usually neglected for pure metals due to the fact that it is two orders smaller than k_e. However, the heat conduction in the lattice should be considered in this model for which it includes solid-liquid and liquid-vapor phase change.

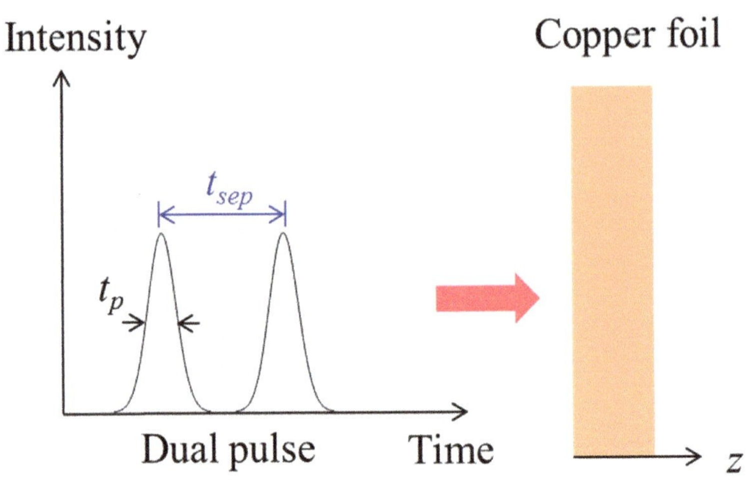

Figure 1. Dual-pulse femtosecond laser irradiated on copper surface.

The laser heat density S for a dual pulse can be expressed in a general form as [13]:

$$S(z,t) = \sum_{i=1}^{2} 0.94 \frac{[1 - R(0,t)]F_i}{t_p} \frac{1}{\delta(z,t) + \delta_b} \exp\left[-\int_0^z \frac{1}{\delta(z,t) + \delta_b} dz - 2.77 \left(\frac{t - (i-1)t_{sep}}{t_p} \right)^2 \right] \tag{3}$$

where $R(0,t)$ is the temperature-dependent surface reflectivity of the material at $z = 0$; F_i is the laser fluence of the i-pulse; $\delta(z, t) = 1/\alpha(z,t)$ is the temperature-dependent optical penetration depth; and $\alpha(z,t)$ is the absorption coefficient. The electron ballistic range δ_b added to the $\delta(z,t)$ here is to take into account the effects of the hot electron diffusion that spreads the absorbed laser energy into a deeper part of electrons. In this work, a constant value of $\delta_b = 15$ nm for copper is used [5].

In previous studies [3], constant surface reflectivity and absorption coefficient are often assumed for laser energy deposition. However, the dynamic optical properties during laser irradiation can significantly alter the irradiated laser energy absorption and influence the distribution of laser heat density, respectively. In this study, a critical point model with three Lorentzian terms is used to calculate the optical reflectivity and the absorption coefficient of copper [24].

The thermophysical properties, *i.e.*, C, k and G, control the thermal transport and temperature distributions in the laser-irradiated material. In this study, polynomial functions are used to describe the C_e and G for the copper material over a range of electron temperatures [14,25]. At a non-equilibrium condition, the k_e used is a function of electron and lattice temperature as [24]:

$$k_e = \chi \frac{(\varphi_e^2 + 0.16)^{5/4}(\varphi_e^2 + 0.44)\varphi_e}{(\varphi_e^2 + 0.092)^{1/2}(\varphi_e^2 + \xi\varphi_l)} \quad (4)$$

where $\phi_e = T_e/T_F$ and $\phi_l = T_l/T_F$. T_F is Fermi temperature; χ and ξ are constants. For copper, $T_F = 8.16 \times 10^4$ K, $\chi = 377$ W/(m·K), and $\xi = 0.139$. The bulk thermal conductivity, specific heat and mass density of copper in solid and liquid phases can be also found in [24].

If the laser fluence of a femtosecond laser pulse is sufficiently high, a solid medium can be melted and ablated through vaporization, and even by phase explosion. To accurately simulate the material removal process, the above 1D TTM is integrated with two phase change models for ultrafast melting/resolidification and evaporation, and a phase explosion model for ejection of metastable liquid and vapor. In the phase change model, the liquid-vapor interfacial velocity can be calculated from the vaporization wave model and shock wave theory [26,27]. The details of these models can be found in reference [28]. They are not described here for brevity.

For metals heated by femtosecond laser pulses, the molten material can be superheated without boiling, because the heating time is too short for the necessary heterogeneous nuclei to form. As the temperature of the superheated liquid approaches the thermodynamic equilibrium critical temperature, T_c, the tensile strength of the liquid rapidly falls to zero. Consequently, homogeneous bubble nucleation occurs at an extremely high rate. The superheated liquid thus relaxes explosively into a mixture of vapor and equilibrium liquid droplets and immediately ejected from the bulk material.

In numerical analyses of ultrafast laser ablation, it is often assumed that phase explosion takes place when the superheated liquid temperature reaches to 90% of T_c [29]. When a volume of the superheated liquid temperature reaches $0.9T_{tc}$, that material is removed under the assumption of phase explosion. The vertical size of the corresponding volume is defined as the ablation depth.

In the numerical simulation, a copper foil with an initial temperature of 300 K is irradiated on the front surface by a dual-pulse femtosecond laser with wavelength of 800 nm, pulse duration of 120 fs and a variable pulse separation time (0.1–150 ps). For each laser pulse, the lasering starts from $t = -2t_p$ and ends at $t = 2t_p$. The laser energy outside this time period is ignored, because it is too small to significantly alter the results. A number of 2500 finite volumes per μm are employed. The thermodynamic equilibrium critical temperature, T_c, is 7696 K for copper [30]. The details of the numerical iteration algorithms for modeling melting/resolidification and vaporization can be found in [31].

3. Results and Discussion

Figure 2 shows the time dependence of the calculated optical reflectivity, electron temperature and lattice temperature on the copper surface for the different laser pulses with the same total laser fluence of 4.0 J/cm^2, *i.e.*, a single pulse with 4.0 J/cm^2 and a dual pulse of 2.0 J/cm^2 per pulse. The pulse separation times of the two dual-pulses, t_{sep}, are 10 ps and 100 ps, respectively. As shown in Figure 2a, the reflectivity decreases quickly during the laser pulse irradiation; the reflectivity at room temperature (t = −0.24 ps) is 0.962, and the minimum values for the three cases are 0.244 (t = 0.12 ps), 0.274 (t = 10.1 ps) and 0.31 (t = 100.1 ps). The time-resolved reflectivity for copper irradiated by a femtosecond laser (806 nm, 100 fs) was also presented previously [32]. The temporal reflectivity decreased from a maximum of 0.85 to a minimum of 0.23, which is similar to our single-pulse case.

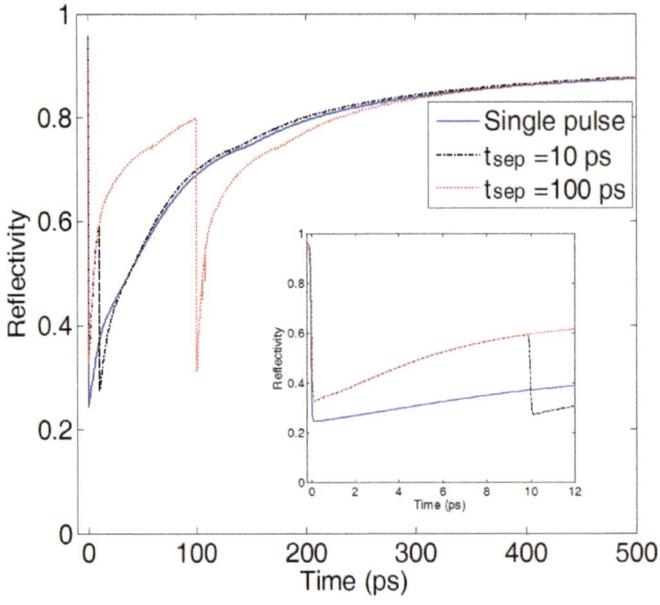

(**a**)

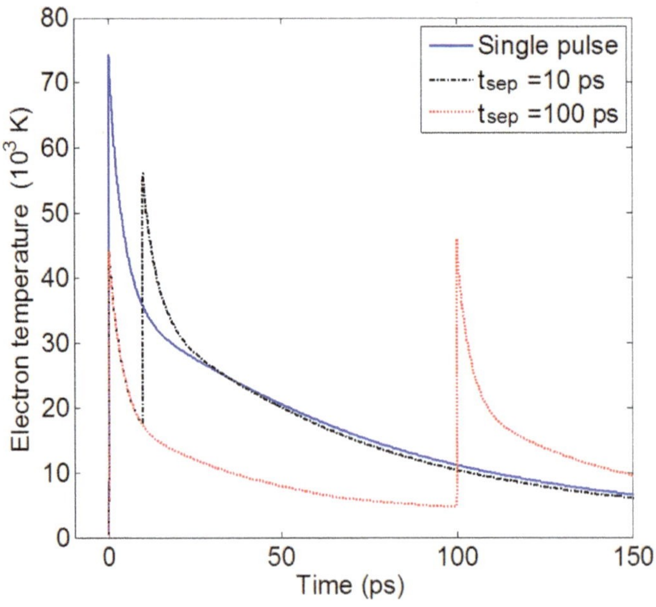

(**b**)

Figure 2. *Cont.*

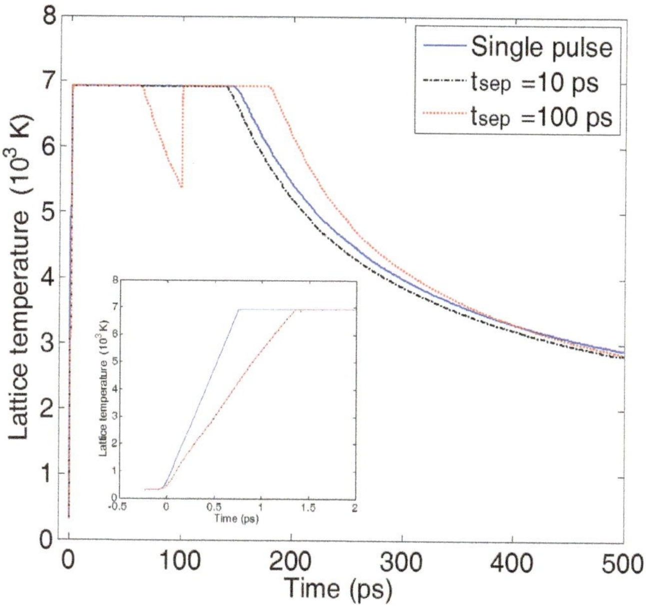

(c)

Figure 2. Time dependence of calculated (**a**) optical reflectivity; (**b**) electron temperature and (**c**) lattice temperature on the copper surface for different laser pulses with the same total laser fluence of 4.0 J/cm^2.

For the dual-pulse cases, the minimum reflectivity after the second pulse irradiation is usually lower than that of the first pulse. For example, in the case of t_{sep} = 10 ps (see inset), the minimum reflectivity for the first and second pulses are 0.327 and 0.274, respectively, since the high electron temperatures (see Figure 2b) excited on the copper surface after the first pulse help decrease the reflectivity for the second pulse. It was evident that these dynamic properties could alter the thermal response.

Figure 2b shows that the highest electron temperature decreases with increases in separation time; the maximum values for the three cases are 74,229 K (t = 0.12 ps), 56,029 K (t = 10.1 ps) and 45,967 K (t = 100.1 ps), respectively. In the dual-pulse cases, however, the peak electron temperature induced by the second pulse irradiation is usually higher than that by the first pulse. For example, in the case of t_{sep} = 100 ps, the peak electron temperatures induced by the first and second pulses are 44,121 K and 45,967 K, respectively. It is worth noting that the surface reflectivity before the second pulse irradiation is lower than that at room temperature (Figure 2a) due to temperature rise in material. The further decrease of the reflectivity by the second pulse increases the amount of laser energy deposited from the second pulse, and thus the electron temperature is increased to a higher level than that by the first pulse.

As shown in Figure 2c, over the timescale of a few picoseconds (see inset), the lattice is heated up via the electron-phonon collisions, and the maximum lattice temperatures after the first pulse for the three cases remain at around 6,926 K, *i.e.*, 90% of the thermodynamic equilibrium critical temperature, T_c (7696 K), which is attributed to the phase explosion. This temperature lasts for about 147 ps for the single pulse and 140 ps for the dual pulse with the pulse separation time 10 ps. However, for the dual-pulse cases with the pulse separation of 100 ps, the lattice temperature starts to drop to below 0.9 T_c at about 63 ps, and quickly reaches 0.9 T_c after the second pulse irradiation. Those time periods of temperature 0.9 T_c shown in Figure 2c indicate the occurrence of phase explosion.

The time histories of material ablation depth resulting from different laser pulses with the same total laser fluence of 4.0 J/cm^2 are shown in Figure 3a. For each case, there are two different ablation rates. The steep ablation rate results mainly from the phase explosion, while the flatter one in from evaporation. For the dual pulse with the separation time of 100 ps, phase explosion stops at 63 ps, followed by a vaportaion, and then re-starts when the second pulse impinges onto the material. This means that a dual pulse with a separation time shorter than 63 ps, phase explosion would continue to a time instant sooner or later than the time (178 ps here) found for the single pulse. It can be found from Figure 3a that the total ablation depths by the dual-pulse with $t_{sep} \leqslant 10$ ps are smaller than that of the single-pulse case. When $t_{sep} \geqslant 30$ ps, the total ablation depth is enhanced, as compared to the single pulse. The enhancement of material ablation by a dual pulse with a longer separation time can be explained as follows. Before the second laser pulse is irradiated, there is more time for the thermal energy to be conducted into the deeper part of material. The spread of energy heats more material to or near to the state of phase explosion, leading to more material ablation by the second pulse. It can be found from Figure 3a that for the case of t_{sep} = 100 ps, the ablation depth resulting from the first and second pulses are 52 nm and 79 nm, respectively.

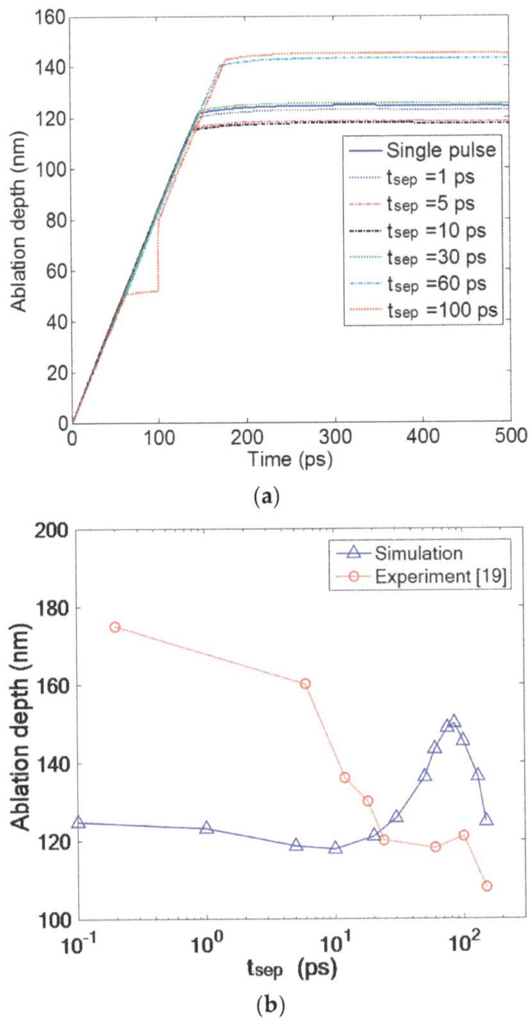

Figure 3. (**a**) Time history of ablation depth resulting from different laser pulses and (**b**) ablation depth as a function of separation time, with a total laser fluence of 4.0 J/cm^2.

The proposed method was used to predict ablation depths, as well as to compare the experimental result by [19] of two delayed femtosecond laser pulses (laser wavelength 800 nm, pulse duration 100 fs). Figure 3b shows the comparison between the simulated and experimental ablation depths with the same total laser fluence of 4.0 J/cm^2. The simulation results from Figure 3b demonstrate that an optimum value of the pulse separation time, t_{sep}, for enhancing the ablation efficiency exists. For the cases studied here with a total fluence 4.0 J/cm^2. The maximum ablation depth is about 150 nm for the dual pulse with the separation time 85 ps. It is noted, however, that the ablation depth by a dual pulse with t_{sep} < 30 ps is smaller than that by the single pulse of 125 nm.

However, the simulation ablation depths are not agreed well with those by experimental result. The simulation result for the dual pulse with t_{sep} < 10 ps is similar to the experimental results by [18,19]; e.g., the ablation depth is similar to that by the single pulse for t_{sep} < 1 ps and smaller for 1ps < t_{sep} < 10 ps. However, for t_{sep} > 10 ps, the experimental result show that the ablation depth decreases with t_{sep} in the range of 10–150 ps. The reason is that when t_{sep} > 10 ps, the plasma starts to shield the target surface and the second laser pulse is spent reheating the plasma, with no additional ablation depth resulting from the second pulse observed.

Interestingly, the study in which a dual-pulse femtosecond laser train was used to groove steel showed that the ablation depth is dramatically increased with an increase in t_{sep}, with a maximum ablation depth in the range of 5–30 ps obtained [21]. Recently, a dual-pulse femtosecond laser for ablation of silver foil in vacuum was presented [33]. It was also found that the ablation depth does not always decrease as the t_{sep} increases longer than 10 ps. For example, with a laser fluence of 106 J/cm^2, the ablation depth decreases for t_{sep} < 10 ps but increases to its maximum value for t_{sep} ~ 60 ps.

In the femtosecond laser material ablation process, it is desirable that the maximum ablation depth be obtained and that the temperature rise could be as low as possible to reduce the heat-affected zone and thermal stresses. Figure 4 shows the variation in the total ablation depth between the single and dual pulse for a total laser fluence of 3–8 J/cm^2. The ablation depth by the dual pulses with a suitable t_{sep}, *i.e.*, t_{sep} = 80, 85, 120, 145, 170 and 190 ps, is higher than that by the single pulse. The straight lines curve-fitted from the calculated data confirm the logarithmic dependence between the ablation depth and the laser fluence. The slope of the fitted line for the dual pulse is found to be higher by about 1.64 times, as compared to that for the single pulse. It is shown that the dual femtosecond pulses with suitable pulse separation time have a potential to increase material removal rates and thereby reducing the undesired thermal effects.

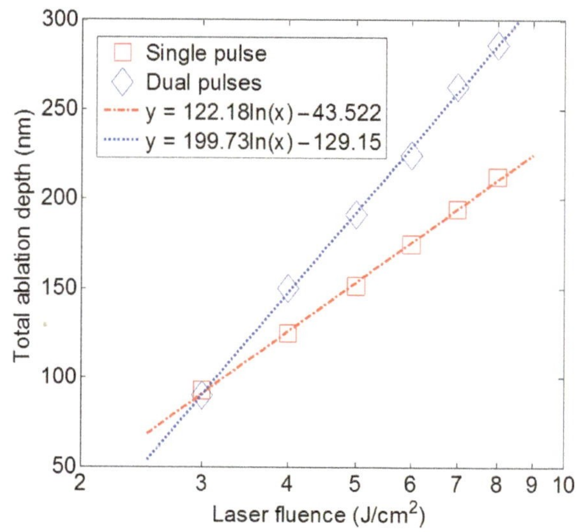

Figure 4. Variation in the ablation depth between single and dual pulses with laser fluence.

4. Conclusions

This paper reports the numerical results of thermal ablation of copper foil by a single- and dual-pulse femtosecond laser for laser fluences of 3–8 J/cm^2. It is found that a dual pulse with a total laser fluence of 4 J/cm^2 and a pulse separation time around 85 ps can increase the amount of material ablated, as compared to a single pulse. It is also demonstrated that a dual pulse with a suitable pulse separation time for different laser fluences can enhance the ablation rate by about 1.6 times. The results show that a dual-pulse femtosecond laser has a potential to improve laser drilling efficiency.

Acknowledgments: The authors would like to thank the Ministry of Science and Technology MOST 103-2218-E-009-025-MY2 for support of this research.

Author Contributions: Chung-Wei Cheng designed and implemented the simulation model, and prepared the manuscript. Jinn-Kuen Chen edited the manuscript, conceptualized the research. All authors have read and approved the final manuscript.

Conflicts of Interest: The authors declare no conflict of interest.

References

1. Nolte, S.; Momma, C.; Jacobs, H.; Tunnermann, A.; Chichkov, B.N.; Wellegehausen, B.; Welling, H. Ablation of metals by ultrashort laser pulses. *J. Opt. Soc. Am. B* **1997**, *14*, 2716–2722. [CrossRef]
2. Gattass, R.R.; Mazur, E. Femtosecond laser micromachining in transparent materials. *Nat. Photonics* **2008**, *2*, 219–225. [CrossRef]
3. Yang, J.; Zhao, Y.; Zhu, X. Theoretical studies of ultrafast ablation of metal targets dominated by phase explosion. *Appl. Phys. A* **2007**, *89*, 571–578. [CrossRef]
4. Wu, B.; Shin, Y.C. A simple model for high fluence ultra-short pulsed laser metal ablation. *Appl. Surf. Sci.* **2007**, *253*, 4079–4084. [CrossRef]
5. Byskov-Nielsen, J.; Savolainen, J.-M.; Christensen, M.S.; Balling, P. Ultra-short pulse laser ablation of copper, silver and tungsten: Experimental data and two-temperature model simulations. *Appl. Phys. A* **2011**, *103*, 447–453. [CrossRef]
6. Colombier, J.P.; Combis, P.; Bonneau, F.; le Harzic, R.; Audouard, E. Hydrodynamic simulations of metal ablation by femtosecond laser irradiation. *Phys. Rev. B* **2005**, *71*, 165406. [CrossRef]
7. Byskov-Nielsen, J.; Savolainen, J.M.; Christensen, M.S.; Balling, P. Ultra-short pulse laser ablation of metals: Threshold fluence, incubation coefficient and ablation rates. *Appl. Phys. A* **2010**, *101*, 97–101. [CrossRef]
8. Hashida, M.; Semerok, A.F.; Gobert, O.; Petite, G.; Izawa, Y.; Wagner, J.F. Ablation threshold dependence on pulse duration for copper. *Appl. Surf. Sci.* **2002**, *197*, 862–867. [CrossRef]
9. Kirkwood, S.E.; van Popta, A.C.; Tsui, Y.Y.; Fedosejevs, R. Single and multiple shot near-infrared femtosecond laser pulse ablation thresholds of copper. *Appl. Phys. A* **2005**, *81*, 729–735. [CrossRef]
10. Le Harzic, R.; Breitling, D.; Weikert, M.; Sommer, S.; Fohl, C.; Valette, S.; Donnet, C.; Audouard, E.; Dausinger, F. Pulse width and energy influence on laser micromachining of metals in a range of 100 fs to 5 ps. *Appl. Surf. Sci.* **2005**, *249*, 322–331. [CrossRef]
11. Weck, A.; Crawford, T.H.R.; Wilkinson, D.S.; Haugen, H.K.; Preston, J.S. Laser drilling of high aspect ratio holes in copper with femtosecond, picosecond and nanosecond pulses. *Appl. Phys. A* **2008**, *90*, 537–543. [CrossRef]
12. Artyukov, I.A.; Zayarniy, D.A.; Ionin, A.A.; Kudryashov, S.I.; Makarov, S.V.; Saltuganov, P.N. Relaxation phenomena in electronic and lattice subsystems on iron surface during its ablation by ultrashort laser pulses. *JETP Lett.* **2014**, *99*, 51–55. [CrossRef]
13. Wang, S.Y.; Ren, Y.; Cheng, C.W.; Chen, J.K.; Tzou, D.Y. Micromachining of copper by femtosecond laser pulses. *Appl. Surf. Sci.* **2013**, *265*, 302–308. [CrossRef]
14. Cheng, C.W.; Wang, S.Y.; Chang, K.P.; Chen, J.K. Femtosecond laser ablation of copper at high laser fluence: Modeling and experimental comparison. *Appl. Surf. Sci.* **2016**, *361*, 41–48. [CrossRef]
15. Lehane, C.; Kwok, H.S. Enhanced drilling using a dual-pulse Nd:YAG laser. *Appl. Phys. A* **2001**, *73*, 45–48. [CrossRef]

16. Forsman, A.C.; Banks, P.S.; Perry, M.D.; Campbell, E.M.; Dodell, A.L.; Armas, M.S. Double-pulse machining as a technique for the enhancement of material removal rates in laser machining of metals. *J. Appl. Phys.* **2005**, *98*, 033302. [CrossRef]
17. Wang, X.D.; Michalowski, A.; Walter, D.; Sommer, S.; Kraus, M.; Liu, J.S.; Dausinger, F. Laser drilling of stainless steel with nanosecond double-pulse. *Opt. Laser Technol.* **2009**, *41*, 148–153. [CrossRef]
18. Semerok, A.; Dutouquet, C. Ultrashort double pulse laser ablation of metals. *Thin Solid Films* **2004**, *453*, 501–505. [CrossRef]
19. Noël, S.; Hermann, J. Reducing nanoparticles in metal ablation plumes produced by two delayed short laser pulses. *Appl. Phys. Lett.* **2009**, *94*, 053120. [CrossRef]
20. Amoruso, S.; Bruzzese, R.; Wang, X.; O'Connell, G.; Lunney, J.G. Multidiagnostic analysis of ultrafast laser ablation of metals with pulse pair irradiation. *J. Appl. Phys.* **2010**, *108*, 113302. [CrossRef]
21. Liebig, C.; Srisungsitthisunti, P.; Weiner, A.M.; Xu, X. Enhanced machining of steel using femtosecond pulse pairs. *Appl. Phys. A* **2010**, *101*, 487–490. [CrossRef]
22. Anisimov, S.I.; Kapeliovich, B.L.; Perel'man, T.L. Electron emission from metal surfaces exposed to ultrashort laser pulses. *J. Exp. Theor. Phys.* **1974**, *39*, 375–377.
23. Chen, J.K.; Beraun, J.E. Numerical study of ultrashort laser pulse interactions with metal films. *Numer. Heat Transf. Part A* **2001**, *40*, 1–20.
24. Ren, Y.; Chen, J.K.; Zhang, Y. Optical properties and thermal response of copper films induced by ultrashort-pulsed lasers. *J. Appl. Phys.* **2011**, *110*, 113102. [CrossRef]
25. Lin, Z.; Zhigilei, L.V.; Celli, V. Electron-phonon coupling and electron heat capacity of metals under conditions of strong electron-phonon nonequilibrium. *Phys. Rev. B* **2008**, *77*, 075133. [CrossRef]
26. Courant, R.; Friedrichs, K.O. *Supersonic Flow and Shock Waves*; Interscience Publishers: New York, NY, USA, 1976.
27. Ionin, A.A.; Kudryashov, S.I.; Makarov, S.V.; Seleznev, L.V.; Sinitsyn, D.V. Generation and detection of superstrong shock waves during ablation of an aluminum surface by intense femtosecond laser pulses. *JETP Lett.* **2011**, *94*, 34–38. [CrossRef]
28. Ren, Y.; Cheng, C.W.; Chen, J.K.; Zhang, Y.; Tzou, D.Y. Thermal ablation of metal films by femtosecond laser bursts. *Int. J. Therm. Sci.* **2013**, *70*, 32–40. [CrossRef]
29. Kelly, R.; Miotello, A. Comments on explosive mechanisms of laser sputtering. *Appl. Surf. Sci.* **1996**, *96–98*, 205–215. [CrossRef]
30. Hess, H. Critical data and vapor pressures of aluminium and copper. *Z. Metallkunde* **1998**, *89*, 388–393.
31. Huang, J.; Zhang, Y.; Chen, J.K. Superheating in liquid and solid phases during femtosecond-laser pulse interaction with thin metal film. *Appl. Phys. A* **2011**, *103*, 113–121. [CrossRef]
32. Sandhu, A.S.; Dharmadhikari, A.K.; Kumar, G.R. Time resolved evolution of structural, electrical, and thermal properties of copper irradiated by an intense ultrashort laser pulse. *J. Appl. Phys.* **2005**, *97*, 023526. [CrossRef]
33. Roberts, D.E.; Plessis, A.d.; Botha, L.R. Femtosecond laser ablation of silver foil with single and double pulses. *Appl. Surf. Sci.* **2010**, *256*, 1784–1792. [CrossRef]

17

Correlation Plenoptic Imaging With Entangled Photons

Francesco V. Pepe [1,2], Francesco Di Lena [2,3], Augusto Garuccio [2,3], Giuliano Scarcelli [4] and Milena D'Angelo [2,3,*]

1 Museo Storico della Fisica e Centro Studi e Ricerche "Enrico Fermi", I-00184 Roma, Italy; francesco.pepe@ba.infn.it
2 Istituto Nazionale di Fisica Nucleare (INFN), Sezione di Bari, I-70126 Bari, Italy; francesco.dilena@uniba.it (F.D.L.); augusto.garuccio@uniba.it (A.G.)
3 Dipartimento Interateneo di Fisica, Università degli studi di Bari, I-70126 Bari, Italy
4 Fischell Department of Bioengineering, University of Maryland, College Park, MD 20742, USA; scarc@umd.edu
* Correspondence: milena.dangelo@uniba.it

Academic Editors: Yanhua Shih and Ronald E. Meyers

Abstract: Plenoptic imaging is a novel optical technique for three-dimensional imaging in a single shot. It is enabled by the simultaneous measurement of both the location and the propagation direction of light in a given scene. In the standard approach, the maximum spatial and angular resolutions are inversely proportional, and so are the resolution and the maximum achievable depth of focus of the 3D image. We have recently proposed a method to overcome such fundamental limits by combining plenoptic imaging with an intriguing correlation remote-imaging technique: ghost imaging. Here, we theoretically demonstrate that correlation plenoptic imaging can be effectively achieved by exploiting the position-momentum entanglement characterizing spontaneous parametric down-conversion (SPDC) photon pairs. As a proof-of-principle demonstration, we shall show that correlation plenoptic imaging with entangled photons may enable the refocusing of an out-of-focus image at the same depth of focus of a standard plenoptic device, but without sacrificing diffraction-limited image resolution.

Keywords: entanglement; ghost imaging; three-dimensional imaging

---

1. Introduction

Plenoptic imaging, also known as light-field or integral imaging, is a novel optical imaging modality [1]. Its key principle is to record the three-dimensional light field of a given scene by measuring both the location and the propagation direction of the incoming light. In particular, several images of the scene, one for each propagation direction of light within the scene, are acquired in a single shot. On one hand, such images correspond to the required viewpoints enabling the three-dimensional reconstruction of the scene. In fact, plenoptic imaging is the simplest method of 3D imaging with the present technological means [2–4]. On the other hand, the available angular information also enables the simplification of low-light shooting: The acquired images can be combined, in post-processing, to give an overall image characterized by the same depth of field of the N original images, but a signal-to-noise ratio N times larger [5].

Plenoptic imaging is currently used in digital cameras enhanced by refocusing capabilities [6–8]; in fact, in photography, plenoptic imaging highly simplifies both auto-focus and low-light shooting [5]. A plethora of innovative applications in 3D imaging and sensing [2,9], stereoscopy [1,10,11] and microscopy [3,12,13] are also being developed. In particular, high-speed large-scale 3D functional imaging of neuronal activity has been demonstrated [4].

However, the potentials of plenoptic imaging are strongly limited by the inherent inverse proportionality between image resolution and maximum achievable depth of field. In fact, plenoptic imaging has so far been implemented by inserting a microlens array in the native image plane, while moving the sensor array behind the microlenses. The image of the scene is reproduced on the microlenses, which thus define the spatial resolution of the acquired image. Each microlens also serves for reproducing, on the sensor array, an image of the camera lens, thus providing the angular information associated with each imaging pixel [5]. As a result, a trade-off between spatial and angular resolution is built in the plenoptic imaging process. To recover the lost resolution, signal processing and deconvolution have been implemented [3,4,14–17].

We have recently proposed a novel approach to plenoptic imaging, named correlation plenoptic imaging (CPI), which exploits the spatio-temporal second-order correlation typical of chaotic light sources to beat the strong coupling between spatial and angular resolution, as imposed to standard plenoptic imaging devices [18,19]. From a fundamental standpoint, the plenoptic application has been the first physical context where the counterintuitive properties of chaotic light (namely, the coexistence of momentum and position correlation [20]) are effectively used to beat intrinsic limits of standard imaging systems. From a practical standpoint, our protocol has been shown to dramatically enhance the potentials of plenoptic imaging. However, in contrast with chaotic light [21], correlation imaging based on entangled photons has been shown to enable sub-shot-noise imaging [22], as required by biomedical and security applications. Hence, in this paper, we investigate the possibility of performing CPI with entangled photons, or twin beams, from spontaneous parametric down-conversion (SPDC) [23]. We show that the peculiar momentum-momentum and position-position correlations typical of such EPR entangled systems [24,25] can be simultaneously exploited to substantially weaken the connection between spatial resolution and depth of field typical of standard plenoptic imaging.

The proposed setup for CPI with entangled photons from SPDC is reported in Figure 1. In view of plenoptic imaging, the setup must enable the parallel acquisition of several images of the given scene, one for each propagation direction of light. In fact, as we shall soon demonstrate, the sensor array S_a retrieves *N coherent ghost images* of the object by means of correlation measurement with each of the pixels of the sensor array S_b. Such images represent different viewpoints of the desired scene. This is quite intuitive considering sensor S_b reproduces the image of the light source. Hence, each coherent ghost image is associated with a different illumination of the object. Interestingly, the single lens L_b replaces the microlens array required in standard plenoptic imaging. In summary, the basic idea of CPI is to replace with a single lens and two separate sensors, the complex system composed of the microlens array followed by a single sensor; spatial and angular measurements are thus physically decoupled, enabling a significant weakening of the inverse proportionality between spatial and angular resolution characterizing standard plenoptic imaging devices.

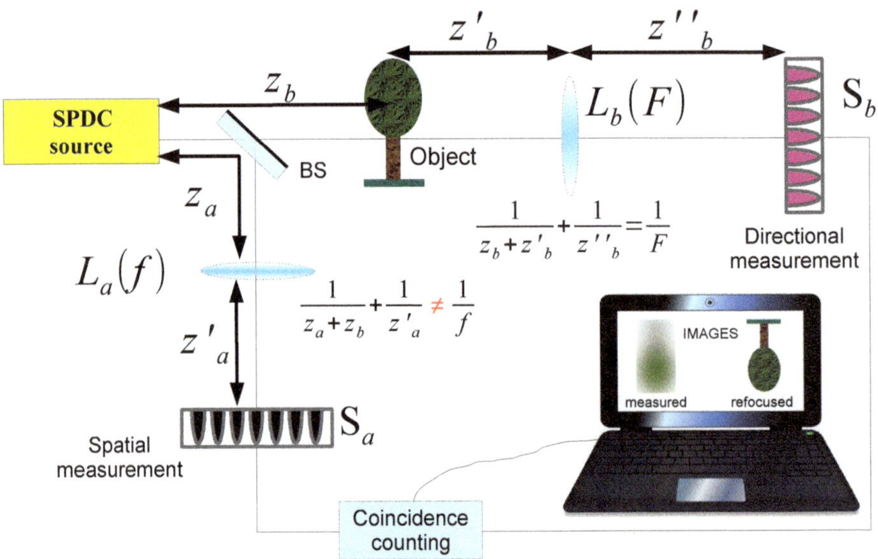

Figure 1. Schematic setup for correlation plenoptic imaging with entangled photons from SPDC. Signal and idler beams emitted from the SPDC source impinge on a beam-splitter (BS). Both beams are split into a reflected path a and a transmitted path b. The reflected beam propagates toward the lens L_a of focal length f and is refracted toward the high resolution sensor array S_a. The transmitted beam propagates through the object, playing the role of the desired scene, and is collected by the lens L_b of focal length F before being detected by the high-resolution sensor array S_b. The two sensors are connected to a coincidence counting circuit. On one hand, distances z_b, z_b', z_b'' are chosen in such a way that the source and the sensor S_b are in conjugate planes of the lens L_b. On the other hand, distances z_a and z_a' are such that, when the two-photon thin-lens equation $1/(z_b + z_a) + 1/z_a' = 1/f$ is satisfied, a ghost image of the object is retrieved on sensor S_a, triggered by sensor S_b.

2. Theoretical Analysis

2.1. Background

The coincidence detection of entangled photons from SPDC is described by the second order Glauber correlation function [26]:

$$G^{(2)}(\boldsymbol{r}_a, \boldsymbol{r}_b; t_a, t_b) = \langle \Psi | E_a^{(-)}(\boldsymbol{r}_a, t_a) E_b^{(-)}(\boldsymbol{r}_b, t_b) E_b^{(+)}(\boldsymbol{r}_b, t_b) E_a^{(+)}(\boldsymbol{r}_a, t_a) | \Psi \rangle, \quad (1)$$

where

$$E_j^{(+)}(\boldsymbol{r}_j, t_j) = \int d\omega \int d\boldsymbol{\kappa}\, a_{\boldsymbol{k}} e^{-i\omega t_j} g_j(\boldsymbol{r}_j, \boldsymbol{k}), \quad (2)$$

is the positive-energy part of the electric field at sensor j (with $j = a, b$), placed in $\boldsymbol{r}_j = (\boldsymbol{\rho}_j, z_j)$, t_j the time of the detection, ω is the frequency and $\boldsymbol{k} = (\boldsymbol{\kappa}, \omega/c)$ the wave vector of the detected radiation, g_j is the Green's function propagating the field mode $\boldsymbol{k}$ from the source to the sensor. The negative-energy part $E_j^{(-)}(\boldsymbol{r}_j, t_j)$ of the electric field is the Hermitian conjugate of the field $E^{(+)}$ of Equation (2). A scalar approximation for the electric field has been assumed, which physically corresponds to considering a fixed polarization of light. The positive and negative-energy parts of the electric field involve the photon annihilation and creation operators ($a_{\boldsymbol{k}}$ and $a_{\boldsymbol{k}}^{\dagger}$), respectively, associated with wave vector $\boldsymbol{k}$. The expectation value in Equation (1) is taken over the two-photon signal-idler state produced by SPDC [27–29]:

$$|\Psi\rangle = \mathcal{N} \int d\nu s(LD\nu) \int d\boldsymbol{\kappa}_i d\boldsymbol{\kappa}_s h_{\mathrm{tr}}(\boldsymbol{\kappa}_i + \boldsymbol{\kappa}_s) a^\dagger_{\boldsymbol{k}_i} a^\dagger_{\boldsymbol{k}_s} |0\rangle, \tag{3}$$

where $\mathcal{N}$ is a normalization constant, ν is the detuning with respect to the central frequency of signal and idler $\Omega_s = \Omega_i = \omega_p/2$, which is linked by phase matching to the central frequency of the pump laser ω_p, L is the length of the SPDC crystal, D is the difference between the inverse group velocities of signal and idler, $s(LD\nu)$ is the spectrum of the SPDC biphoton [30,31], and h_{tr} is the Fourier transform of the pump transverse profile:

$$\mathcal{F}(\boldsymbol{\rho}) = \int d\boldsymbol{\kappa} e^{i\boldsymbol{\kappa}\cdot\boldsymbol{\rho}} h_{\mathrm{tr}}(\boldsymbol{\kappa}). \tag{4}$$

We have assumed, for simplicity, degenerate SPDC radiation, but the result can be easily generalized to the non-degenerate situation [32,33]. Without loss of generality, we shall further assume the source to be monochromatic, in such a way that the time dependence of the correlation function will not be relevant. By employing the canonical commutation relations $[a_{\boldsymbol{k}}, a_{\boldsymbol{k}'}] = 0$ and $[a_{\boldsymbol{k}}, a^\dagger_{\boldsymbol{k}'}] = \delta(\boldsymbol{k} - \boldsymbol{k}')$, with δ the Dirac delta distribution, and the inversion symmetry of the Fourier transform of the transverse pump profile $h_{\mathrm{tr}}(\boldsymbol{\kappa}) = h_{\mathrm{tr}}(-\boldsymbol{\kappa})$, the spatial part of the two-photon correlation function reads:

$$\Gamma(\boldsymbol{\rho}_a, \boldsymbol{\rho}_b) = \left| \int d\boldsymbol{\kappa}_a \int d\boldsymbol{\kappa}_b g_a(\boldsymbol{\rho}_a, \boldsymbol{\kappa}_a) g_b(\boldsymbol{\rho}_b, \boldsymbol{\kappa}_b) h_{\mathrm{tr}}(\boldsymbol{\kappa}_a + \boldsymbol{\kappa}_b) \right|^2, \tag{5}$$

up to irrelevant constants. This result indicates the strong coupling between the two remote sensors, as enabled by the momentum-momentum entanglement characterizing SPDC biphotons.

Let us now evaluate the propagators in the two arms of the setup depicted in Figure 1; we shall assume for simplicity the lenses to be diffraction-limited. In arm a, light propagates in free space for a distance z_a from the source to the lens L_a and is then detected by the sensor S_a, placed at a distance z'_a from the lens. In the paraxial approximation, propagation of a field with frequency $\Omega \simeq ck_z$ in free space from $(\boldsymbol{\rho}_1, z_1)$ to $(\boldsymbol{\rho}_2, z_2)$ is described by the function [34]:

$$\mathcal{G}(\boldsymbol{\rho}_2 - \boldsymbol{\rho}_1, z_2 - z_1) = \frac{-i\Omega e^{i\frac{\Omega}{c}(z_2 - z_1)}}{2\pi c(z_2 - z_1)} G(\boldsymbol{\rho}_2 - \boldsymbol{\rho}_1)_{\left[\frac{\Omega}{c(z_2 - z_1)}\right]} \tag{6}$$

with $G(\boldsymbol{x})_{[y]} = e^{iy|\boldsymbol{x}|^2/2}$. Knowing the initial field $E(\boldsymbol{\rho}_1)$, one can determine the final field $E(\boldsymbol{\rho}_2) = \int d\boldsymbol{\rho}_1 E(\boldsymbol{\rho}_1) \mathcal{G}(\boldsymbol{\rho}_2 - \boldsymbol{\rho}_1, z_2 - z_1)$. Propagation through a lens of focal length f is described by $G(\boldsymbol{\rho}_l)_{[-\Omega/(cf)]}$. Hence, the propagator associated with arm a of the setup reads:

$$\begin{aligned} g_a(\boldsymbol{\rho}_a, \boldsymbol{\kappa}_a) &= C_a(z_a, z'_a) \int d\boldsymbol{\rho}_s \int d\boldsymbol{\rho}_\ell e^{i\boldsymbol{\kappa}_a\cdot\boldsymbol{\rho}_s} G(\boldsymbol{\rho}_\ell - \boldsymbol{\rho}_s)_{\left[\frac{\Omega}{cz_a}\right]} G(\boldsymbol{\rho}_\ell)_{\left[-\frac{\Omega}{cf}\right]} G(\boldsymbol{\rho}_a - \boldsymbol{\rho}_\ell)_{\left[\frac{\Omega}{cz'_a}\right]} \\ &= C'_a(z_a, z'_a) G(\boldsymbol{\rho}_a)_{\left[\frac{\Omega}{c}\left(\frac{1}{z_a} - \frac{\zeta(z_a, z'_a)}{z'^2_a}\right)\right]} \int d\boldsymbol{\rho}_s e^{i\boldsymbol{\kappa}_a\cdot\boldsymbol{\rho}_s} G(\boldsymbol{\rho}_s)_{\left[\frac{\Omega}{cz_a}\left(1 - \frac{\zeta(z_a, z'_a)}{z_a}\right)\right]} e^{-\frac{i\Omega\zeta(z_a, z'_a)}{cz_a z'_a}\boldsymbol{\rho}_s\cdot\boldsymbol{\rho}_a}, \end{aligned} \tag{7}$$

where

$$\zeta(z_a, z'_a) = \left(\frac{1}{z_a} + \frac{1}{z'_a} - \frac{1}{f}\right)^{-1}, \tag{8}$$

$\boldsymbol{\rho}_s$ and $\boldsymbol{\rho}_\ell$ are transverse coordinate on the source and the lens L_a plane, respectively, and C_a, C'_a contain irrelevant constants. In arm b, light propagates for a distance z_b from the source to the object which represents the desired scene to image, then for a distance z'_b from the object to lens L_b, and a further distance z''_b before being detected by the sensor S_b. By indicating with A the aperture function of the object, and assuming the focusing condition $1/(z_b' + z_b'') + 1/z_b = 1/F$ to be satisfied, the propagator associated with arm b of the setup reads:

$$\begin{aligned} g_b(\boldsymbol{\rho}_b, \boldsymbol{\kappa}_b) =& C_b(z_b, z_b') \int d\boldsymbol{\rho}_s \int d\boldsymbol{\rho}_o \int d\boldsymbol{\rho}_\ell' e^{i\boldsymbol{\kappa}_a \cdot \boldsymbol{\rho}_s} A(\boldsymbol{\rho}_o) G(\boldsymbol{\rho}_o - \boldsymbol{\rho}_s)_{\left[\frac{\Omega}{cz_b}\right]} G(\boldsymbol{\rho}_\ell' - \boldsymbol{\rho}_o)_{\left[\frac{\Omega}{cz_b'}\right]} \\ & \times G(\boldsymbol{\rho}_\ell)_{\left[-\frac{\Omega}{cF}\right]} G(\boldsymbol{\rho}_b - \boldsymbol{\rho}_\ell)_{\left[\frac{\Omega}{cz_b''}\right]} \\ =& C_b'(z_b, z_b') G(\boldsymbol{\rho}_b)_{\left[\frac{\Omega}{cz_b''}\left(1 - \frac{1}{z_b''}\left(\frac{1}{z_b'} + \frac{1}{z_b''} - \frac{1}{F}\right)^{-1}\right)\right]} \int d\boldsymbol{\rho}_s d\boldsymbol{\rho}_o e^{i\boldsymbol{\kappa}_a \cdot \boldsymbol{\rho}_s} G(\boldsymbol{\rho}_s)_{\left[\frac{\Omega}{cz_b}\right]} A(\boldsymbol{\rho}_o) e^{-\frac{i\Omega}{cz_b}\left(\boldsymbol{\rho}_s + \frac{\boldsymbol{\rho}_b}{M}\right) \cdot \boldsymbol{\rho}_o}, \end{aligned} \tag{9}$$

where $\boldsymbol{\rho}_o$ and $\boldsymbol{\rho}_\ell'$ are transverse coordinate on the object and the lens L_b planes, respectively, $M = z_b''/(z_b + z_b')$ is the magnification of the image of the source on the sensor array S_b, and C_b, C'_b contain irrelevant constants.

By inserting in Equation (5) the Green's function given by Equations (7) and (9), and the laser pump profile on the SPDC crystal, as defined in Equation (4), one finds that the second order correlation function associated with signal-idler pairs from SPDC is given by the plenoptic correlation function:

$$\begin{aligned} \Gamma(\boldsymbol{\rho}_a, \boldsymbol{\rho}_b) = \mathcal{K}(z_a, z_a', z_b, z_b') \Bigg| & \int d\boldsymbol{\rho}_o A(\boldsymbol{\rho}_o) \int d\boldsymbol{\rho}_s \mathcal{F}(\boldsymbol{\rho}_s) G(\boldsymbol{\rho}_s)_{\left[\frac{\Omega}{c}\left[\frac{1}{z_b} + \frac{1}{z_a}\left(1 - \frac{\zeta(z_a, z_a')}{z_a}\right)\right]\right]} \\ & e^{-\frac{i\Omega\zeta(z_a, z_a')}{cz_a z_a'} \boldsymbol{\rho}_s \cdot \boldsymbol{\rho}_a} e^{-\frac{i\Omega}{cz_b}\left(\boldsymbol{\rho}_s + \frac{\boldsymbol{\rho}_b}{M}\right) \cdot \boldsymbol{\rho}_o} \Bigg|^2, \end{aligned} \tag{10}$$

where $\mathcal{K}$ contains irrelevant constants.

2.2. *Plenoptic Properties of the Correlation Function and Refocusing Capability*

As shown in Equation (10), the proposed CPI protocol is theoretically described by a second order correlation function encoding both spatial and angular information, hence, characterized by the key re-focusing capability typical of plenoptic imaging.

To develop an intuition about the result of Equation (10), we consider the simple case in which the distance between the object and the source $z_b = z_{bF}$ satisfies the two-photon thin lens equation [25,35]:

$$\frac{1}{z_a + z_{bF}} + \frac{1}{z_a'} = \frac{1}{f}. \tag{11}$$

In this case, by integrating the result of Equation (10) over the whole sensor array S_b, one gets the standard (*incoherent*) ghost image of the object, magnified by a factor of $m = \frac{z_a'}{z_a + z_{bF}}$, namely [25,35],

$$\Sigma_F(\boldsymbol{\rho}_a) = \int d\boldsymbol{\rho}_b \Gamma(\boldsymbol{\rho}_a, \boldsymbol{\rho}_b) \propto \int d\boldsymbol{\rho}_o |A(\boldsymbol{\rho}_o)|^2 \left| h_{\text{tr}} \left[\frac{\Omega}{cz_{bF}} \left(\boldsymbol{\rho}_o + \frac{\boldsymbol{\rho}_a}{m} \right) \right] \right|^2, \tag{12}$$

where h_{tr} is the Fourier transform of the laser pump profile, as defined in Equation (4). The above result is valid in the hypothesis that h_{tr} is similar to or narrower than the Fourier transform of the imaging lens L_a. In fact, such incoherent ghost image is formally equivalent to the incoherent image one would obtain in a standard imaging system characterized by a point-spread function h_{tr} given by the Fourier transform of the imaging lens aperture function.

However, the second order correlation function of Equation (10) can do much better than standard ghost imaging: The deep physical difference arises from the coherent nature of the ghost image it describes,

$$\Gamma_F(\rho_a, \rho_b) = \mathcal{K}(z_a, z'_a, z_{bF}, z'_b) \left| \int d\rho_o A(\rho_o) h_{\mathrm{tr}} \left[\frac{\Omega}{c z_{bF}} \left(\rho_o + \frac{\rho_a}{m} \right) \right] e^{-\frac{i\Omega}{c z_{bF}} \frac{\rho_b}{M} \cdot \rho_o} \right|^2, \tag{13}$$

that is obtained from the general expression (10), when the focusing condition in Equation (11) holds.

The coherence of such ghost image is the immediate consequence of measuring coincidences between the *spatial* sensor S_a and any single pixel of the *angular* sensor S_b. This can be better understood in terms of the Klyshko picture [35] reported in Figure 2: The light illuminating the object and contributing to the coincidence detection between any two pairs of pixels ρ_a and ρ_b has a well defined propagation direction (*i.e.*, it is spatially coherent). As made clear from Figure 2, the Klyshko picture also enables the interpretation of the proposed setup for CPI with entangled photons as a sort of *correlation* pinhole camera. Such a perspective helps developing an intuition about the analogy between the proposed scheme and standard plenoptic imaging, as well as understanding the role played by the sensor S_b in retrieving the angular information about the two-photon light field. In fact, due to the quasi one-to-one correspondence between points on the sensor S_b and points on the source, one can trace, in post-processing, the geometrical ray connecting each point of the source with each point of the object. This leads to the peculiar refocusing and 3D imaging capabilities of plenoptic imaging.

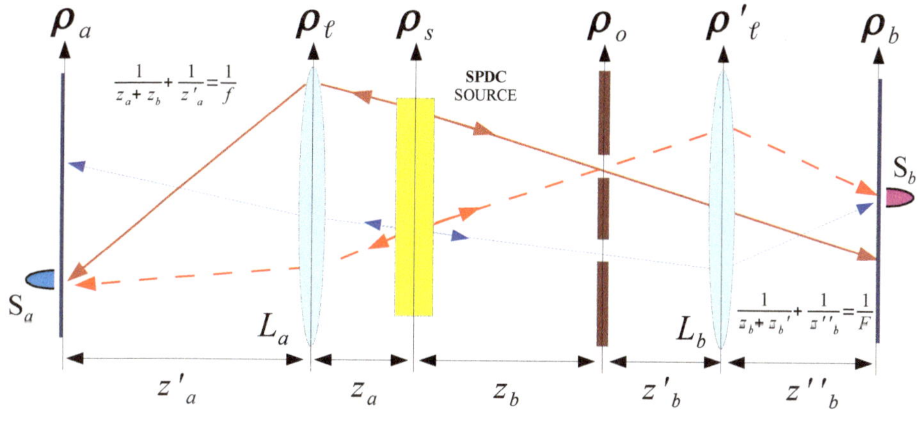

Figure 2. Unfolded version, or Klyshko picture, of the schematic setup reported in Figure 1, in the case in which the ghost image of the object is focused on the sensor S_a. On one hand, by means of coincidence detection, the lens L_a reproduces on the sensor S_a the ghost image of the object. On the other hand, the source and the sensor S_b are in conjugate planes of the lens L_b. However, based on the advanced-wave model proposed by Klyshko, the effect can be understood by treating sensor S_b as the light source and the SPDC source as a simple mirror. The solid and the dashed lines represent two-photon amplitudes that pass through the same slit; hence, at second order, they are focused in the same point of sensor S_a. The dashed and the dotted two-photon amplitudes are emitted by the same source point and are thus focused in the same point of sensor S_b.

Now, to explicitly demonstrate this last point and better highlight the plenoptic properties of the second-order correlation function of Equation (10), we shall consider the more general out-of-focus situation ($z_b \neq z_{bF}$) and rewrite it as a product of the pump profile $\mathcal{F}$ and the object aperture function A with the phase factor $e^{i\frac{\Omega}{c}\varphi(\rho_o,\rho_s;\rho_a,\rho_b)}$, with:

$$\varphi(\rho_o,\rho_s;\rho_a,\rho_b) = \left[\frac{1}{z_b} + \frac{1}{z_a}\left(1 - \frac{\zeta(z_a,z_a')}{z_a}\right)\right]\frac{|\rho_s|^2}{2} - \frac{\zeta(z_a,z_a')}{z_a z_a'}\rho_s\cdot\rho_a - \frac{1}{z_b}\left(\rho_s + \frac{\rho_b}{M}\right)\cdot\rho_o, \tag{14}$$

namely

$$\Gamma(\rho_a,\rho_b) \propto \left|\int d\rho_o A(\rho_o)\int d\rho_s \mathcal{F}(\rho_s)e^{i\frac{\Omega}{c}\varphi(\rho_o,\rho_s;\rho_a,\rho_b)}\right|^2. \tag{15}$$

The stationary points of the phase defined in Equation (14) enable us to determine the geometrical correspondence between points on the object and the source with points on the sensors S_a and S_b, respectively. In particular, the stationarity of φ with respect to ρ_s determines the object point that gives the predominant contribution to the integral of Equation (15), that is:

$$\rho_o = -\frac{z_b}{z_{bF}}\frac{\rho_a}{m} - \frac{\rho_b}{M}\left(1 - \frac{z_b}{z_{bF}}\right), \tag{16}$$

where the identity $\zeta(z_a,z_a') = (z_{bF} + z_a)z_a/z_{bF}$ has been used. When the focusing condition of Equation (11) is satisfied, this object point becomes independent of the specific sensor pixel ρ_b. Hence, the focused ghost image is not sensitive to the change of perspective enabled by the high resolution of the angular sensor S_b. On the other hand, the stationarity of φ with respect to ρ_o yields the focusing of the source on the sensor S_b:

$$\rho_s = -\frac{\rho_b}{M}. \tag{17}$$

Thus, in the geometrical optics limit, the second order correlation function of Equation (15) reduces to the product of the tilted and rescaled geometrical image of the object and the source profile:

$$\Gamma_G(\rho_a,\rho_b) \sim \left|A\left[-\frac{z_b}{z_{bF}}\frac{\rho_a}{m} - \frac{\rho_b}{M}\left(1 - \frac{z_b}{z_{bF}}\right)\right]\right|^2 \left|\mathcal{F}\left(-\frac{\rho_b}{M}\right)\right|^2. \tag{18}$$

Interestingly, by properly rescaling the variable ρ_a, the object can be completely decoupled from the source; in fact, the rescaled second order correlation function

$$\Gamma_G^{\mathrm{ref}}\left[\frac{z_{bF}}{z_b}\rho_a + \frac{\rho_b}{M}m\left(1 - \frac{z_{bF}}{z_b}\right), \rho_b\right] \sim \left|\mathcal{F}\left(-\frac{\rho_b}{M}\right)\right|^2 \left|A\left(-\frac{\rho_a}{m}\right)\right|^2, \tag{19}$$

gives the perfect geometrical image of the desired scene. Such rescaling is formally identical to the one employed both in standard plenoptic imaging [5] and in correlation plenoptic imaging with chaotic light [18,19].

Similar to standard plenoptic imaging, the signal to noise ratio of the refocused image can be improved by integrating the result of Equation (19) over the whole sensor array ρ_b, thus employing light coming from the whole light source:

$$\Sigma^{\mathrm{ref}}(\rho_a) = \int d\rho_b \Gamma^{\mathrm{ref}}\left[\frac{z_{bF}}{z_b}\rho_a + \frac{\rho_b}{M}m\left(1 - \frac{z_{bF}}{z_b}\right), \rho_b\right]. \tag{20}$$

This result represents the refocused incoherent ghost image of an object placed at a generic distance z_b from the source, and is thus the central result of the present paper.

The possibility of reconstructing the light field and refocusing an out-of-focus image, as reported in Equation (20), lies on the accuracy with which both object and source points are in a one-to-one correspondence with points on sensors S_a and S_b, respectively. We have already demonstrated that the Fourier transform of the transverse pump profile determines the object point spread function (see Equation (12)), with a spot size $\Delta\rho_a \sim mcz_{bF}/(\Omega D_s)$, where D_s is the diameter of the pump profile.

On the other hand, it is easy to check that the source is imaged with a point spread function given by the Fourier transform of the object aperture function. From Equation (10), one can infer that a point on the source corresponds to a spot of width $\Delta\rho_b \sim Mcz_b/(\Omega d)$ on the sensor S_b, with d the smallest length scale of the aperture function of the object. Thus, as far as the pixel sizes lie above the resolution limits, the spatial and angular resolution are decoupled. The structure of a standard plenoptic device, instead, entails an inverse proportionality relation between the angular resolution and the spatial resolution of the focused image, also in the geometrical-optics regime [1,5]. Thus, our protocol of plenoptic imaging with entangled photons enables us to beat this intrinsic limitation and achieve a larger depth of field (depending on the angular resolution), by leaving unchanged the resolution on the focused image and the total number of pixels.

3. Simulation of CPI With Entangled Photons From SPDC

In Figure 3, we show the enhanced depth of field induced by the refocusing capability of the SPDC correlation plenoptic protocol. A mask with a transparent letter E, whose thickness is $d = 0.2\,\text{mm}$, is placed in a setup with $z_a = 10\,\text{mm}$, $z'_a = 30\,\text{mm}$, and $f = 12\,\text{mm}$, which would give a focused ghost image magnified by $m = 1.5$. The object mask is illuminated by SPDC photons with $\lambda = 1\,\mu\text{m}$, generated by a pump whose Gaussian transverse profile has width $\sigma = 0.6\,\text{mm}$. With respect to the source, the object is placed at a distance $z_b = 3\,\text{mm}$, which is less than one third of the focused plane distance $z_{bF} = 10\,\text{mm}$. The ghost image of such an object would be focused at $z'_{aF} = 5z'_a$. The widths of the sensors S_a and S_b are fixed to $W_a = 6md = 1.8\,\text{mm}$ and $W_b = 4M\sigma = 1.9\,\text{mm}$, with $M = 0.8$ the magnification of the source image reproduced on S_b. Their pixel size $\delta = 6\,\mu\text{m}$ is close to both resolution limits, as defined by the source and the object's aperture. The results reported in Figure 3 clearly indicate that the refocusing procedure enables the recovery of the information on the aperture function of the object, which is completely lost in the misfocused ghost image.

We shall now compare the above results with the one achievable by a standard plenoptic camera having the same pixel size and total number of pixels per side ($N_{\text{tot}} = N_a + N_b = 620$). To this end, we introduce the parameter $\alpha = S_i/S'_i$, given by the ratio between the distance S_i from the focusing element to the image plane, and the actual distance S'_i between the focusing element (imaging lens) and the detector. Generally, perfect refocusing is possible if [5]

$$\left|1 - \frac{1}{\alpha}\right| < \frac{\Delta x}{\Delta u}, \tag{21}$$

where Δx is the minimum distance that can be resolved on the image plane, and Δu the minimum distance that can be resolved on the imaging lens. In a standard plenoptic camera, if the sensor have pixels of size δ, the image resolution is given by $\Delta x^{(\text{p})} = 2\delta N_u^{(\text{p})}$, while each pixel δ coincides with an area of width $\Delta u^{(\text{p})} = 2D_s/N_u^{(\text{p})}$ on the lens, with D_s the lens diameter. Hence,

$$\left(\frac{\Delta x}{\Delta u}\right)^{(\text{p})} = \frac{\delta}{D_s}\left(N_u^{(\text{p})}\right)^2. \tag{22}$$

In CPI instead, $\Delta x^{(\text{c})} = 2\delta$, since pixels of width δ can be used also to retrieve the image. On the other hand, the resolution on the imaging lens is given by $\Delta u^{(\text{c})} = 2D_s/N_b$, where D_s is the effective diameter of the lens L_a, that can be obtained by properly scaling the size D'_s of the pump profile:

$$D_s = D'_s\left(1 + \frac{z_a}{z_b}\right). \tag{23}$$

In this case, the right-hand side of the perfect refocusing condition given in Equation (21) reads

$$\left(\frac{\Delta x}{\Delta u}\right)^{(c)} = \frac{\delta}{D_s} N_b. \tag{24}$$

Hence, the maximum achievable depth of focus, in the setup employed for the simulation reported in Figure 3, is $|1 - 1/\alpha| < 0.26$. A standard plenoptic camera with the same pixel size and total number of pixels per side would enable us to achieve this same depth of focus provided $N_u^{(p)} = 18$ pixels are employed for the angular resolution; this condition imposes a loss of spatial resolution by a factor 18 ($\Delta x^{(p)} = 0.1$ mm) with respect to the one of the CPI protocol considered above.

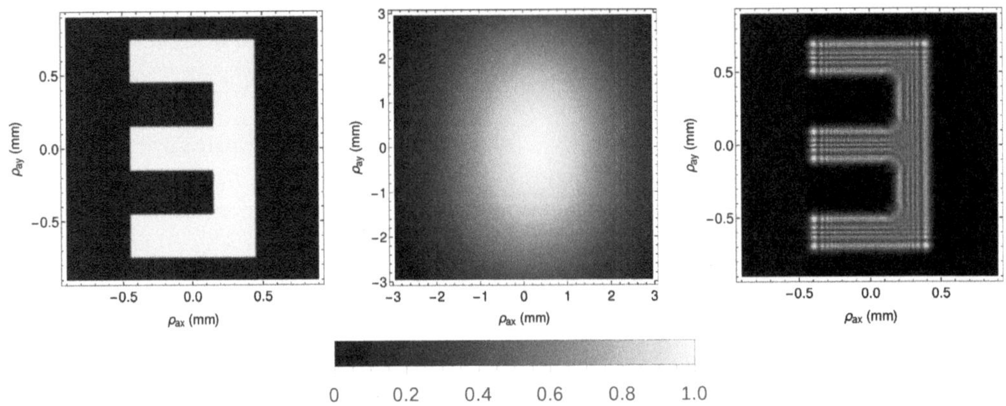

Figure 3. Comparison between focused (left), misfocused (center) and refocused (right) images of a two-dimensional object. Intensities are normalized to their maximum value in all panels. The out-of-focus and the refocused images are taken in the *same* setup, in which $z_b \approx z_{bF}/3$.

4. Discussion

At the heart of the refocusing capability of the second order correlation function of Equation (10), is the larger depth of focus of the coherent ghost image (Equation (13)), with respect to the incoherent ghost image (Equation (12)), as reported in Figure 4. In fact, the maximum achievable depth of focus of the proposed CPI scheme is the result of the increased depth of focus of coherent ghost imaging, with respect to incoherent ghost imaging.

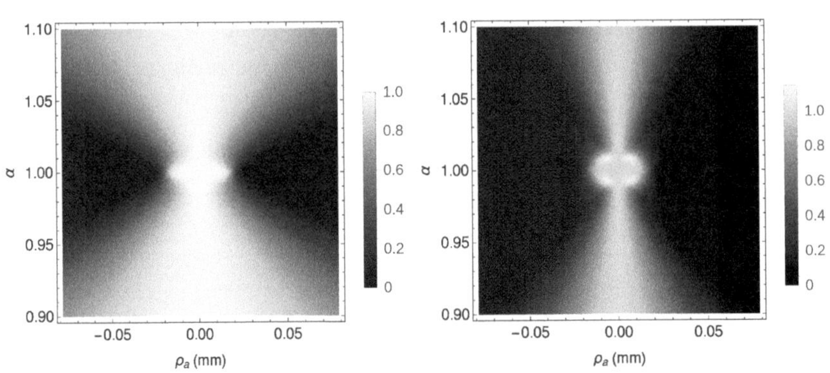

Figure 4. Comparison between the coherent and the incoherent ghost image of a single slit of width $a = 26\,\mu$m, as given by Equations (12) and (13), respectively, in the same setup described in Section 3. Both functions have been normalized to their value in $\rho_a = 0$ for any value of α.

This can be better understood by considering the origin of both the out-of-focus and the refocused image: The first one is obtained by integrating the out-of-focus coherent image (Equations (10), (15), or (18)) over the whole sensor S_b, exactly as it would do a bucket detector of standard ghost imaging; the second one is obtained by integrating, over the same sensor S_b, the rescaled version of such out-of-focus coherent image, as indicated in Equation (20). Now, as shown in Figure 5, the out-of-focus coherent image is a projection of the focused image (hence, it is either enlarged or reduced with respect to it) as seen by the *viewpoint* defined by the specific value of ρ_b. The integration all such coherent images over the whole sensor S_b implies the overlap of all the projections taken from the different viewpoints ρ_b; the resulting incoherent image is thus characterized by a loss of resolution, namely, it appears out of focus. The rescaled coherent image restores the correct size of the focused image and, most important, *tilts* the image in such a way to cancel the specific viewpoint from which it was taken. As a consequence, the integration of all such rescaled coherent images over the whole sensor S_b has no more detrimental effect on the resolution of the resulting incoherent image; the post-processed image thus appears refocused.

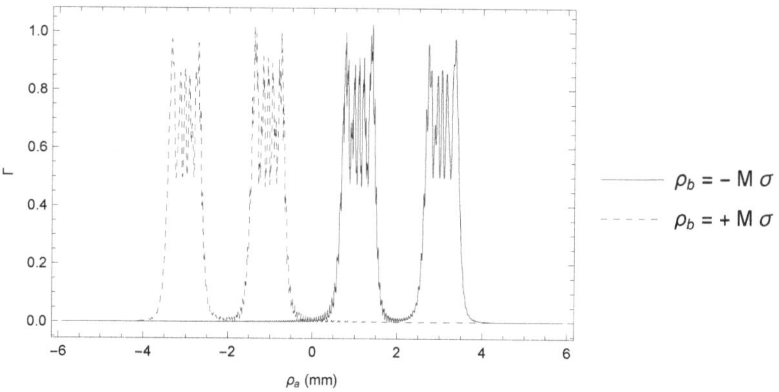

Figure 5. Observation of a double slit of width $a = 0.2$ mm and center-to-center distance $2a$ from two different points of view. Here, the setting is one-dimensional, with the same parameters as the setup described in Section 3. The coherent ghost images of Equation (10) enable us to change the point of view on any out-of-focus plane by selecting the point ρ_b on the sensor S_b, corresponding to a source point $\rho_s = -\rho_b/M$. In this case, the axis of the double slit coincides with the optical axis, and the chosen points on S_b are $\rho_b = -M\sigma$ (solid line, on the right) and $\rho_b = +M\sigma$ (dashed line, on the left).

5. Conclusions and Outlook

In view of practical applications, it is worth mentioning that all the above results apply to both reflective and transmitting objects. In addition, in contrast with chaotic light, entangled photons from SPDC enable us to employ different wavelengths in the two arms of the setup: Light illuminating the object is not required to have the same spectrum as light being remotely detected by S_a to retrieve the desired image [32,33]. This is quite interesting in view of applications requiring specific illumination wavelenghts for the object. In this scenario, one may choose two different sensors for maximizing the detection efficiency.

As plenoptic imaging is being broadly adopted in diverse fields such as digital photography [6–8], microscopy [3,4], 3D imaging, sensing and rendering [2], our proposed scheme has direct applications in several biomedical and engineering fields. Interestingly, the coherent nature of the correlation plenoptic imaging technique may lead to innovative coherent microscopy modality.

Acknowledgments: This work has been supported by the MIUR project P.O.N. RICERCA E COMPETITIVITA' 2007-2013 - Avviso n. 713/Ric. del 29/10/2010, Titolo II - "Sviluppo/Potenziamento di DAT e di LPP" (project n. PON02-00576-3333585), the INFN through the project "QUANTUM", the UMD Tier 1 program and the Ministry of Science of Korea, under the "ICT Consilience Creative Program" (IITP-2015-R0346-15-1007).

Author Contributions: Milena D'Angelo, Giuliano Scarcelli and Augusto Garuccio conceived and designed the proposed scheme for CPI; Francesco V. Pepe performed the theoretical calculation; Francesco Di Lena performed the simulation; Milena D'Angelo and Francesco V. Pepe wrote the paper. All authors have read and approved the final manuscript.

Conflicts of Interest: The authors declare no conflict of interest. The founding sponsors had no role in the design of the study; in the collection, analyses, or interpretation of data; in the writing of the manuscript, and in the decision to publish the results.

References

1. Adelson, E.H.; Wang, J.Y.A. Single Lens Stereo with a Plenoptic Camera. *IEEE Trans. Pattern Anal. Mach. Intell.* **1992**, *14*, 99–106.
2. Xiao, X.; Javidi, B.; Martinez-Corral, M.; Stern, A. Advances in three-dimensional integral imaging: Sensing, display, and applications. *Appl. Opt.* **2013**, *52*, 546–560.
3. Broxton, M.; Grosenick, L.; Yang, S.; Cohen, N.; Andalman, A.; Deisseroth, K.; Levoy, M. Wave optics theory and 3-D deconvolution for the light field microscope. *Opt. Express.* **2013**, *21*, 25418–25439.
4. Prevedel, R.; Yoon, Y.-G.; Hoffmann, M.; Pak, N.; Wetzstein, G.; Kato, S.; Schrödel, T.; Raskar, R.; Zimmer, M.; Boyden, E.S.; *et al.* Simultaneous whole-animal 3D imaging of neuronal activity using light-field microscopy. *Nat. Methods* **2014**, *11*, 727–730.
5. Ng, R.; Levoy, M.; Brédif, M.; Duval, G.; Horowitz, M.; Hanrahan, P. *Light Field Photography with a Hand-Held Plenoptic Camera*; Tech Report CSTR 2005-02; Stanford University Computer Science: Stanford, CA, USA, 2005.
6. Lytro ILLUM. Available online: https://www.lytro.com/illum (accessed on 2 June 2016).
7. Raytrix. Available online: http://www.raytrix.de/ (accessed on 2 June 2016).
8. 3D capture for the next generation. Available online: http://www.pelicanimaging.com (accessed on 2 June 2016).
9. Liu, H.; Jonas, E.; Tian, L.; Jingshan, Z.; Recht, B.; Waller, L. 3D imaging in volumetric scattering media using phase-space measurements. *Opt. Express.* **2015**, *23*, 14461–14471.
10. Muenzel, S.; Fleischer, J.W. Enhancing layered 3D displays with a lens. *Appl. Opt.* **2013**, *52*, D97–D101.
11. Levoy, M.; Hanrahan, P. Light field rendering. In *Computer Graphics Annual Conference Series*, Proceedings of the SIGGRAPH 1996, New Orleans, LA, USA, 4–9 August 1996; ACM SIGGRAPH: New York, NY, USA, 1996; pp. 31–42.
12. Levoy, M.; Ng, R.; Adams, A.; Footer, M.; Horowitz, M. Light field microscopy. *ACM Trans. Graph.* **2006**, *25*, 924–934.
13. Glastre, W.; Hugon, O.; Jacquin, O.; de Chatellus, H.G.; Lacot, E. Demonstration of a plenoptic microscope based on laser optical feedback imaging. *Opt. Express* **2013**, *21*, 7294–7303.
14. Waller, L.; Situ, G.; Fleischer, J.W. Phase-space measurement and coherence synthesis of optical beams. *Nat. Photonics* **2012**, *6*, 474–479.
15. Georgiev, T.; Zheng, K.C.; Curless, B.; Salesin, D.; Nayar, S.; Intwala, C. Spatio-Angular Resolution Tradeoff in Integral Photography. In *Eurographics Symposium on Redering (2006)*; Akenine-Möller, T., Heidrich, W., Eds.; The Eurographics Association: Geneva, Switzerland, 2006.
16. Schroff, S.A.; Berkner, K. Image formation analysis and high resolution image reconstruction for plenoptic imaging systems. *Appl. Opt.* **2013**, *52*, D22–D31.
17. Pérez, J.; Magdaleno, E.; Pérez, F.; Rodríguez, M.; Hernández, D.; Corrales, J. Super-Resolution in Plenoptic Cameras Using FPGAs. *Sensors* **2014**, *14*, 8669–8685.
18. D'Angelo, M.; Pepe, F.V.; Garuccio, A.; Scarcelli, G. Correlation Plenoptic Imaging. *Phys. Rev. Lett.* **2016**, *116*, 223602.
19. Pepe, F.V.; Scarcelli, G.; Garuccio, A.; D'Angelo, M. Plenoptic imaging with second-order correlations of light. *Quantum Meas. Quantum Metrol.* **2016**, *3*, 20–26.
20. Ferri, F.; Magatti, D.; Gatti, A.; Bache, M.; Brambilla, E.; Lugiato, L.A. High-Resolution Ghost Image and Ghost Diffraction Experiments with Thermal Light. *Phys. Rev. Lett.* **2005**, *94*, 183602.
21. Brida, G.; Chekhova, M.V.; Fornaro, G.A.; Genovese, M.; Lopaeva, L.; Ruo Berchera, I. Systematic analysis of signal-to-noise ratio in bipartite ghost imaging with classical and quantum light. *Phys. Rev. A* **2011**, *83*, 063807.

22. Brida, G.; Genovese, M.; Ruo Berchera, I. Experimental realization of sub-shot-noise quantum imaging. *Nat. Photonics* **2010**, *4*, 227–230.
23. Klyshko, D.N. *Photons and Nonlinear Optics*; CRC Press: Boca Raton, FL, USA, 1988.
24. D'Angelo, M.; Valencia, A.; Rubin, M.H.; Shih, Y.H. Resolution of quantum and classical ghost imaging. *Phys. Rev. A* **2005**, *72*, 013810.
25. D'Angelo, M.; Shih, Y.H. Quantum Imaging. *Laser Phys. Lett.* **2005**, *2*, 567–596.
26. Scully, M.O.; Zubairy M.S. *Quantum Optics*, 1st ed.; Cambridge University Press: Cambridge, UK, 1997.
27. Rubin, M.H.; Klyshko, D.N.; Shih, Y.H.; Sergienko, A.V. Theory of two-photon entanglement in type-II optical parametric down-conversion. *Phys. Rev. A* **1994**, *50*, doi:10.1103/PhysRevA.50.5122.
28. Rubin, M.H. Transverse correlation in optical spontaneous parametric down-conversion. *Phys. Rev. A* **1996**, *54*, doi:10.1103/PhysRevA.54.5349.
29. Burlakov, A.V.; Chekhova, M.V.; Klyshko, D.N.; Kulik, S.P.; Penin, A.N.; Shih, Y.H.; Strekalov, D.V. Interference effects in spontaneous two-photon parametric scattering from two macroscopic regions. *Phys. Rev. A* **1997**, *56*, doi:10.1103/PhysRevA.56.3214.
30. Kim, Y.-H. Measurement of one-photon and two-photon wave packets in spontaneous parametric downconversion. *J. Opt. Soc. Am. B* **2003**, *20*, 1959–1966.
31. Baek, S.-Y.; Kim, Y.-H. Spectral properties of entangled photon pairs generated via frequency-degenerate type-I spontaneous parametric down-conversion. *Phys. Rev. A* **2008**, *77*, 043807.
32. Rubin, M.H.; Shih, Y. Resolution of ghost imaging for nondegenerate spontaneous parametric down-conversion. *Phys. Rev. A* **2008**, *78*, 033836.
33. Karmakar, S.; Shih, Y. Two-color ghost imaging with enhanced angular resolving power. *Phys. Rev. A* **2010**, *81*, 033845.
34. Goodman, J.W.; *Introduction to Fourier Optics*, 2nd ed.; McGraw-Hill Science/Engineering/Math: New York, NY, USA, 1996.
35. Pittman, T.B.; Shih, Y.H.; Strekalov, D.V.; Sergienko, A.V. Optical imaging by means of two-photon quantum entanglement. *Phys. Rev. A* **1995** *52*, R3429.

Permissions

All chapters in this book were first published in Technologies, by MDPI; hereby published with permission under the Creative Commons Attribution License or equivalent. Every chapter published in this book has been scrutinized by our experts. Their significance has been extensively debated. The topics covered herein carry significant findings which will fuel the growth of the discipline. They may even be implemented as practical applications or may be referred to as a beginning point for another development.

The contributors of this book come from diverse backgrounds, making this book a truly international effort. This book will bring forth new frontiers with its revolutionizing research information and detailed analysis of the nascent developments around the world.

We would like to thank all the contributing authors for lending their expertise to make the book truly unique. They have played a crucial role in the development of this book. Without their invaluable contributions this book wouldn't have been possible. They have made vital efforts to compile up to date information on the varied aspects of this subject to make this book a valuable addition to the collection of many professionals and students.

This book was conceptualized with the vision of imparting up-to-date information and advanced data in this field. To ensure the same, a matchless editorial board was set up. Every individual on the board went through rigorous rounds of assessment to prove their worth. After which they invested a large part of their time researching and compiling the most relevant data for our readers.

The editorial board has been involved in producing this book since its inception. They have spent rigorous hours researching and exploring the diverse topics which have resulted in the successful publishing of this book. They have passed on their knowledge of decades through this book. To expedite this challenging task, the publisher supported the team at every step. A small team of assistant editors was also appointed to further simplify the editing procedure and attain best results for the readers.

Apart from the editorial board, the designing team has also invested a significant amount of their time in understanding the subject and creating the most relevant covers. They scrutinized every image to scout for the most suitable representation of the subject and create an appropriate cover for the book.

The publishing team has been an ardent support to the editorial, designing and production team. Their endless efforts to recruit the best for this project, has resulted in the accomplishment of this book. They are a veteran in the field of academics and their pool of knowledge is as vast as their experience in printing. Their expertise and guidance has proved useful at every step. Their uncompromising quality standards have made this book an exceptional effort. Their encouragement from time to time has been an inspiration for everyone.

The publisher and the editorial board hope that this book will prove to be a valuable piece of knowledge for researchers, students, practitioners and scholars across the globe.

List of Contributors

Peter Schnauber, Alexander Thoma, Christoph V. Heine, Alexander Schlehahn, Manuel Gschrey, Ronny Schmidt, Caspar Hopfmann, Jan-Hindrick Schulze, André Strittmatter, Tobias Heindel, Sven Rodt and Stephan Reitzenstein
Institut für Festkörperphysik, Technische Universität Berlin, Hardenbergstraße 36, 10623 Berlin, Germany

Liron Gantz and David Gershoni
The Physics Department, Technion—Israel Institute of Technology, Haifa 32000, Israel

Benjamin Wohlfeil
Zuse-Institut Berlin (ZIB), Takustraße 7, 14195 Berlin, Germany

Ulrike Woggon
Institut für Optik und Atomare Physik, Technische Universität Berlin, Straße des 17. Juni 135, 10623 Berlin, Germany

Gautam Kumar
Electrical and Systems Engineering, Washington University in St. Louis, 1 Brookings Drive, St. Louis, MO 63130, USA

Mayuresh V. Kothare
Chemical and Biomolecular Engineering, Lehigh University, 111 Research Drive, Bethlehem, PA 18015, USA

Nitish V. Thakor
Biomedical Engineering, Johns Hopkins University, Baltimore, MD 21218, USA

Marc H. Schieber
Neurobiology and Anatomy, University of Rochester Medical Center, Rochester, NY 14642, USA

Hongguang Pan
School of Electrical and Control Engineering, Xi'an University of Science and Technology, Xi'an 710071, China

Baocang Ding
Department of Automation, School of Electronic and Information Engineering, Xi'an Jiaotong University, Xi'an 710049, China

Weimin Zhong
Key Laboratory of Advanced Control and Optimization for Chemical Processes, Ministry of Education, East China University of Science and Technology, Shanghai 200237, China

Yuan-Jen Chang, Chao-Hsuan Chang, Jin-Chen Hsu and Chia-Lung Kuo
Department of Mechanical Engineering, National Yunlin University of Science and Technology, Douliou, Yunlin 640, Taiwan

Chao-Ching Ho
Department of Mechanical Engineering, National Taipei University of Technology, Taipei 106, Taiwan

Zeyang Liao and M. Suhail Zubairy
Institute for Quantum Science and Engineering (IQSE) and Department of Physics and Astronomy, Texas A&M University, College Station, TX 77843-4242, USA

Fahad AlGhannam
Institute for Quantum Science and Engineering (IQSE) and Department of Physics and Astronomy, Texas A&M University, College Station, TX 77843-4242, USA
The National Center for Applied Physics, KACST, P.O.Box 6086, Riyadh 11442, Saudi Arabia

Philip Hemmer
Department of Electrical & Computer Engineering, Texas A&M University, College Station, TX 77843-4242, USA

Colin A. Scholes
Department of Chemical and Biomolecular Engineering, The University of Melbourne, Melbourne 3010, Australia

Minh T. Ho and Dianne E. Wiley
School of Chemical Engineering, University of New South Wales, Kensington 2052, Australia

Yin Hsien Fung, Pimonpan Sompet and Mikkel F. Andersen
Dodd-Walls Centre for Photonics and Quantum Technologies, Department of Physics, University of Otago, 9016 Dunedin, New Zealand

Samuel P. Santos and João C. Bordado
Centro de Recursos Naturais e Ambiente (CERENA), Instituto Superior Técnico, Universidade de Lisboa, Av. Rovisco Pais, s/n, Lisboa 1049-001, Portugal

João F. Gomes
Centro de Recursos Naturais e Ambiente (CERENA), Instituto Superior Técnico, Universidade de Lisboa, Av. Rovisco Pais, s/n, Lisboa 1049-001, Portugal
Área Departamental de Engenharia Química (ADEQ), Instituto Superior de Engenharia de Lisboa, Instituto Politécnico de Lisboa, R. Conselheiro Emídio Navarro 1, Lisboa 1959-007, Portugal

Rohaya Mohamed Halim, Che Rahmat Che Mat and Nu'man Abdul Hadi
Milling & Processing Unit, Engineering & Processing Research Division, Malaysian Palm Oil Board, Ministry of Plantation Industries & Commodities, 6, Persiaran Institusi, Bandar Baru Bangi, 43000 Kajang, Selangor, Malaysia

Ridzuan Ramli
Biomass Technology Unit, Engineering & Processing Research Division, Malaysian Palm Oil Board, Ministry of Plantation Industries & Commodities, 6, Persiaran Institusi, Bandar Baru Bangi, 43000 Kajang, Selangor, Malaysia

Choo Yuen May
Director General's Office, Malaysian Palm Oil Board, Ministry of Plantation Industries & Commodities, 6, Persiaran Institusi, Bandar Baru Bangi, 43000 Kajang, Selangor, Malaysia

Nasrin Abu Bakar
Energy & Environment Unit, Engineering & Processing Research Division, Malaysian Palm Oil Board, Ministry of Plantation Industries & Commodities, 6, Persiaran Institusi, Bandar Baru Bangi, 43000 Kajang, Selangor, Malaysia

Antonino Cosentino
Cultural Heritage Science Open Source (CHSOS), Viagrande 95029, Italy

Lakkana Piewkhaow, Christine W. Chan and Paitoon Tontiwachwuthikul
Faculty of Engineering and Applied Science, University of Regina, Regina, SK S4S 0A2, Canada

Anastassia Manuilova and MalcolmWilson
ArticCan Energy Services, Regina, SK S4P 3N6, Canada

Prabhakar Sharma
School of Ecology and Environment Studies, Nalanda University, Rajgir, Nalanda, Bihar 803116, India
Department of Earth Sciences, Uppsala University, Uppsala 75236, Sweden

Fritjof Fagerlund, Ulrika Iverfeldt and Amanda Eskebaek
Department of Earth Sciences, Uppsala University, Uppsala 75236, Sweden

Alvaro Recio, Sip Chen Liew and Josephine M. Hill
Department of Chemical and Petroleum Engineering, University of Calgary, 2500 University Drive NW, Calgary, AB T2N 1N4, Canada

Dennis Lu
CanmetENERGY, Natural Resources Canada, 1 Haanel Drive, Ottawa, ON K1A 1M1, Canada

Ryad Rahman and Arturo Macchi
Department of Chemical and Biological Engineering, University of Ottawa, 161 Louis Pasteur Street, Ottawa, ON K1N 6N5, Canada

Sylvain Barraud, Romain Lavieville, Louis Hutin, Heorhii Bohuslavskyi and Maud Vinet
Laboratoire d'électronique des technologies de l'information (LETI), Université Grenoble Alpes and CEA, Minatec Campus, F-38054 Grenoble, France

Andrea Corna, Paul Clapera, Marc Sanquer and Xavier Jehl
Institut Nanosciences et Cryogénie (INAC), Université Grenoble Alpes and CEA, F-38054 Grenoble, France

Mario D'Aniello, Francesco Portioli and Raffaele Landolfo
Department of Structures for Engineering and Architecture, University of Naples "Federico II", Naples 80134, Italy

Abdussalam El-Suleiman
Department of Mechanical Engineering Technology, Federal Polytechnic, Nasarawa 962101, Nigeria

Nnamdi B. Anosike
Department of Mechanical Engineering, Nnamdi Azikiwe University, Awka 420107, Nigeria

Pericles Pilidis
Centre for Propulsion Engineering, Cranfield University, Cranfield MK43 0AL, UK

Chung-Wei Cheng
Department of Mechanical Engineering, National Chiao Tung University, No. 1001, Ta Hsueh Road, Hsinchu 30010, Taiwan

Jinn-Kuen Chen
Department of Mechanical and Aerospace Engineering, University of Missouri, Columbia, MO 65211, USA

Francesco V. Pepe
Museo Storico della Fisica e Centro Studi e Ricerche "Enrico Fermi", I-00184 Roma, Italy
Istituto Nazionale di Fisica Nucleare (INFN), Sezione di Bari, I-70126 Bari, Italy

Francesco Di Lena, Augusto Garuccio and Milena D'Angelo
Istituto Nazionale di Fisica Nucleare (INFN), Sezione di Bari, I-70126 Bari, Italy
Dipartimento Interateneo di Fisica, Università degli studi di Bari, I-70126 Bari, Italy

Giuliano Scarcelli
Fischell Department of Bioengineering, University of Maryland, College Park, MD 20742, USA

Index

A

Aermod, 125, 131, 133, 144-145, 148, 151

Air Dispersion Modeling, 125, 127-128, 131-133, 142, 148-149

Air Separation Unit (ASU), 133

Anti-reflection (AR), 1-2

Anti-reflection Coating, 1, 8

Aqueous Amine Solutions, 87

Ar-coated Microlenses, 1, 4

Art Conservation, 112, 114

Artificial Sensory Feedback, 9-11, 21-22, 24-25, 27-29, 32

Atomic Physics, 64, 178

B

Blue-detuned, 66-67, 69, 73-79, 81-83, 86

Bonded Splices, 185, 192-193

Brain-machine Interfaces, 9, 31-32

C

Calcium Looping

Combustion, 166

Calcium Oxide, 166, 176

Canvas Paintings, 118

Carbon Capture, 54, 56, 62, 90, 126-127, 150-152, 166, 176, 196, 203-204

Carbon Dioxide, 61-63, 87, 95, 125, 135, 138, 151, 166, 196, 203-204

Charge-coupled Device (CCD), 35

Chemical Absorption, 62, 87, 90, 92

Closed-loop Bmi, 10-12, 19-20, 22-25, 28

Co2 Compression, 138, 196-197

Co2 Dosing, 89

Collision Beam Parameters, 76-77, 82

Collisional Blockade, 64-65, 78-83, 85

Combined Cycle Gas Turbine (CCGT), 196, 203

Conservation Science, 112, 122

Cooling Light Parameters, 77, 81

Copper Oxide, 166

Cortical Circuit Model, 9, 11-12, 29

Cryogenic, 50, 52, 54, 62, 134, 137, 179, 184, 197, 203-204

Cultural Heritage Science, 112, 120

D

Data Inventory Modeling, 133

Deterministic Devices, 1

Dipole Trap, 64, 85

Double Strap Test, 185

Dry Separation System, 97

Dual Pulse, 205-206, 208, 210-212

E

Electrical Quantum Metrology, 178

Electron Pumps, 178-180, 182-184

Extraction Efficiency, 1-2, 4

F

Femtosecond Laser, 205, 207, 211-213

Few-body Physics, 64

Fire Particles, 153, 162

Fire-born Particles, 153-155, 157-158, 160-163

Frequency-domain, 114

I

Intracortical Micro-stimulation, 25

K

Kalman Filter Based Decoder Design, 17

Kernel Losses, 97-98, 101, 104-105, 107, 110

Kernel Recovery, 97, 101, 104-105, 110

Kernel Recovery Rate, 97, 101, 105, 110

L

Lap Shear Test, 185

Life-cycle Assessment (LCA), 125, 127

Light-assisted Collisions, 64-65, 68-70, 72-83, 85

M

Magnetic Resonance Imaging, 44, 50, 52

Membrane Gas Separation, 54, 61-62

Methyldiethanolamine (MDEA), 87-88, 93

Microelectronics, 178-179, 181

Model Predictive Control, 9, 11, 21, 23, 29

Model Predictive Control (MPC), 21, 23
Mosfets, 178

N
Nanoscale, 34, 43, 49, 86
Narrow Bandwidth, 116
Nmr Imaging, 44
Nut and Kernel Size Distribution, 102

O
Optical Opaque Materials, 113
Optically Detected Magnetic Resonance, 44, 49, 51

P
Palm Kernel, 97-98, 101, 109-111
Palm Oil Mill, 97-98, 110
Palm Shell, 97-98, 109-111
Panel Paintings, 116, 123
Photon-extraction Efficiency (PEE), 1-2
Pilot Unit Scale, 90, 92-94
Porous Media, 153-154, 156-165
Post-combustion, 54, 56-58, 61-62, 125, 127-128, 132-133, 138-140, 142, 146, 148-149, 151, 166, 176, 204
Power Consumption, 196
Pre-combustion Co2 Capture Processes, 125
Prime Mover, 196-197, 199, 203

Q
Quantum Dots, 1, 7-8, 184
Quantum Information Processing, 64, 83-84
Quantum Optics, 1, 52, 225
Qubits, 52, 64, 84, 86

R
Retention, 153, 159-163

S
Saskatchewan Lignite, 125, 127-128, 134, 149-150
Separation System, 97-98, 101, 105-106
Single Atom, 64-65, 69, 73, 75-83, 85-86
Single-photon Source, 1, 7
Steam Carbonation, 166
Straining, 153, 159-163
Subwavelength Lithography, 44
Sulfur Recovery Process, 135
Superresolution, 44

T
Terahertz (THZ), 112, 124
Thermal Treatment, 166, 171-173
Thz Range, 114
Thz Spectroscopy, 115, 118
Thz Tomography, 116, 120
Time-domain, 114, 116, 122-124
Total Dirt Content, 101
Transport, 128, 153-156, 160-162, 164-165, 196-197, 203-204, 207
Triggered Single Photons, 6
Two-temperature Model, 205, 212

W
Wall Paintings, 114, 118
Wiener Filter Based Decoder Design, 15, 17-18
Winnowing Column, 97-98, 100-101, 105, 110-111

Lecture Notes in Computer Science 4669

Commenced Publication in 1973
Founding and Former Series Editors:
Gerhard Goos, Juris Hartmanis, and Jan van Leeuwen